Nanotechnology, the Brain, and the Future

# Yearbook of Nanotechnology in Society

Volume 3

*Series Editor*

David H. Guston, *Arizona State University*

For further volumes:
http://www.springer.com/series/7583

Sean A. Hays • Jason Scott Robert
Clark A. Miller • Ira Bennett
Editors

# Nanotechnology, the Brain, and the Future

Springer

*Editors*
Sean A. Hays
The Center for Nanotechnology in Society
Arizona State University
Tempe, AZ, USA

Jason Scott Robert
The Center for Nanotechnology in Society
Arizona State University
Tempe, AZ, USA

Clark A. Miller
The Center for Nanotechnology in Society
Arizona State University
Tempe, AZ, USA

Ira Bennett
The Center for Nanotechnology in Society
Arizona State University
Tempe, AZ, USA

Chapters 11and 15 has been published in *Nanoethics*, Vol. 2, 2008, on pp. 241–249 and pp. 305–316 respectively.

Chapter 14 has been published in *Nature*, Vol. 456 (No. 7223), 2008 on pp. 702–705.

Chapter 18 has been published online in *Journal of the Royal Society Interface* on 2 June 2010, doi: 10.1098/rsif.2010.0158.focus, on pp. 1–12

Chapter 19 has been published in *Recommendations for a Municipal Health & Safety Policy for Nanomaterials: A Report to the Cambridge City Manager*, 2008, on pp. 1–14.

Chapter 20 has been published in the conference *Nanotechnology in Cambridge: What Do You Think?* (May 22, 2008)

ISBN 978-94-007-1786-2 ISBN 978-94-007-1787-9 (eBook)
DOI 10.1007/978-94-007-1787-9
Springer Dordrecht Heidelberg New York London

Library of Congress Control Number: 2012937038



Printed on acid-free paper

Springer is part of Springer Science+Business Media (www.springer.com)

# Preface

When I was a young television watcher, a ubiquitous and now much spoofed public service announcement ran like this:

> Image of an egg held in a hand. Voice: "This is your brain." Image of a hand cracking the egg and dumping it into sizzling fry pan. Voice: "This is your brain on drugs. Any questions?"

In this third volume of the *Yearbook of Nanotechnology in Society*, we begin to explore the question, "Will this be your brain on nano?"

Most of the time when writers have mentioned the nouns "nanotechnology" and "brain" together in the same sentence, the discussion has veered toward speculative accounts of the enhancement of human cognition or other capacities. Indeed, an exploration early in the history of nanotechnology in the United States – the volume on *Converging Technologies for Improving Human Performance* (Roco and Bainbridge 2003) – set a tone, if not a trend, that the US National Nanotechnology Initiative had human enhancement as an important institutional goal. While this current continued, it has over time become more submerged beneath criticisms that it was too speculative (e.g., Nordmann 2007) as well as beneath more, arguably, immediate concerns like environmental health and safety of nanomaterials.

The Center for Nanotechnology in Society at Arizona State University (CNS-ASU), which is the home of this *Yearbook* series, inserted itself into the issue of nanotechnology and human enhancement through establishing a research program in "Human Identity, Enhancement, and Biology." While originally as broad and, perhaps, as inchoate, as that string of nouns, the research program eventually coalesced around a specific interest in the human brain. What this meant for the research at CNS-ASU was that – in addition to a relatively autonomous inquiry into the societal aspects of nanotechnology and the brain – a broad set of other research programs at the Center would orient some of their work to include such concerns as well. Thus, bibliometric analysis, public opinion polling, large-scale deliberation and public engagement, historical and analogical inquiry, and other techniques of "real-time technology assessment" (Guston and Sarewitz 2002) were brought to bear on this one relatively narrow slice of nanotechnology. As described in detail in the editors' introduction, this volume is the fruit of that effort.

What stands out in this effort to understand nanotechnology and the brain in an "end-to-end" effort across a set of empirical inquiries is, in fact, how necessary that empirical work is to get a good handle on the phenomena of interest. Bibliometric analysis revealed a vast substrate of research related to nanotechnology and the brain – although not much of it immediately related to questions of enhancement. Public opinion polling and large-scale deliberation revealed a public quite uneasy with plausible applications of nanotechnology for human enhancement, and yet still quite committed to applications for therapy. Inquiries among those with visual or hearing impairments even suggested that therapies for such "target populations" might be received at best with great ambivalence. And these anticipatory discussions are set against a backdrop of an emerging technical literature that shows greater facility with visualizing, understanding, and manipulating the brain, as well as credible, precautionary findings that nanomaterials in the environment, in addition to those in therapies or enhancements, could in fact influence the brain.

This is not to say that all anticipatory research needs to be empirical. The *Yearbook* contains efforts at the development of theory and concepts – for example, attempting to explain the relationship between our often-misguided popular understandings of intelligence and our beliefs about enhancement. It also contains documentation of some of the political and policy action that has been motivated in part by environmental health and safety concerns that include the understanding that nanoparticles can cross the blood–brain barrier.

At least for the 400 years since Hamlet pondered the relationship between uncertain knowledge and potentially rash action, we have been challenged publicly to discern when the quality and quantity of what is known is sufficient for the task. This discernment is part of the strategic vision of CNS-ASU, what we call anticipatory governance (Barben et al. 2008), which links the capacities of foresight into plausible futures, engagement of lay publics, and integration of social science and humanistic perspectives with ongoing natural science and engineering. The first volume of the *Yearbook* demonstrated the great variety of approaches to anticipating the futures of nanotechnology. The second *Yearbook* explored what we can know about the consequences for equity, equality, and human development of acting through nanotechnology – even at this early stage in the game.

The contents of this third volume demonstrate quite readily that there are important questions to be attended to, now, if our brains on nano are to be happy and healthy.

Tempe, Arizona David H. Guston

## References

Barben, D., E. Fisher, C. Selin, and D.H. Guston. 2008. Anticipatory governance of nanotechnology: Foresight, engagement, and integration. In *The handbook of science and technology studies*, ed. E.J. Hackett, O. Amsterdamska, M. Lynch and J. Wajcman, 979–1000. Cambridge: MIT Press.

Guston, D.H., and D. Sarewitz. 2002. Real-time technology assessment. *Technology in Society* 24(1–2): 93–109.

Nordmann, A. 2007. If and then: A critique of speculative nanoethics. *NanoEthics* 1(1): 31–46.

Roco, M., and W.S. Bainbridge. 2003. *Converging technologies for improving human performance*. New York: Springer.

# Contents

1 **Introduction: Ethics and Anticipatory Governance of Nano-Neurotechnological Convergence** ... 1
Jason Scott Robert, Clark A. Miller, and Valerye Milleson

**Part I Introduction to RTTA**

2 **Applications of Nanotechnology to the Brain and Central Nervous System** ... 21
Christina Nulle, Clark A. Miller, Alan Porter, and Harmeet Singh Gandhi

3 **Public Attitudes Towards Nanotechnology-Enabled Cognitive Enhancement in the United States** ... 43
Sean A. Hays, Clark A. Miller, and Michael D. Cobb

4 **U.S. News Coverage of Neuroscience Nanotechnology: How U.S. Newspapers Have Covered Neuroscience Nanotechnology During the Last Decade** ... 67
Doo-Hun Choi, Anthony Dudo, and Dietram A. Scheufele

5 **Nanotechnology, the Brain, and the Future: Ethical Considerations** ... 79
Valerye Milleson

6 **A New Model for Public Engagement: The Dialogue on Nanotechnology and Religion** ... 97
Richard Milford and Jameson M. Wetmore

**Part II Brain Repair and Brain-Machine Implants**

7 **The Age of Neuroelectronics** ... 115
Adam Keiper

8 **The Cochlear Implant Controversy: Lessons Learned for Using Anticipatory Governance to Address Societal Concerns of Nano-scale Neural Interface Technologies** ........ 147
Derrick Anderson

9 **Healing the Blind: Perspectives of Blind Persons on Methods to Restore Sight** ........ 159
Arielle Silverman

10 **Nanotechnology, the Brain, and Personal Identity** ........ 167
Stephanie Naufel

11 **Ethical, Legal and Social Aspects of Brain-Implants Using Nano-Scale Materials and Techniques** ........ 179
Francois Berger, Sjef Gevers, Ludwig Siep, and Klaus-Michael Weltring

**Part III Enhancing the Brain and Cognition**

12 **The Complex Cognitive Systems Manifesto** ........ 195
Richard P.W. Loosemore

13 **Narratives of Intelligence: The Sociotechnical Context of Cognitive Enhancement in American Political Culture** ........ 219
Sean A. Hays

14 **Towards responsible use of cognitive-enhancing drugs by the healthy** ........ 235
Henry Greely, Barbara Sahakian, John Harris, Ronald C. Kessler, Michael Gazzaniga, Philip Campbell, and Martha J. Farah

15 **The Opposite of Human Enhancement: Nanotechnology and the Blind Chicken Problem** ........ 247
Paul B. Thompson

16 **National Citizens' Technology Forum: Nanotechnologies and Human Enhancement** ........ 265
Patrick Hamlett, Michael D. Cobb, and David H. Guston

17 **Panelists' Reports by State: Arizona, California, Colorado, Georgia, New Hampshire, and Wisconsin (a–f)** ........ 285

**Part IV Nanoparticle Toxicity and the Brain**

18 **A Review of Nanoparticle Functionality and Toxicity on the Central Nervous System** ........ 313
Z. Yang, Z.W. Liu, R.P. Allaker, P. Reip, J. Oxford, Z. Ahmad, and G. Reng

**19 Recommendations for a Municipal Health & Safety Policy for Nanomaterials** ........ 333
Cambridge Nanomaterials Advisory Committee, Cambridge Public Health Department

**20 Nanotechnology in Cambridge: What Do You Think?** ........ 357

**21 Anticipatory Governance in Practice? Nanotechnology Policy in Cambridge, Massachusetts** ........ 373
Shannon N. Conley

**Index** ........ 393

# Chapter 1
# Introduction: Ethics and Anticipatory Governance of Nano-Neurotechnological Convergence

**Jason Scott Robert, Clark A. Miller, and Valerye Milleson**

The brain is the center of what makes us human. Art, culture, science, democracy, religion, technology – products of human thought and ideas, products of the unique capabilities of the human brain. Yet, for all but the last few decades, the human brain has been essentially a mystery, a biological organ whose functioning we could at best guess at from observations of human behavior and mental pathologies but about which we knew exceedingly little. All of that has changed, however. The rise of neuroscience as an interdisciplinary field of inquiry, deep investments during the 1990s – the Decade of the Brain – in new technologies, such as functional magnetic resonance imaging, and long persistence on the part of pioneering scientists have given us a wealth of new insights into the behavior and function of the human central nervous system.

Now, neuroscience is about to change again, as is the human brain. Already, scientists have begun to attach neural probes to the human brain, creating partial and highly preliminary, but nonetheless real, interfaces between the brain and computers. Adaptive machines are being integrated into the nervous system in laboratories with the design of creating radically more capable prosthetic devices. Brain repair is increasingly common and effective, even in the face of severe gunshot wounds to the head, such as that experienced by US Congresswoman Gabrielle Giffords. Yet, progress to date is only the beginning. What lies ahead will involve the ability not only to even more fully understand and repair the brain but also to extensively interact with and manipulate the brain and, especially, to upgrade and enhance its capabilities through technology, not just learning.

Driving these changes is the rapid convergence of neuroscience with nanoscale science and engineering (NSE). As recently as a few years ago, prominent neuroscientists continued to express deep skepticism that nanotechnology would have much

J.S. Robert (✉) • C.A. Miller • V. Milleson
The Center for Nanotechnology in Society, Arizona State University,
P.O. Box 875603, AZ 85287-5603, Tempe, USA
e-mail: jason.robert@asu.edu; clark.miller@asu.edu

S.A. Hays et al. (eds.), *Nanotechnology, the Brain, and the Future*,
Yearbook of Nanotechnology in Society 3, DOI 10.1007/978-94-007-1787-9_1,

of an impact on their field for the near future. Yet, the two fields had already begun to converge by the late 1990s and that convergence has only accelerated since 2000. As Nulle et al. document in their chapter in this volume, over 10,000 scientific papers involving the application of nanotechnology to neuroscience were published between 1991 and 2007, with roughly 1,600 articles published in 2007, the final year for which complete data were available in their study. These studies reflect the beginnings of partnership that seems inevitably to put into human hands the ability to understand, repair, and intervene in the biology of the brain at molecular scales – and to design nano-biotechnological devices that allow machines and computers to interface seamlessly with human neurological and cognitive processes. It may not happen today, maybe not next year, but it will almost certainly happen sooner than anyone might have imagined only a few years ago.

So what does this convergence of neuroscience and nanotechnology portend? How will people react to nano-neurotechnologies? How will our values, behaviors, relationships, institutions, and communities change as we integrate new and emerging brain technologies created through NSE into our lives, our economies, and our societies? What ethical challenges do these technologies pose? How will their risks and benefits be distributed, in the United States and around the globe?

This book offers a first step toward beginning to answer some of these questions. It does so through an integration of three distinct forms of humanistic and social science analysis: ethics, anticipatory research, and real-time technology assessment. Ethics is broadly understood here to subsume not just the narrow subfield of ethical analysis within philosophy but the broader normative analysis of the meaning, value, and import of new and emerging fields of science and technology to individuals, to communities, and to society as a whole. Anticipatory research seeks to understand not simply the ethical and societal dimensions of what scientists and engineers can do today but to be proactive in grappling with what science and technology may bring tomorrow. Almost all innovative science and engineering work is done with some vision of society's future in mind; anticipatory research, as we practice it here, strives to bring a robust social and ethical analysis to an understanding of what that future might look like. We do this in an attempt to help avoid deeply unethical or otherwise problematic outcomes in society that might have been anticipated. Finally, real-time technology assessment seeks to provide a systematic form of technology assessment that operates at the cutting edge of science, in parallel and in partnership with the science and engineering enterprise, in real-time. Its design is to provide a capacity to inform society and policy decision makers with robust technology assessment in time to provide "upstream" analysis and governance of major scientific and technological transitions in society. We describe these three research approaches at greater length below.

## 1.1 Why 'Nanoethics'?

Why care about the ethics of nanotechnology and its convergence with neuroscience? Indeed, why care about ethics at all, particularly as it pertains to science and technology (Grunwald 2000)? Some would claim that all that matters from an ethical perspective is already tackled by the engineer, the researcher, the policy-maker, the practitioner; the process safeguards itself against nefarious intrusions, so why should the ethicist – philosopher or social scientist – gadfly interfere (Grunwald 2005; Litton 2007)?

Let us count the ways. First, skills involved in ethical and societal issue identification, analysis, and resolution are only rarely, if ever, incorporated into the professional education of engineers, scientists, and policy-makers. Where they are incorporated, they are often added on to, rather than fully integrated into, the curriculum. Therefore, there is an abundant lack of relevant ethical expertise within these specific professions, as in most professions. Second, the moral imagination of researchers and practitioners may be insufficiently broad and deep to reflect upon ethical issues, even if for no other reason than that these folks are preoccupied with their professional tasks and so are, at best, unaware, (and, at worst, actively ignorant), of wider ethical and societal considerations. Third, whether moral values are ever, let alone frequently, on the mind of political decision-makers has never been well-established; indeed, given the complexity of policy-making, there is every reason to believe that contextual political and pragmatic considerations overpopulate their values landscape, crowding out ethical reflections. For these reasons, and many more, ethics and ethicists matter to good science and technology (see also Resnik 1998).

Even so, naysayers might contend that ethics and ethicists have, or should have, no pride of place concerning nanoscience and nanotechnology. One complaint occasionally rendered is that "nanotechnology" is either ill defined or unable to point to a specific field (Alpert 2008, 57; Hodge et al. 2007, 10), such that any potential concept of "nanoethics" will be just as incoherent a field as nanotechnology (Allhoff 2008, 4). A more common complaint is that the ethical issues surrounding nanotechnology are not unique to nanotechnology (Allhoff 2008; Godman 2008; Grunwald 2005; Holm 2005; Keiper 2007; Litton 2007; Parr 2005). As Paul Litton proclaims, "None of the ethical concerns associated with nanotechnology is unprecedented, and none raises novel ethical issues or demands new ethical principles" (Litton 2007, 23). Other critics are concerned about the fact that much of ethical reflection in nanotechnology seems to focus on "extreme" hypotheticals like Drexler's "gray goo" scenario (Drexler 1986) and the like; such reflection, detractors maintain, is a waste of time and financial resources, ultimately stalling progress in ethical discussion (Grunwald 2005; Litton 2007; Nordmann 2007). There is also an occasional appeal to pragmatism, claiming that by focusing on an ethics of nanotechnology we risk "reinventing the bioethical wheel," a waste of precious ethics resources (Alpert 2008; Litton 2007; Parens and Johnston 2007, S61). Ultimately, much of the argument against nanoethical analysis can be summed up as follows: any ethical issues relating to nanotechnology – assuming one can adequately confine and define them – will not be sufficiently unique and/or realistic so as to be worthy of the associated resource expenditure. This is so particularly considering the presence of risk analysis and related fields of bioethics and other genres of applied ethics.

These complaints notwithstanding, there are many sound reasons to think that there is good value to be found in ethical analysis of nanotechnology. One is the fact that many emerging technologies, (or will have), a 'nano' component. As former US

Undersecretary of Commerce Philip Bond said in the 2003 National Nanotechnology Initiative Workshop, "Nanotechnology is coming and it won't be stopped" (Bond 2003, 17). Thus, the putative problem some detractors cite about nanotechnology not being field specific actually *supports* the need for *greater* ethical analysis. So long as the presence of nanoscale components pervade several areas of science and engineering, the need to engage in meaningful ethical dialogue will only increase rather than decrease. Moreover, given that different fields, including neuroscience, already raise their own ethical issues, which may in turn be compounded with the dispersion of nanoscale science and engineering (NSE), the intersections and convergences between these several fields will almost undoubtedly compound and complicate their ethics even further (Roco 2003; Roco and Bainbridge 2003; van de Poel 2008).

Additionally, there is good reason to believe that risk analysis, bioethics, and applied ethics are actually ill equipped to adequately deal with these issues. According to some, traditional risk analysis is not sufficient to deal with understanding risk in nanotechnology in particular (Hansson 2004), especially since nanotechnology already poses novel environmental, health, and safety risks (Lin and Allhoff 2007, 9). Moreover, risk analysis generally misses some kinds of risks, such as various social risks, that are nonetheless real; and even if they were covered by risk analysis, studies show that risk is not all that matters to citizens, ethically speaking, so the creation and implementation of any ethical public policy must be more broadly informed (Pidgeon and Rogers-Hayden 2007).

Similar problems are present with respect to bioethics and applied ethics, as we know them, as Mark Meaney summarizes well:

> [W]ith the advent of the development of nanotechnologies, it is not at all clear that the standard set of principles and rules in bioethics will help in a complete ethical analysis and evaluation of nanotechnologies. Although traditional approaches to bioethics may prove appropriate to some aspects of the subject matter, developments in nanotechnology are of such import with such extensive ethical and social implications that we will have to develop alternative approaches to facilitate the study of the conditions for the responsible development of nanotechnologies (Meaney 2006, 687).

This position is seconded by Philip Bond: "The technologies under development today – especially the converging technologies of nanotechnology, biotechnology, information technology, and cognitive sciences – are so powerful and revolutionary, their applications are likely to create ethical societal challenges beyond our current framework" (Bond 2003, 19). While there are some who try to translate nanoethical issues into the frameworks of bioethics and risk analysis, whether these efforts will suffice remains to be seen.[1]

The third reason that we ought to engage in ethical analysis of nanotechnology is that the dispersion of NSE has the potential to reshape, refine, or make more urgent many extant ethical concerns. Even nanoethics opponent Armin Grunwald admits

---

[1] In addition, even if the ethical issues of nanotechnology could be successfully translated into bioethics and risk analysis frameworks, we would have the metaethical question of whether or not those frameworks are in themselves sufficient.

that "Nanotechnological innovations can accelerate or facilitate the realization of certain technical possibilities" (Grunwald 2005, 194), thereby making ethical analysis of their development and consequences more pressing. For instance, the very scale of NSE may pose increased or novel risks to human health. For example, the potential for nanoparticles to cross the blood-brain barrier brings the possibility of not only beneficial applications but also inadvertent exposure of the brain to toxic chemicals. Indeed, nanoparticles have been shown to travel to the brain through other pathways as well, including along the olfactory nerve. Likewise, even though there is already ethical discussion of the issue of transhumanism without the incorporation of NSE, NSE may potentially make more feasible the types of neurological enhancements and human-machine interfaces that transhumanists hope will take humanity to the next level. Thus, even if the ethical issues that arise in nanotechnology may not be unique to one field, "if nanotech is as revolutionary as many expect then it makes these questions much more important and urgent than in other areas" (Parr 2005, 395; Robert 2008).

Moreover, there are strong pragmatic reasons to engage in ethical analysis of nanoscale science and engineering. Again, we appeal to Philip Bond, who concludes that addressing social and ethical issues "is the necessary thing to do because it is essential for speeding technology adoption, broadening the economic and societal benefits, and accelerating and increasing our return on investment" (Bond 2003, 21). Despite the concern that ethical analysis focuses too much on hypothetical situations, nanoethics proponents maintain that such analysis is necessary for pragmatic reasons. One reason is that public perception both strongly affects the course of technology development and is, in turn, influenced by not only current scientific realities but also potential technological futures. Ethical analysis must thus grapple with those visions – however hypothetical – to fully assess and inform public dialogue and deliberation (Kaiser 2006, 669). Moreover, without assessing and dismissing the extreme hypotheticals, we will not know which issues really warrant resource expenditure (Bond 2003). Thus, "[while] nanoethics lacks any metaphysical autonomy (from other areas of applied ethics)...the field can receive a pragmatic justification" (Allhoff 2008, 4).

Finally, we offer a humanitarian consideration: "Good science is ethical science" (Robert 2008, 234; Resnik 1998). Therefore, if – ultimately – we want good science, then ethics must be part of the process. Nanotechnology scholar Rosalyn Berne underscores this position in her work *Nanotalk*: "My assertion is that ethics particular to nanotechnology is needed in order to guide nanotechnology development towards humanitarian aims" (Berne 2006, 75). As with many other fields with an ethics component, there is the danger of nanoethics being co-opted, through ethicists for hire (Johnson 2007), so "[a]ddressing societal and ethical issues is the right thing to do and the necessary thing to do. It is the right thing to do because as ethically responsible leaders we must ensure that technology advances human well-being and does not detract from it" (Bond 2003, 21).

Accordingly, despite the claims that NSE lacks the requisite uniqueness to warrant associated resource expenditure in ethics, there are good reasons to believe ethical reflection with respect to nanotechnology, whether we opt to call it 'nanoethics' or not, is imperative. Given the wide range of fields that will likely be affected by NSE and their diverse range of resulting ethical issues, and given that ethical analysis is necessary for NSE to develop pragmatically and in humanitarian fashion, ethical reflection on NSE is indeed both the right course of action and a good and necessary course of action, too.

## 1.2 Why Nanotechnology and the Brain?

Although there may be some ethical, societal, or policy issues that are best explored with regard to nanoscale science and engineering as a whole, this volume of *The Yearbook of Nanotechnology in Society* has a narrower focus: the application of NSE to neuroscience and the human brain. There are many reasons for this particular choice. First, NSE has the potential to advance neuroscience in brain imaging, diagnosis, drug development, drug delivery, neurosurgery, and neural prosthetic design in ways that have only recently come to be recognized.[2] Second, the application of NSE to the human brain – potentially leading to treatments for debilitating diseases or to cognitive enhancement – has a high probability of important, long-term moral, ethical, political, and societal implications that call for substantive social science and humanities research. Third, the relatively early stage of NSE application to neuroscience enables the development of real-time technology assessment (Guston and Sarewitz 2002; Guston 2008) capabilities in parallel with the emergence of new research directions, which is key to any adequate anticipatory deliberation and governance enterprise.

Consider this example: the NSE-enabled development and refinement of implantable neural prosthetic devices.[3] It as long been assumed that NSE should yield solutions to a fundamental technological challenge in the design of neural prosthetics devices, namely the development of tiny, flexible, reliable, chronic, multi-electrode recording and signaling methods for the cerebral neocortex. Such solutions may depend on miniaturization strategies, localization strategies, or strategies for harnessing scale-dependent properties of nanomaterials as coatings for implantable devices or their components. All of these solutions raise a wide range of ethical, societal, and policy issues, from considerations about demonstrating the safety and efficacy of these devices in preclinical and clinical studies to determining the perspectives of intended consumers.[4] Including, also, the allocation of scarce research dollars to such high-tech interventions with limited clinical target

[2] For an overview of the state of the science, see, e.g., the *Journal of Nanoneuroscience* (http://www.aspbs.com/jns.htm) and a 2010 report on nanobiomedicine that focuses particularly on nanoneurotechnology (http://pharmabiotech.ch/reports/nanobiotechnology/)

[3] This paragraph is drawn, with slight modification, from Robert (2008).

[4] Especially people with disabilities – and especially given the controversy within deaf communities about an early neural prosthetic, the cochlear implant.

audience to worries about potential misuses of neural implants for surveillance or even 'substituted decisional authority' (behavioral control or the influencing of decisions). While these issues are by no means unique to nanoscale science and engineering, as we noted above, they are nonetheless worthy of transdisciplinary analysis. Because the NSE-neuroscience research connection is still in its infancy – or perhaps even earlier in development – it is especially well suited to the task of anticipatory research.

## 1.3 Why Focus on the Future via Anticipatory Research?[5]

The most fundamental reason to undertake anticipatory, prospective research in technology assessment is to demonstrate that affecting future outcomes in science and technology is not only possible, but socially and morally desirable. The future does not just happen to us; it is made to happen. The processes by which the future is made to happen are central and proper focal points for ethicists, scientists, and anyone else who is interested in the well being of future societies. The point of anticipatory technology assessment is therefore not to let imaginations run wild about what the future will be like. Leave that work to pundits, bloggers, and transhumanists. Our task, instead, is to undertake upstream and midstream assessment of NSE technological trajectories, grounded in collaborative engagement with scientists and engineers, bibliometric and workforce analysis, data about public perceptions, outputs from public deliberation, and whatever other sources of serious social science and humanistic evidence and analysis can and should be brought to bear.

The point of engaging the future in an anticipatory mode is to affect the world we will ourselves inhabit and/or that we will bestow upon future generations. Anticipatory technology assessment thus has a practical ambition: to influence the developments in science and technology that will help to produce tomorrow's tomorrows. Given the time lag between discovery and application, there is both time and opportunity to work upstream and to guide the development of particular applications over others and along particular pathways over others (Wilsdon and Willis 2004, 47).

Not just any impact on science and technology research and development will do; the point of anticipatory technology assessment is to have a positive formative impact, based on relevant values, and negotiated visions of the good that may be integrated upstream into research and development activities. What, precisely, these values and visions are (or should be) is often an open question, and so it is critical to discover, describe, evaluate, and deliberate about key individual and societal values that will or should guide scientific research and technological development. Because of the relative novelty of NSE research, (despite a long history), there is a distinct opportunity at present to begin to embed societal and ethical considerations

[5] This section is largely drawn, with some modification, from Robert et al. (under review).

into the design parameters at an early stage of research and development in nanotechnology. In short, the convergence of nano- and neurotechnologies has only just begun. Given societies' deep interests in the outcomes of this convergence, our goal is to bring additional voices and insights to bear on the shaping and governance of that convergence than would otherwise likely be present within the neuroscientific and nanotechnological science, engineering, business, and policy communities.

## 1.4 End-to-End, Real-Time Technology Assessment

In May 2007, under the co-direction of Jason Scott Robert and Clark A. Miller, the Center for Nanotechnology in Society at Arizona State University (CNS-ASU) launched an important new initiative to pilot-test a novel research approach to anticipatory governance. The initiative created what has come to be known within the project as the first "end-to-end" (E2E), real-time technology assessment (RTTA) of nanotechnology and its implications for neuroscience and the future of the human brain. What does that mean?

Real-time technology assessment (Guston and Sarewitz 2002) is an approach to the assessment of new and emerging technologies that presupposes the possibility of – and seeks to create the capacity for – conducting technology assessment in an ongoing, systematic fashion "upstream" in the innovation process (Wilsdon and Willis 2004). The goal of RTTA is to provide assessment of the social and ethical dimensions of new and emerging technologies as these technologies emerge, rather than after, and to provide feedback from the assessment into the innovation process to help "steer" the pathways taken by socio-technological developments.

As implemented by CNS-ASU, RTTA involves four key elements:

1. *Research and innovation systems mapping* – the design of tools and instruments that enable assessors to track and map developments in scientific research and technological innovation and, therefore, to understand and assess what is happening at the cutting edge of scientific activity, to highlight potential opportunities for intervention, and to facilitate anticipation of future research and innovation trajectories. At CNS-ASU, this work has been carried out with a team of researchers at Georgia Institute of Technology.
2. *Public value mapping* – the design of tools and instruments that enable assessors to track and map emerging public opinion and values that are forming around new technologies, and therefore to understand and bring to bear in their evaluations of new technologies emerging public sentiments and to identify potential areas of emerging public concern. At CNS-ASU, this work has been carried out primarily by researchers at the University of Wisconsin-Madison, with collaboration from colleagues at Arizona State University.
3. *Public dialogue and deliberation about the future* – the design of tools and instruments that enable assessors to effectively engage diverse publics in dialogue and deliberation about potential future technologies; the social, political, and economic formations that may get created in, around, and through those technologies; and

the potential futures that citizens would like to see emerge. At CNS-ASU, this work has been carried out with researchers at Arizona State University and North Carolina State University, and with collaboration from a national network of researchers at other universities.

4. *Reflexive engagement and integration* – the design of tools and instruments that enable assessors to engage effectively with scientists, engineers, and others engaged in the innovation process and, therefore, to provide feedback and successfully integrate assessment insights into scientific research and technological innovation. At CNS-ASU, this research has been carried out at Arizona State University.

The vision of the initiative imagined and implemented in this project was to draw on the resources of the entire CNS-ASU team, including all four of the major research thrusts described above, to pilot test an "end-to-end" (E2E) RTTA of the application of nanotechnology to neuroscience and the human brain. The effort began by establishing a collaborative group of researchers across CNS-ASU to plan a series of crosscutting studies that would draw on the capacities of each of the four research thrusts. Over the subsequent 3 years, those studies have resulted in a wealth of new quantitative and qualitative data and significant new insights into what nanotechnology may mean for the human central nervous system and, as a consequence, human society in the twenty-first century.

## 1.5 Overview of the Volume

This volume is divided into four parts. Part I provides an introduction to and overview of real-time technology assessment as a method for studying the application of nanotechnology to neuroscience and the brain. Chapter 2, by Christina Nulle, Clark A. Miller, Alan Porter, and Harmeet Singh Gandhi, opens this introduction with an overview of the current state of nanoscale science and engineering research applied to the field of neuroscience. Delving into a database of one-million-plus scientific articles in the field of nanotechnology, Nulle, and colleagues used research and innovation systems analysis and mapping techniques to identify and classify over 10,000 articles focused on the brain and central nervous system. They found that annual publication rates were growing from approximately 200 articles per year in the early 1990s to 1,500 articles per year by the mid-2000s. Roughly, 40% of these articles were published by researchers from the US, 40% from Europe, 10% from Japan, and 5% from China. Their study shows that the bulk of this research involves the application of NSE tools and methods to the study of the brain. Roughly, 10% of the research represents the development of biosensors for the brain, and a small but growing fraction of research is focused on nanoscale structures in the brain and the design of nanoscale interfaces between engineered systems and neurobiological systems. Articles in the database identified close to 20 distinct diseases as targets for NSE research, with Alzheimer's, brain cancer, and Parkinson's as the top three. Overall, Nulle and colleagues identified the five most important domains of nano-neuro research as: (1) visualizing nerve and brain structure and dynamics at the nanoscale; (2) visualizing nerve growth and regeneration; (3) developing scaffolding for nerve regeneration experiments; (4) visualizing and structuring neural prosthetic interfaces; and (5) improved cancer detection and identification.

Chapter 3, by Sean A. Hays, Clark A. Miller, and Michael Cobb, presents a second major source of data for real-time technology assessment, the mapping of public opinion, attitudes, and values regarding new and emerging applications of nanotechnology to the human brain and cognition. The study reports on a nationally representative public opinion survey regarding nanotechnology-enabled cognitive enhancement, conducted in 2008, as well as a second, follow-up survey, conducted in 2010. Consistent with previous surveys regarding nanotechnology and other emerging technologies, respondents expressed little knowledge of these technologies but nonetheless offered strong opinions regarding their use. Overall, respondents broadly opposed the use of technology to enhance human cognition and abilities. Technologies designed for therapeutic application scored much better, however, even when they contributed to ancillary non-medical enhancements of patient abilities. Opposition also varied by gender, with women expressing significantly greater opposition than men to most enhancement technologies. Other interesting findings included that, generally, respondents expect human enhancement technologies to be expensive and available only to the wealthy, want government to secure equal access to these technologies, but would not want insurance to pay for them.

Chapter 4, by Doo-Hun Choi, Anthony Dudo, and Dietram Scheufele presents data from a survey of media coverage of nanotechnology and its application to neuroscience and the brain. As with the survey data presented in the prior chapter, Chap. 4 seeks to understand the ways in which emerging technologies are framed in public dialogue and deliberation. For emerging technologies like nanotechnology, about which the public has little knowledge, media reporting can play a critical role in framing the technology for publics. To understand how media contribute to opinion formation about neuroscience nanotechnology, this study tracks the evolution of US newspaper coverage of this issue over time and presents a content analysis to examine the amount of coverage, authorship patterns, and thematic emphases in journalistic accounts of this specific application of nanotechnology. The findings suggest that US newspaper coverage of nanotechnology applications in neuroscience has grown since its first appearance in the early 1980s, accounting during the 2000s for roughly one-third of all articles on nanotechnology and health and 20% of all nanotechnology articles. At the same time, patterns of authorship show a steep decline in authors who write more than one article on the topic. This is consistent with the event-driven character of the coverage, as non-specialist journalists respond to releases of new research results or other events in their local coverage area. Finally, the findings indicate a general shift over time from a focus on benefits to equal coverage of the benefits and risks of these technologies to society.

Chapters 5 and 6 shift from the data intensive elements of real-time technology assessment toward broader discussions of the normative considerations that surround nanotechnology and its application to the brain. Chapter 5, by Valerye

Milleson, presents a review of the ethical literatures in nanoethics and neuroethics designed to begin to explore the ethical challenges that may arise at the intersection of nanotechnology and neuroscience. Using a combination of approaches, the study generated a systematic delineation of key ethical issues as they apply to nanotechnology alone and nanotechnology in occurrence with neural applications, identifying three major areas of concern: risk, enhancement, and human-machine interaction. While these three areas are well known from both nanoethics and neuroethics, the study concludes that the convergence of nano- and neurotechnologies creates novel or heightened ethical concerns across all three domains that deserve detailed inquiries but so far have received little to no attention from scholars in either ethical field. The results also highlight – as has also been noted for the field of nanoethics more broadly – the overly narrow formulation of ethical analysis on individual risk that dominates current research, to the neglect of broader concerns about social risk and social justice.

Chapter 6, by Richard Milford and Jameson M. Wetmore, presents the results of a novel approach to upstream public engagement designed to illuminate a broad and deep sense of how new nanotechnology applications intersect with the values, goals, concerns, wishes, and identities of individuals and communities. While public engagement has become a focal point of efforts to bring greater democracy to the governance of new and emerging technologies, too often an overly broad definition of "public" smears out the very differences among individuals and communities out of which real meanings of new technologies emerge. This study presents the result of an alternative approach that seeks specifically to highlight strong identities and belief systems as the focus for selection and dialogue, in this case through a focused dialogue on religious perspectives on nanotechnology and its application to the brain. The results demonstrate the importance of this kind of tailored approach for getting beyond the least common denominator kinds of meanings, (e.g., individual risks and benefits, as noted in Chap. 5), often attached to new technologies. By taking a more tailored approach, the study opened up new avenues for understanding how individuals and communities make sense of changing technological capacities, fostered nuanced discussions that engaged complex ethical perspectives, elicited the expression of deep personal beliefs, and engaged human capacities for creativity and imagination in grappling with what new and emerging technologies may mean for societies' futures.

Following this introductory part, subsequent parts explore a diverse range of RTTA-based analyses focused on three distinct domains of nanotechnology application and impact. Part II focuses on the use of nanotechnology for brain repair and the design of brain-machine interfaces. Part III focuses on the use of nanotechnology for cognitive enhancement. Finally, Part IV focuses on the toxicity of nanoparticles to the human brain and efforts to govern nanoparticle risks in an anticipatory fashion.

Beginning Part II, on brain-machine interfaces, Adam Keiper provides a detailed overview of both the twenty-first century's "age of neuroelectronics" and its historical context in Chap. 7. Situating today's expanding research into brain-machine interfaces against earlier backdrops, Keiper describes the multiple historical pathways through which scientists have inquired into the human mind and the foundations they have laid for the halting, if nonetheless sometimes remarkable advances of the last few decades. Keiper highlights both the tremendous potential benefits to patients of ongoing advances in brain-machine interface research, as well as the severe limits of current knowledge. Today's capabilities are, Keiper observes, far from the fancies of science fiction or visionaries and unlikely to lead anytime soon to transhuman transcendence. Yet, the chapter also warns, research in the next few decades is likely to lead to advances that raise very hard questions for scientists, patients, and society. It will raise questions about how much we are able to accomplish by looking at the brain as a complex but ultimately electro-mechanical system, and how much we may lose of our humanity if we come to see the brain only as a machine.

Chapter 8, by Derrick Anderson, continues the trend of situating research on brain repair and brain-machine interfaces within a larger historical context. Exploring the specific case of cochlear implants (the first brain-machine implant in widespread use), the chapter uses historical comparison as a tool for anticipatory thinking about future implants and the social and ethical challenges to which they may give rise. In particular, Anderson challenges the classical model of innovation and diffusion of new technologies, in which scientific discoveries lead to technological applications that are adopted by individuals over time as they diffuse through various populations. By contrast, Anderson observes for the case of cochlear implants, a range of additional factors are critical to how a new technology is received. These include regulatory frameworks, distributions of wealth (and health insurance), and cultural factors that contribute to highly divergent framings of technology and its benefits and costs.

Chapter 9, by Arielle Silverman, extends the study of brain repair and brain-machine interfaces to the study of blindness and emerging technologies for "curing" what is, in the United States, the third most feared disease, behind cancer and AIDS. Building on a description of efforts to cure blindness, many of which are often portrayed by researchers and activists as not only medically but socially imperative to return the damaged individual to normal functioning, the chapter explores the perceptions of blind persons of both blindness and its potential technologies of repair. Presenting the results of a web survey of 281 self-identifying blind individuals, the chapter observes that both blindness and its potential cures are the subject of divergent opinions among survey respondents. Major determinants of these divergences include degree of blindness, age of blindness onset, skill with alternative approaches, such as Braille, level of potential repair, kind of technological intervention, and attendant risks of repair. At one end of a spectrum, individuals who are and have always been completely blind tended to most embrace blindness as an identity rather than a debilitation and therefore to be most skeptical of repair technologies. At the other end, individuals who have become partially blind during or after childhood tended to be supportive of repair technologies. In another example of divergence, respondents tended to give "biologically based" therapies (with real risks and limitations) greater support than electro-mechanical therapies (like artificial sensors, also with real risks and limitations). Perhaps most significant, given current and likely medium-term capabilities, respondents were typically only interested in visual prosthetics or therapies if they could create enough visual perception to enable reading.

Exploring still another important dimension of the ethical terrain surrounding brain-machine interfaces, Chap. 10, by Stephanie Naufel, examines the potential implications of these technologies for sensibilities of human personhood. Observing that personhood is as much or more a legal and moral construct as it is a reflection of any underlying reality, Naufel inquires into how divergent characteristics of brain-machine technologies may influence how individuals see themselves and are seen by others. Especially, given the centrality of the physical brain to the perception of mind, (and the centrality of mind to perceptions of identity, responsibility, blame, and personality), the chapter analyzes how technological characteristics, such as the permanence of implants, their invasiveness, and their impact on personality, may shape the degree to which brain-machine interfaces drive a reconsideration of human values and identities. The chapter applies its analysis to a range of technologies, including cochlear implants, deep brain stimulation, and BrainGate interfaces, to provide a context for thinking about future nanotechnology-enabled neurological interfaces.

Part II concludes with Chap. 11, by Francois Berger, Sjef Gevers, Ludwig Siep, and Klaus-Michael Weltring, which provides a synthesis and overview of the potential applications of nanotechnology to brain-machine implants and the social, ethical, and legal challenges that these applications present. The chapter assesses these challenges in the context of the ethics of short-term developments in the testing of devices in clinical trials; the risks associated with medium-term developments in the implementation of these devices in medical practice; and the longer-term social, ethical, and legal questions raised by human enhancements. The chapter thus also provides an important bridge to Part III, which examines brain enhancement.

Part III offers a series of studies that illuminate key aspects of the emerging debate over cognitive enhancement of the brain. The part begins with Chap. 12, by Richard Loosemore, which offers a provocative look into the prospect that the brain and, especially, human cognition need to be understood as at least partially complex systems. Complex systems, Loosemore offers, are those in which we cannot mathematically relate the behavior of the individual elements of the system, (in this case, neurons), to the higher order behavior of the system as a whole, (in this case, cognition and psychological behavior). Loosemore's conclusion that cognitive systems risk being at least partially complex implies, most importantly, that current paradigms for studying cognition within the scientific community may simply produce a series of workable but ultimately poor models of how the brain works. Instead, Loosemore suggests, scientists should be looking for complex models of the brain and cognition that transcend locally optimal solutions and instead seek to model overall system behavior.

The other side of Loosemore's analysis is that while it may be possible to optimize models of very narrow applicability to cognitive function, it may not be possible to do so for generalized cognition. This point is made in Chap. 13, by Sean A. Hays, in discussing the problem of cognitive enhancement. While many people talk about cognitive enhancement as if referring to the generalized enhancement of human intelligence, Hays argues that the more likely result is a range of enhancement technologies that focus on optimizing or enhancing specific facets of cognition.

Thus, for example, the impact of Ritalin on cognition is not one of generalized increase in intelligence but of specific enhancement of the ability to concentrate on a task. Engaging closely with the work of Stephen Jay Gould and John Carson, as well as the work of contemporary neuroscience, Hays demonstrates the ways in which a theory of unitary, generalized inheritance have resonated in the competitive political culture of the United States, to the neglect of more distributed models of multiple intelligences. Hays goes on to argue that this framing of intelligence has had profound implications for both proponents and opponents of human cognitive enhancement. He also argues that an alternative framing that focuses instead on the theory of distributed and multiple intelligences would radically transform the debate. It would eliminate both its over-exuberant, exaggerated claims, as well as the deep concerns of many of its opponents.

Chapter 14, by Henry Greeley, John Harris, Ronald C. Kessler, Michael Gazzaniga, Philip Campbell, and Martha J. Farah, shifts the discussion of human enhancement in a different direction and focuses on the need for rules to govern the use of enhancement technologies. The authors observe that cognitive enhancement drugs are already in use in society, especially on university campuses; they argue that these drugs are beneficial; and they insist that rules are needed to balance these benefits against potential risks in order to set the stage for the responsible and widespread adoption of cognitive enhancement technologies in society. In the process, the authors make a series of highly provocative claims. They claim that enhancement drugs should be seen as being in the same category as education, tutors, good nutrition, and a laptop computer connected to the Internet. They also claim that the difference between therapy and enhancement is morally and ethically meaningless. Finally, they claim that society should adopt a "presumption that mentally competent adults should be able to engage in cognitive enhancement using drugs" (this volume, p. 240). While not everyone will agree with these positions – indeed, the survey results presented by Hays, Miller, and Cobb in Chap. 3 suggest that most Americans disagree strongly with them – the importance of grappling seriously with both the technologies of cognitive enhancement and the policy proposals put forward in this chapter is undeniable. Humanity stands at the cusp of innovative new technologies that may radical alter the human brain and its capabilities, and it is critical that societies engender deep public discussions of the social and ethical consequences of potential changes.

Paul Thompson offers an alternative approach to considering the ethics of human enhancement in Chap. 15. Thompson observes that ethical arguments about the use of nanotechnology for human enhancement are similar to arguments about the ethics of disenhancing animals using technology to prevent animal discomfort or disease in agricultural settings. The problem, however, Thompson notes, is that long debates about the ethics of deliberately creating "blind chickens" have not resolved seemingly inherent tensions between our intuitive reactions to disenhancement – "That's wrong!" – and an array of ethical arguments that can be made in support of such technologies. Worse, Thompson posits, ethical jargons that have emerged to explain what is wrong with disenhancement technologies in fact do little to explain why it is wrong. Most importantly, Thompson concludes, ethical debates about human enhancement seem likely to encounter similar quagmires between stubborn intuitions and moral concerns. Despite the wishes of Greeley and his colleagues in Chap. 14, right answers to the treatment of human enhancement technologies may not emerge from rational conversations among medical, ethical, and educational professionals. It may be the case that broader publics must engage with difficult questions about the kinds of societies they wish for themselves and their progeny.

Chapter 16 presents the results of one such experiment in public engagement about human enhancement. The Center for Nanotechnology in Society at Arizona State University conducted in 2008 a National Citizens Technology Forum (NCTF) to facilitate citizen dialogue about the use of nanotechnologies for human enhancement. The NCTF exercise involved citizens from six cities working independently and together over the course of a month to explore the convergence of nanotechnology, biotechnology, information technology, and cognitive science and its social and ethical implications. Chapter 17 presents both an overview of the NCTF effort and a synthesis of its conclusions, as well as the individual citizens' reports drafted at each site.

The final part of the book continues an emphasis of public engagement with new technologies, but focusing on the potential risks posed by nanotechnology to the brain and central nervous system. Chapter 18, by Z. Yang Z.W. Liu, R.P. Allaker, P. Reip, J. Oxford, Z. Ahmad, and G. Reng, opens the part with a recent review of the both the functionality and toxicity of nanoparticles in relation to the central nervous system. Driving many potential biomedical applications of nanoparticles are their unique abilities to interact with cellular structures and processes, as well as larger biophysical domains. Yet, these same abilities raise potentially troublesome concerns regarding exposure to nanoparticles in the environment, especially given the growing use of nanoparticles in research and manufacturing. Yang et al. highlight risks from exposure to nanoparticles through a range of exposure pathways, from the breaking off from materials implanted in the body, such as hip replacements, to inhalation of airborne particulates. In particular, the authors review concerns about the risks posed to human neural cells, which are particularly vulnerable cells. Although brain and other neural cells are typically protected from chemical exposures, nanoparticles have the potential to easily transit biological membranes, such as the blood-brain barrier. Studies have shown that nanoparticles can be transported from the environment into the brain and can cause neural damage there.

Chapters 19, 20 and 21 subsequently examine a case study of anticipatory governance of nanotechnology that took place in the City of Cambridge, MA, in 2008. Beginning in the summer of 2007, the City Manager of Cambridge convened a novel Cambridge Nanomaterials Advisory Committee to provide advice to the city regarding the regulation of nanotechnology in the city. As one of the few cities to directly regulate biotechnology research within its boundaries, Cambridge has a legacy of programs designed to provide regulation and oversight for new and emerging technologies. Working with the Cambridge Public Health Department, the Cambridge Nanomaterials Advisory Committee produced a report to the City Manager recommending the development of a series of policy initiatives regarding nanotechnology. These recommendations include developing a database of the use of nanomaterials in the city, assistance to companies and others handling nanomaterials to better understand nanoparticle risks, development of public information about nanoparticle risks, tracking of research and regulation in other jurisdictions, and annual reporting to the City Council to provide ongoing updates and insights to Council members. The report does not recommend new regulation. This report is included here as Chap. 19.

As part of generating the report, the Cambridge Public Health Department solicited public input through the development of a citizens' dialogue and forum, carried out with the help of the Boston Museum of Science. This event provided insights to city residents on nanotechnology and its use in consumer products as well as the review process underway by the Cambridge Nanotechnology Advisory Council. Residents then took part in a small group discussion exercise in which they role-played City Council members and were asked to make recommendations regarding city responses to a variety of new nanotechnology-based consumer products. Responses were then collected from each discussion group and provided to the Cambridge Public Health Department. Chapter 20 displays the materials used in this exercise, as well as the responses of the discussion groups.

Finally, in Chap. 21, Shannon Conley, analyzes the work of the Cambridge Public Health Department, Cambridge Nanotechnology Advisory Council, and Boston Museum of Science against the model of anticipatory governance. The chapter asks whether the work of these organizations can be modeled as anticipatory governance and, if so, what anticipatory governance looks like in practice and how it compares to other approaches to regulating science and technology. Drawing on observations of the process and interviews with key participants, Conley concludes that important aspects of the activities of local regulatory agencies function in similar ways to anticipatory governance and can therefore serve as a site for exploring the possibilities and limitations of anticipatory governance mechanisms. At the same time, she argues that anticipatory governance theories can, in turn, help advance the practice of science and technology regulation.

## References

Allhoff, Fritz. 2008. On the autonomy and justification of nanoethics. In *Nanotechnology & society: Current and emerging ethical issues*, ed. Fritz Allhoff and Patrick Lin, 3–38. New York: Springer Science and Business Media.

Alpert, Sheri. 2008. Neuroethics and nanoethics: Do we risk ethical myopia? *Neuroethics* 1: 55–68.

Berne, Rosalyn W. 2006. *Nanotalk: Conversations with scientists and engineers about ethics, meaning, and belief in the development of nanotechnology*. Mahwah: Lawrence Erlbaum Associates.

Bond, Phillip J. 2003. Preparing the path for nanotechnology. In *Nanotechnology: Societal implications – Maximizing benefits for humanity: Report of the national nanotechnology initiative workshop, December 2–3, 2003*, ed. Mihail C. Roco and Williams Sims Bainbridge, 16–21. Arlington: National Science Foundation.

Drexler, K.Eric. 1986. *Engines of creation: The coming era of nanotechnology*. New York: Anchor Books.
Godman, Marion. 2008. But is it unique to nanotechnology? *Science and Engineering Ethics* 14: 391–403.
Grunwald, Armin. 2000. Against over-estimating the role of ethics in technology development. *Science and Engineering Ethics* 6: 181–196.
Grunwald, Armin. 2005. Nanotechnology – A new field of ethical inquiry? *Science and Engineering Ethics* 11: 187–201.
Guston, David H. 2008. Innovation policy: Not just a jumbo shrimp. *Nature* 454: 940–941.
Guston, David H., and Daniel Sarewitz. 2002. Real-time technology assessment. *Technology in Society* 24: 93–109.
Hansson, Sven Ove. 2004. Great uncertainty about small things. *Technè* 8(2): 26–35.
Hodge, Graeme, Diana Bowman, and Karinne Ludlow. 2007. Introduction: Big questions for small technologies. In *New global frontiers in regulation: The age of nanotechnology*, ed. Graeme Hodge, Diana Bowman, and Karinne Ludlow, 3–26. Northampton: Edward Elgar Publishing, Inc.
Holm, Sören. 2005. *Does nanotechnology require a new 'nanoethics'?* London: Cardiff Centre for Ethics, Law & Society.
Johnson, Deborah G. 2007. Ethics and technology 'in the making': An essay on the challenge of nanoethics. *NanoEthics* 1: 21–30.
Kaiser, Mario. 2006. Drawing boundaries of nanoscience – Rationalizing the concerns? *The Journal of Law, Medicine & Ethics* 34: 667–674.
Keiper, Adam. 2007. Nanoethics as a discipline? *The New Atlantis* 16: 55–67.
Lin, Patrick, and Fritz Allhoff. 2007. Nanoscience and nanoethics: Defining the discipline. In *Nanoethics: The ethical and social implications of nanotechnology*, ed. Fritz Allhoff, Patrick Lin, James Moor, and John Weckert, 3–16. Hoboken: Wiley-Interscience.
Litton, Paul. 2007. "Nanoethics"? What's new? *The Hastings Center Report* 37: 22–25.
Meaney, Mark E. 2006. Lessons from the sustainability movement: Toward an integrative decision-making framework for nanotechnology. *The Journal of Law, Medicine & Ethics* 34: 682–688.
Nordmann, Alfred. 2007. If and then: A critique of speculative nanoethics. *NanoEthics* 1: 31–46.
Parens, Erik, and Josaphine Johnston. 2007. Does it make sense to speak of neuroethics? *European Molecular Biology Organization* 8: S61–S64.
Parr, Douglas. 2005. Will nanotechnology make the world a better place? *Trends in Biotechnology* 23(8): 395–398.
Pidgeon, Nick, and Tee Rogers-Hayden. 2007. Opening up nanotechnology dialogue with the public: Risk communication or 'upstream engagement'? *Health, Risk and Society* 9(2): 191–210.
Resnik, David B. 1998. *The ethics of science: An introduction*. New York: Routledge.
Robert, Jason Scott. 2008. Nanoscience, nanoscientists, and controversy. In *Nanotechnology & society: Current and emerging ethical issues*, ed. Fritz Allhoff and Patrick Lin, 225–239. New York: Springer Science and Business Media.
Robert, Jason Scott, Ira Bennett, and Jennifer Brian. Forthcoming. *Just scenarios? Cultivating anticipatory assessment of novel nanotechnologies*.
Roco, Mihail C. 2003. Broader societal issues of nanotechnology. *Journal of Nanoparticle Research* 5: 181–189.
Roco, Mihail C., and W.S. Bainbridge (eds.). 2003. *Converging technologies for improving human performance: Nanotechnology, biotechnology, information technology and cognitive science*. Dordrecht/Boston: Kluwer Academic Publishers.
van de Poel, Ibo. 2008. How should we do nanoethics? A network approach for discerning ethical issues in nanotechnology. *NanoEthics* 2: 25–38.
Wilsdon, J., and R. Willis. 2004. *See-through science: Why public engagement needs to move upstream*. London: Demos.

# Part I
# Introduction to RTTA

# Chapter 2
# Applications of Nanotechnology to the Brain and Central Nervous System

**Christina Nulle, Clark A. Miller, Alan Porter, and Harmeet Singh Gandhi**

> [Leo Zonneveld]: '*US and European governments are introducing programmes for the convergence of nanobio-and cognitive technologies all aiming to improve human performance (Roco and Bainbridge 2003). How do you feel about these new trends, which connect studies in neuroscience to technological aims?*'
> "Professor the Lord Winston: 'Well, I don't think we should necessarily trouble ourselves with nanotechnology in this regard. Its possible impact is probably not yet relevant to mainstream neuroscience and therefore difficult to judge.'" – *Reshaping the Human Condition*

## 2.1 Introduction

In *Reshaping the Human Condition: Exploring Human Enhancement*, a recent report by the Rathenau Institute, in collaboration with the British Embassy in the Hague and the UK Parliamentary Office of Science and Technology, Professor the Lord Winston, Professor of Science and Society at Imperial College dismisses any concern about the application of nanotechnology research to the field of neuroscience as too nascent and inconclusive (Zonneveld et al. 2008). No other mention of nanotechnology occurs in the report. This view is common. In informal conversations with a number of leading researchers in the field of neural prosthetics, our colleague found little knowledge of nanotechnology or expectation that it would have any significant impact on the field for the near future. Current neural prosthetics technologies operate at the micrometer scale range, at the smallest, and this was deemed as sufficient for the design of neural implant devices (Robert, personal communication). Imagine

C. Nulle • C.A. Miller (✉)
The Center for Nanotechnology in Society, Arizona State University,
P.O. Box 875603, AZ 85287-5603, Tempe, USA
e-mail: clark.miller@asu.edu

A. Porter • H.S. Gandhi
School of Public Policy, Georgia Institute of Technology, 685 Cherry Street,
GA 30332-0345, Atlanta

S.A. Hays et al. (eds.), *Nanotechnology, the Brain, and the Future*,
Yearbook of Nanotechnology in Society 3, DOI 10.1007/978-94-007-1787-9_2,

our surprise, then, when a search of Web of Science generated over 10,000 research articles at the intersection of nanotechnology and neuroscience.

The theory of anticipatory governance is built on the idea of integrating deliberations about the social and ethical dimensions of science and technology upstream in the innovation process (Barben 2008). By grappling with the possible meanings of new and emerging technologies for society as early as possible in their design and development, this theoretical framework suggests that societies can position themselves both to anticipate potential risks and benefits of new technologies and to help shape their design and application (Wilsdon and Willis 2004). Ultimately, such work aims to help build the necessary capacity throughout society to reflexively and democratically govern the fashioning of future technological societies (Voss et al. 2006; Sclove 1995).

One key capacity required for such work is the ability to detect emerging trends in new and emerging technologies. The Center for Nanotechnology in Society at Arizona State University (CNS-ASU) has built one such technology, in collaboration with researchers at Georgia Institute of Technology, designed to facilitate the detection, mapping, and potentially even forecasting of new and emerging science and technology in the field of nanotechnology. At the core of this technology is an extensive database of over one million research articles from 1990 to 2008 captured by a detailed definition of the emerging field of nanotechnology (Porter et al. 2008). With this tool, researchers are able to develop nuanced and sophisticated analyses of emerging domains of nanotechnology research, helping to facilitate more systematic deliberations of the possible societal and ethical outcomes of nanotechnology research and innovation.

In this chapter, we used the Georgia Tech/CNS-ASU database to identify and explore the main concepts, trends, and trajectories of nanotechnology research applied to neuroscience, the brain, and the central nervous system. In distinct contrast to commentators who have downplayed the importance of such work, we found extensive research being carried out in this rapidly expanding field. Indeed, we found work occurring with potentially transformative implications for the ability of scientists to: (a) understand the human brain; (b) identify, diagnose, and repair brain and nerve disease and damage; (c) provide effective interfaces between the brain and the central nervous system and human technologies; and (d) enhance human cognitive and nerve function. Our findings thus agree with Silva, who conducted a more limited review of the field:

> Applications of nanotechnology to neuroscience are already having significant effects, which will continue in the near future. Short-term progress has benefited in vitro and ex vivo studies of neural cells, often supporting or augmenting standard technologies. These advances contribute to both our basic understanding of cellular neurobiology and neurophysiology, and to our understanding and interpretation of neuropathology. Although the development of nanotechnologies designed to interact with the nervous system in vivo is slow and challenging, they will have significant, direct clinical implications. Nanotechnologies targeted at supporting cellular or pharmacological therapies or facilitating direct physiological effects in vivo will make significant contributions to clinical care and prevention. (Silva 2006)

However, we would also go further. Now is the time to begin a serious dialogue among neuroscientists, nanotechnology researchers, ethicists, social scientists, regulators, policymakers, and the public about this burgeoning field of work and the social and ethical dilemmas that are likely to accompany its growing technological accomplishments.

## 2.2 Overview of Nano-Neuro Research

To analyze research trends in the application of nanotechnology to neuroscience, (which we term nano-neuro research), we began with a database of Science Citation Index (SCI) records for all nanotechnology publications from 1990 to mid-2008 contained in the Web of Science (WOS). Our search based on an algorithm developed at Georgia Institute of Technology in collaboration with CNS-ASU (Porter et al. 2008). We subsequently applied specific search terms to extract from this database those articles that focused on the brain or central nervous system. To do this, we used the following search terms:

brain, neur*, fmri, "functional mri", synap*, myelin, axon*, serotonin, spinal, paralysis, paralyzed, prosthe*, nerv*, plexus, gangli*, olfact*, cortex, cornea*, cerebell*, cerebr*, parietal, broca, parkinson*, alzheimer*, blind*, deaf*, parapleg*, quadripleg*, glial*, glioblastoma, retin*, epilep*, aneurysm, stroke, amnesia, middle ear, visual, vision, cortic*

This search yielded 10,763 records of journal article/presentation abstracts, which form the basis for the following analysis. This database was cleaned and analyzed using VantagePoint text-mining software.[1]

We begin this chapter with an overview of the nano-neuro material. We observe in Fig. 2.1 that the number of nanotechnology publications identified by our nano-neuro search terms has grown substantially in absolute annual publication numbers

**Fig. 2.1** Publication trend of nano-neuro articles (2008 records are adjusted to reflect end-of-year projections)

[1] http://www.thevantagepoint.com/

**Table 2.1** Top publication countries

| Rank | Countries | Records | Percent (%) |
|---|---|---|---|
| 1 | US | 4,570 | 42.46 |
| 2 | Germany | 1,234 | 11.47 |
| 3 | Japan | 985 | 9.15 |
| 4 | UK | 823 | 7.65 |
| 5 | France | 586 | 5.44 |
| 6 | China | 496 | 4.61 |
| 7 | Italy | 483 | 4.49 |
| 8 | Canada | 450 | 4.18 |
| 9 | Spain | 261 | 2.42 |
| 10 | Switzerland | 259 | 2.41 |
| 11 | Sweden | 253 | 2.35 |
| 12 | Australia | 215 | 2.00 |
| 13 | S. Korea | 198 | 1.84 |
| 14 | Netherlands | 188 | 1.75 |
| 15 | India | 180 | 1.67 |
| 16 | Taiwan | 157 | 1.46 |
| 17 | Israel | 153 | 1.42 |
| 18 | Brazil | 148 | 1.38 |
| 19 | Russia | 146 | 1.36 |
| 20 | Belgium | 143 | 1.33 |
| 21 | Austria | 122 | 1.13 |
| 22 | Poland | 119 | 1.11 |
| 23 | Denmark | 108 | 1.00 |
| 24 | Singapore | 76 | 0.71 |
| 25 | Turkey | 66 | 0.61 |
| | All Europe | 4,660 | 43.30 |

from 1990 through 2007 (full data for 2008 were not yet available). Overall, from 1991 to 2007, the field grew almost eightfold, from approximately 200 publications per year in 1991–1992 to just under 1,600 publications per year in 2007. Growth was steady but relatively slow during the first decade however, up until 2001, after which the number of publications per year began to expand much more rapidly. Indeed, the number of publications per year triples between 2001 and 2007. Additionally, as a percentage of all SCI publications in the overall nanotechnology database, the share of articles published in the field of nano-neuro research also increased significantly between 1990 and 2007.

Following these overall trends, Table 2.1 displays the top 25 countries by number of published nano-neuro articles. Note that this table involves multiple counts for some articles, as researchers from multiple countries may be included as authors on a single article. In cases such as these, each article has been counted once for each country involved. Among the top 25 countries, European countries account for 4,660 articles, or almost exactly the same number as the United States. Nonetheless, while the total number of articles published outside of the United States and Europe is relatively small, it is significant that over 100 articles have been published in nano-neuro research by authors in 23 separate countries including China, South

**Table 2.2** Top 25 Universities pursuing nano-neuro research

| Rank | University | Number of records |
|---|---|---|
| 1 | Harvard University | 177 |
| 2 | University of Illinois | 104 |
| 3 | University of Texas | 104 |
| 4 | Johns Hopkins University | 98 |
| 5 | University of Wisconsin | 97 |
| 6 | University of Tokyo | 90 |
| 7 | University of Michigan | 89 |
| 8 | UC San Diego | 88 |
| 9 | UC Los Angeles | 83 |
| 10 | UC San Francisco | 82 |
| 11 | Stanford University | 79 |
| 12 | University of Southern California | 76 |
| 13 | Chinese Academy of Science | 75 |
| 14 | Northwestern University | 73 |
| 15 | University of Pennsylvania | 72 |
| 16 | University of Toronto | 72 |
| 17 | CNRS | 71 |
| 18 | MIT | 71 |
| 19 | University of Washington | 70 |
| 20 | University of California, Berkeley | 68 |
| 21 | Duke University | 67 |
| 22 | University of Frankfurt | 61 |
| 23 | Cornell University | 60 |
| 24 | University of Cambridge | 59 |
| 25 | University of Munich | 58 |

Korea, India, Taiwan, Israel, Brazil, Russia, and Poland, and authors in Turkey have published an additional 66 articles. Nano-neuro research is thus expanding globally and is increasingly widely distributed.

The global distribution of nano-neuro research is also apparent among both the top universities and top researchers in the field. Table 2.2 displays the top 25 institutional affiliations of the publications captured in the database. While 18 are in the United States, including the top five, two are in Germany, and one each are in Japan, China, Canada, France, and England.

It is also interesting to examine the distribution of nano-neuro research across scientific fields. Here, too, we observe a very broad distribution of research across a wide range of basic scientific fields, ranging from the neurosciences (17%), to biochemistry (13.5%), chemistry (11.5%), and engineering and materials (10%). Clinical applications include pharmacology (7%), neurology (5.5%), ophthamology (5%), surgery (2.5%), radiology (1.75%), and pathology (1.67%). We present subject category data in Fig. 2.2 as an overlay map of subject categories in nano-neuro research, in comparison to the full range of subject categories found in World of

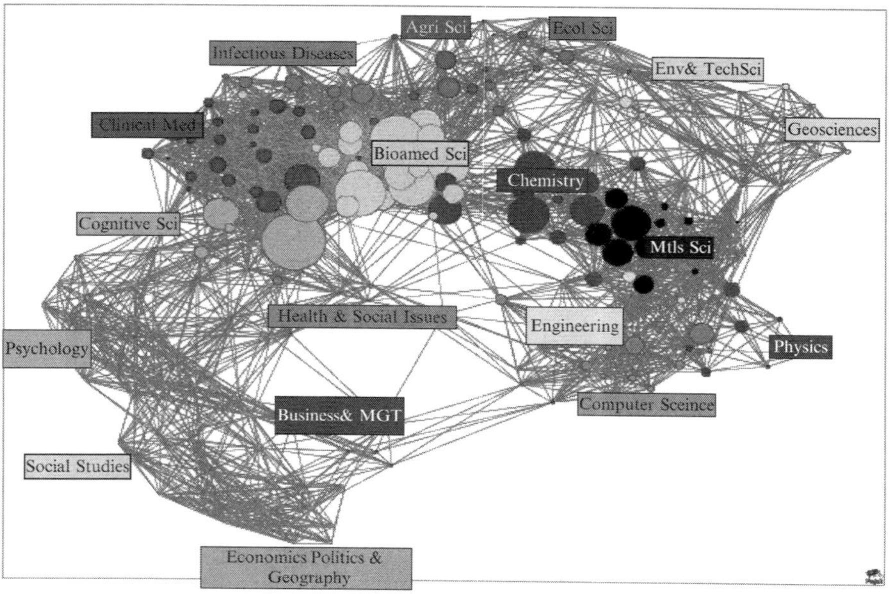


**Fig. 2.2** Nano-neuro subject category overlay map (By contrast, the map indicates little or no work being done in areas that might examine the potential implications for nano-neuro research in fields such as psychology, health and social issues, social studies, business and management, or economics)

Science. This map was generated following the methods of Porter and Youtie (2009). The map shows circles for subsets of subject categories, with larger circles indicating more articles citing that subject category. Colors indicate general domains of scientific research, as indicated in the color labels on the map. Consistent with Silva (2006), the map indicates that the bulk of work remains in core areas of basic research. The bulk of nano-neuro research is found, not surprisingly, in cognitive science and the biomedical sciences, indicating its focus in neuroscience, and in chemistry and material sciences, from its nanotechnology elements. At the next level down, the map indicates work in clinical medicine, computer science, physics, and infectious diseases. An unexpected element of this work was the intersection of infectious diseases with nano-neuro, which we explore later in this chapter.

## 2.3 Thematic Breakdown of Nano-Neuro Research

In this section, we begin to analyze the content of nano-neuro research. Overall, the top 10% of data suggests the major keywords in nano-neuro are microscopy, biosensors, neurodegenerative diseases, biological techniques, biocompatibility, Parkinson's disease, in vitro/in vivo, neurite outgrowth, drug delivery, polymers, endothelial

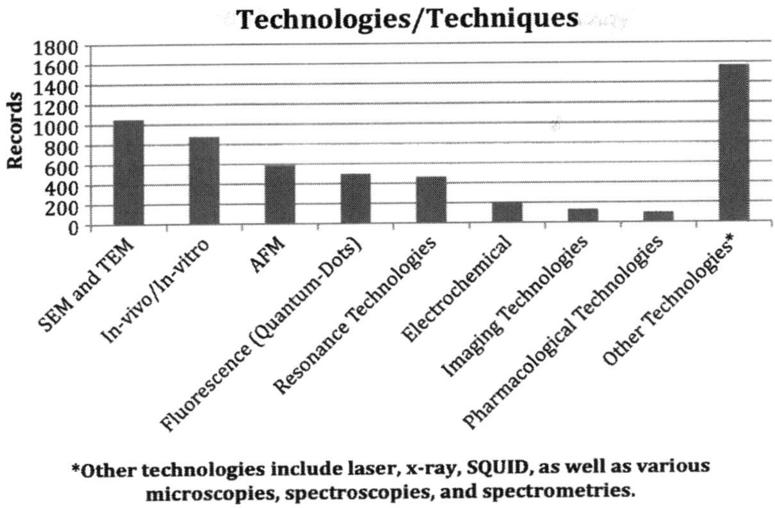


**Fig. 2.3** Distribution of articles mentioning each technology or technique

cells, gene expression, apoptosis, Alzheimer's disease, microtubules, dopamine, calcium, and microspheres. Based on these keywords and the overview analyses above, we decided to focus on three overarching themes. The first examines the technologies and techniques used in nano-neuro research. The second examines diseases under investigation in nano-neuro research. The third examines applications targeted in nano-neuro research.

### *2.3.1 Technologies and Techniques*

Our research identified 5,692 records citing various technologies and techniques used in nano-neuro research. Figure 2.3 offers the breakdown of the top eight technologies identified in the database. In-vitro and in-vivo are, we recognize, not precisely techniques; nonetheless, we included them to show the relative distribution of in-vitro and in-vivo research in comparison to other research approaches.

Perhaps most obvious from Fig. 2.3 is the broad dominance of microscopy and spectroscopy technologies such as scanning electron microscopy (SEM), transmission electron microscopy (TEM), and atomic-force microscopy (AFM) as well as other imaging technologies such as fluorescence, x-ray, and laser technologies. As in other areas of nanotechnology research, the capacity image at molecular scales not only forms a substantial fraction of the research, but is also opening up broad new research horizons in visualizing the basic structural foundations of materials and their properties. Other technologies identified indicate research focusing on biosensors; superconducting quantum interface devices (SQUID); pharmacokinetic

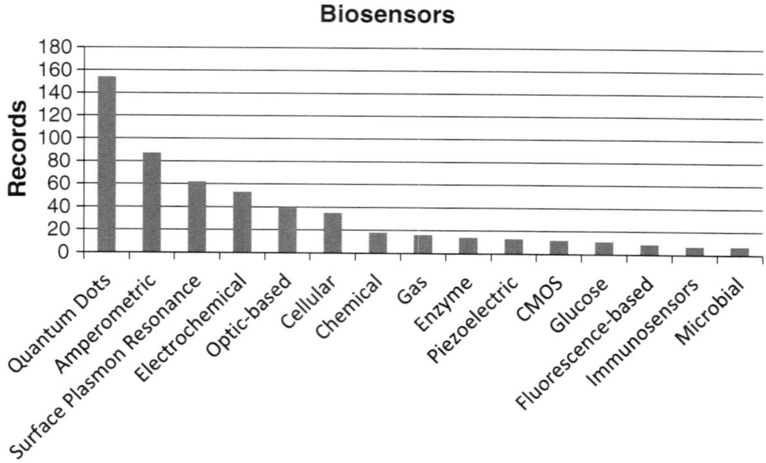


**Fig. 2.4** Top 15 areas of biosensors research in nano-neuro

technologies; gold nanoparticles; drug targeting; and in situ hybridization, which involve evaluations and manipulations on the chromosomal level.

Biosensors incorporate at least one biological component in the process of detecting or analyzing the materials sensed. Our search yielded approximately 800 records referring to biosensors, 50% of which represent the top five biosensors in Fig. 2.4, which displays the top 15 biosensor related keywords in the database.

Most of the biosensors discovered in the database fall into the main categories of either electrochemical or optical, but there are of course various others such as piezoelectric biosensors. Optical sensors are the most common of these main divisions and total nearly half of all biosensors in the database. Within optical sensors, the most common are surface plasmon resonance (SPR) and quantum dots. SPR sensors have been used as "lab-on-a-chip" technologies and are mentioned in conjunction with self-assembled monolayers, Alzheimer's disease, gold nanoparticles, and to a lesser extent silver nanoparticles. SPR sensors are also referenced as a tool for dual detection, for instance of highly toxic nerve-agent analogs, as well as for their monitoring and recording capabilities. Quantum dot sensors are growing in popularity due to the breadth of their capabilities. For instance, quantum dots are mentioned as being a better alternative to organic dyes in providing contrast in imaging. Additionally, they are used for optical detection, tagging brain tumor cells, and labeling cell surface proteins. Quantum dots are also mentioned in conjunction with in-vivo, nanocrystals, fluorescence, brain cancer/tumor, Alzheimer's disease, and to a lesser extent, cytotoxicity and glycine receptors.

Electrochemical sensors are the next most prominent biosensor in the literature and seem to aid more in brain injuries and strokes and in Alzheimer's disease detection. These sensors are mentioned in conjunction with voltammetry, nerve agents, brain tissue, electropolymerized films, glucose, and organophosphate pesticides. Amperometric biosensors are used in tandem with an electrochemical element and

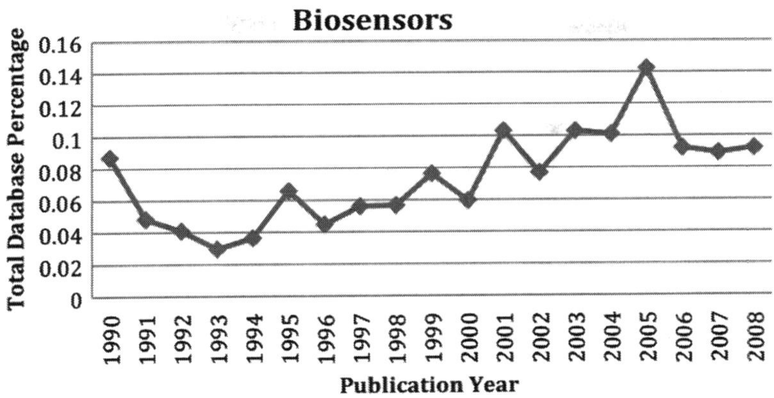


**Fig. 2.5** Percentage of articles pursuing research with biosensors

are mentioned in reference to Alzheimer's disease, brain injury, epilepsy, stroke, and vascular disease. Other keywords associated with amperometric sensors include enzyme immobilization, liquid-chromatography, carbon nanotubes, choline, dopamine, glutamate, pesticides, and voltammetry.

The line graph above magnifies the research activity in biosensors, allowing us to see the growing significance of biosensor research from 1990 to 2008, with biosensors accounting for 10% of all nano-neuro research after 2001 (Fig. 2.5).

### 2.3.2 *Diseases*

Our search identified 3,776 records citing various diseases. We expressly searched for over 300 diseases, and of those, we found hits on over 200 diseases and ailments searched for within the database. Figure 2.6 shows the breakdown of the top ten diseases. In addition to those diseases represented in the graph, we found research focusing on a variety of mental health conditions, cancers (breast and prostate), retinopathy, deafness, spiral atrophy, seizures, epilepsy, and a large number of infectious diseases, among others.

Figure 2.6 offers interesting insight into nano-neuro research. First, it is useful to note that nano-neuro research covers the full spectrum of brain and nervous system diseases, including Alzheimer's and Parkinson's, epilepsy and seizure, stroke, spinal cord injury, schizophrenia, and brain cancer. Particularly striking is the broad appearance of infectious diseases in nano-neuro research. Cumulatively, infectious diseases comprised 14% of the articles relating to diseases, which is both significant and unexpected.

Our research identified 75 infectious diseases out of 115 such diseases searched for within the database. The references are inclusive of the five main divisions of

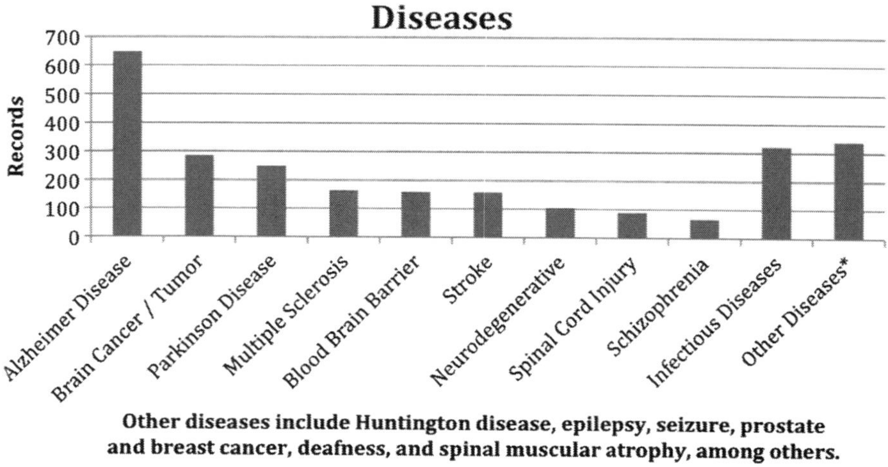


**Fig. 2.6** Diseases involved in nano-neuro research

infectious diseases: viral, bacterial, parasitic, fungal, and prion diseases, totaling 533 records in each of the 75 infectious diseases identified. Overall, infectious disease research nearly consistently averaged 5% over the 19 years represented in this study.

In general, infectious disease research on the nano-neuro level appears to focus on cellular and molecular recognition and reconstruction. Infectious diseases offer an opportunity to study, observe, and even manipulate how organisms and toxins bind to the cell surface and potentially reach the brain. Observing allows researchers to test various techniques and technologies such as in-vitro/in-vivo or imaging techniques on the molecular level to determine effectiveness, to improve delivery of natural or synthetic elements, or to develop receptors for new therapies. In most cases, research in this area appears to concentrate on observing molecular changes, with or without manipulation, that have potential to affect the brain or nervous system, or that mimic conditions that do.

Table 2.3 shows the infectious diseases identified that appeared in five or more records within the database. In addition to the 29 infectious diseases represented in the table below, we found research focusing on 44 other infectious diseases, including anthrax.

## 2.4 Applications

Our research identified 1,707 records referring to anticipated applications of nano-neuro research. Figure 2.7 shows the breakdown of the top ten applications:

The top three categories (drug delivery, prosthetics/implants, and transplants) each have specialties of focus. In drug delivery, the primary application foci are for

**Table 2.3** Distribution of nano-neuro research across infectious diseases

| Infectious disease | Number of records | Infection type |
|---|---|---|
| HIV/AIDS | 87 | Viral |
| Encephalitis | 49 | Viral |
| Herpes | 44 | Viral |
| Creutzfeldt-Jakob disease | 39 | Prion |
| Cholera | 37 | Bacterial |
| Pneumonia | 34 | Viral |
| Meningitis | 29 | Viral |
| Influenza | 23 | Viral |
| Tetanus | 23 | Viral |
| Cytomegalovirus | 22 | Viral |
| Toxoplasmosis | 21 | Parasitic |
| Hepatitis | 15 | Viral |
| Aspergillosis | 14 | Fungal |
| Enterobacteria | 13 | Parasitic |
| Malaria | 13 | Viral |
| Rabies | 13 | Viral |
| Urinary tract | 10 | Bacterial |
| Leishmaniasis | 9 | Parasitic |
| Schistosomiasis | 9 | Parasitic |
| Ascariasis | 8 | Parasitic |
| Candida | 8 | Fungal |
| Taeniasis | 7 | Parasitic |
| Trypanosomiasis | 7 | Parasitic |
| Tuberculosis | 7 | Viral |
| Leukoenephalopathy | 6 | Viral |
| Polio | 5 | Viral |
| Campylobacter | 5 | Bacterial |
| Pertussis (whooping cough) | 5 | Bacterial |
| Gerstmann-Straussler-Scheinker Syndrome | 5 | Prion |

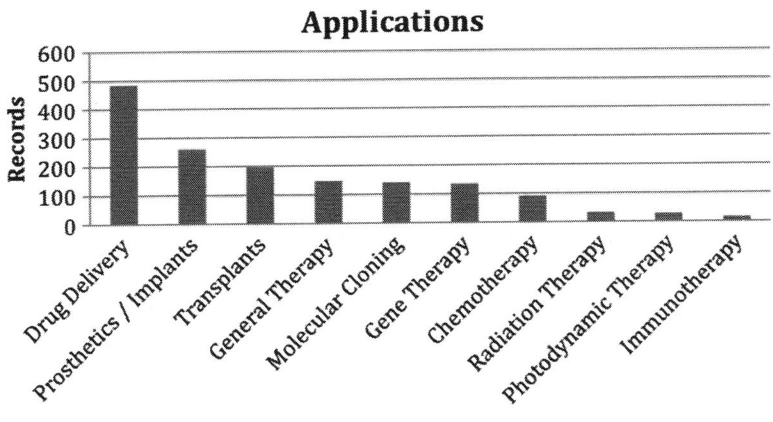


**Fig. 2.7** Nano-neuro research applications

ocular and gene therapy. For prosthetics, the primary applications deal with neural, hand, and abdominal prosthetics; within transplants, the primary applications are for neural, brain, cell, and corneal transplants; and in the category of implants, primary uses are for cochlear, brain, and neural implants.

### *2.4.1 Domains of Nanotechnology Application in Neuroscience*

In the first two-thirds of this paper, we offered an overview of research in nanotechnology applied to neuroscience and neurotechnology. In this second part of the paper, we offer a somewhat more detailed analysis of specific areas of application, identifying with greater focus and insight the rapidly growing impact of nanotechnology on our ability to understand, create interfaces with, and manipulate neural and cognitive systems. We focus, in particular, on the six fields: quantum dots, paralysis, epilepsy, stroke, brain-machine interface, and cochlear research. Collectively, these areas reflect approximately 600 total records or 6% of the overall research abstracts collected in the database. These six fields were selected to provide a reasonable cross-section of fields categorized by diverse orientations, including nanotechnology-related terms, (especially quantum dots), diseases, (paralysis, epilepsy), and applications, (brain-machine interfaces and cochlear research).

For this portion of the study, we examined in detail bibliographic records drawn from the larger database of nano-neuro publications. Within each record, qualitative textual analysis was carried out on the title, abstract, and keywords to identify the specific research represented by the record. Individual research studies were then sorted into an emergent classification to identify six broad categories of nano-neuro research. These categories are not exhaustive, but rather represent major domains of research. For each, we describe below the broad research area and offer a few examples of interesting research projects.

#### 2.4.1.1 Visualizing Brain and Nerve Structure and Dynamics at the Nanoscale

One of the central contributions of nanotechnology to many fields of science has been improved visualization tools, such as electron microscopes and atomic force microscopes that enable materials to be visualized on the scale of nanometers, thousands of times smaller than the best light microscopes. Indeed, these instruments have become so ubiquitous in many fields that they are taken for granted as the necessary tools for modern scientific research. It is not particularly surprising, therefore, that a very large fraction of the nanotechnology work being done in neuroscience involves nanoscale microscopy (see Fig. 2.3 above).

In addition to nanoscale microscopy and quantum dots, other nanoparticles, like nano-diamond crystals, are also increasingly being used as tools for visualizing and probing the biological structure and functioning of the brain (Liu et al. 2008).

Quantum dots are nanoscale structures that fluoresce and offer flexible chemical structures that can be used to target the attachment of quantum dots to specific molecules. Quantum dots can thus be used as valuable contrast agents for enabling the visualization of biological structures and processes and can be designed to bind to specific molecular binding sites, allowing them to tag particular biological reactions or locations within complex molecules (Wang et al. 2005).

Finally, nanoelectric arrays are also being designed that allow for measurement of electrical signals from individual neurons within a group of cells and from different locations within a single nerve cell. In this case, cells are cultured on top of arrays of nanofibers that are individually addressed, allowing for detection and monitoring of signals at each point in the array. These arrays can be used to develop a better understanding of cell signaling and transmission pathways, providing new insights into cell behavior and functioning (Wickramanayake et al. 2005; Mazzatenta et al. 2007; Greve et al. 2007).

Collectively, these and other new approaches to visualization are opening up significant new capacities for understanding the brain and nervous system at molecular scales and rapidly altering the capacity of neuroscience research to examine and understand cellular structures and dynamics. One illustration of the potential impacts of nanotechnology for scientific understanding of normal brain and nerve functioning, the causes of disease, and possibilities for clinical therapies for nerve and brain repair and regeneration, comes from a review of potential applications of nanotechnology in Alzheimer's research:

> In this report, we present the promises that nanotechnology brings in research on the AD diagnosis and therapy. They include its potential for the better understanding of the AD root cause molecular mechanisms, AD's early diagnoses, and effective treatment. The advances in AD research offered by the atomic force microscopy, single molecule fluorescence microscopy and NanoSIMS microscopy are examined here. In addition, the recently proposed applications of nanotechnology for the early diagnosis of AD including bio-barcode assay, localized surface plasmon resonance nanosensor, quantum dot, and nanomechanical cantilever arrays are analyzed. Applications of nanotechnology in AD therapy including neuroprotections against oxidative stress and anti-amyloid therapeutics, neuroregeneration, and drug delivery beyond the blood brain barrier (BBB) are discussed and analyzed. All of these applications could improve the treatment approach of AD and other neurodegenerative diseases (Nazem and Mansoori 2008).

More broadly, both quantum dots and electron microscopy analyses have been used to examine detailed aspects of cellular structure and behavior in nerve cells. Examples of the use of quantum dots as imaging include studies of ion channels, protein receptor sites, neurochemical flows, molecular morphology, RNA, cell tracking, clinical pathology of degenerative eye diseases, and many other applications (O'Connell et al. 2006; Howarth et al. 2008; Chapman et al. 2008; Liang et al. 2005; Ji et al. 2006; Tomlinson 2006; Tsai et al. 2008; Yamamoto et al. 2007). Examples of electron microscopy use included examinations of: the molecular and cellular level processes that contribute to damage to nerve cells after spinal cord injuries or in hereditary spinal cord diseases, enabling a better understanding of the full effects of spinal cord degeneration and helping explain certain observed therapeutic outcomes (Werner et al. 2007; Tator 1995); and the molecular basis of nerve degeneration in paraplegia and muscle degeneration from denervation (Biral et al. 2008). In addition to imaging, quantum dots are also being tested for use in disease therapies, such as degenerative eye diseases (Christie and Kompella 2008).

#### 2.4.1.2 Visualizing Nerve Growth and Regeneration

One specific area of nanoscale visualization research is that of nerve growth and regeneration. Nerve regeneration is crucial to efforts to find new therapies for both nerve damage, (e.g., due to spinal cord injury or other trauma that impacts the nervous system), and degenerative nerve diseases, especially in the brain, (e.g., Parkinson's, Alzheimer's, and many others). A wide array of medical research is seeking to promote nerve regeneration, either through pharmaceutical development or, for example, the use of stem cells to regrow damaged sections of nerves or the brain. Efforts using stem cells include implanting neural stem cells to act as progenitor cells for regrowth or as producers of neurotrophic factors that encourage other cells to grow and replicate.

Nano-imaging techniques can be used to visualize the process of nerve growth and regeneration at molecular scales, thus significantly enhancing researchers' ability to both understand and manipulate molecular level processes involved. There are several potential goals here. Nano-imaging can help researchers understand underlying biological processes of nerve growth and can also help evaluate the success of experimental or clinical efforts to use nerve regeneration in therapeutic applications. One study using electron microscopy, for example, examined neurite growth on dorsal root ganglia when cultured on different surfaces, using microscopy to study the detailed growth patterns observed (Chakrabortty et al. 2000). Other studies used quantum dot imaging to study how growing nerve cells respond to external stimuli that direct their growth (Bouzigues et al. 2004, 2007; Rajan and Vu 2006; Cui et al. 2007; Echarte et al. 2007).

#### 2.4.1.3 Developing Scaffolding for Nerve Regeneration Experiments

Beyond visualization of nanoscale phenomena, nanotechnology is also emerging as a potentially valuable tool for enhancing the outcomes of clinical applications of tissue engineering and cell transplantation therapies by enabling understanding, visualization, and control of cellular interactions at nanoscales. One important area where this work is going on is in the field of spinal cord injuries and other neurodegenerative diseases. Here, in addition to improving scientific understanding of nerve growth and regeneration through novel nano-imaging technologies, nanotechnologies are also emerging as potentially important elements in the design of scaffolding to help facilitate nerve regeneration.

Work in this area has focused on experimental assessments of the potential for nanomaterials to help structure and guide nerve growth and regeneration, such as the use of nanotechnology substrates for nerve regeneration and growth, including single-walled carbon nanotubes. Research has shown that carbon nanotubes have good electrical conductivity, corrosion resistance, and strength, as well as chemical functionalization, cell adhesion, and biocompatibility characteristics that make them promising materials on which to grow cells. Experiments have shown, in particular, that nerve growth on carbon nanotube-based materials is similar in terms of cell growth and differentiation, neurite growth, and biocompatibility to other commonly used materials (Liopo et al. 2006; Rochkind et al. 2006). The use of various tools capable of developing nanoscale patterns on solid materials is also being explored as a potential approach to nerve regeneration, where axonal growth can be encouraged to follow certain scalar patterns (Johansson et al. 2006).

Building on these results, a mouse model was used to test the use of nanomaterials to help fashion viable scaffolding for nerve regeneration in the case of spinal cord injury. Nanofibers were placed inside a tubular scaffold that was then used to culture human embryonic spinal cord stem cells and subsequently sutured into a 4 mm gap cut in the spinal cord. After 3 months, the results showed partial recovery of function in one or two limbs of all of the mice using human embryonic stem cells within the scaffold, highlighting the potential value of such scaffolds in clinical cell transplantation for spinal cord injury and paraplegia (Rochkind et al. 2006).

Like the discussion of Alzheimer's above, researchers see nerve injury as a key site where nanotechnology may have enormous implications:

> Cell transplantation to treat diseases characterized by tissue and cell dysfunction, ranging from diabetes to spinal cord injury, has made great strides preclinically and towards clinical efficacy. In order to enhance clinical outcomes, research needs to continue in areas including the development of a universal cell source that can be differentiated into specific cellular phenotypes, methods to protect the transplanted allogeneic or xenogeneic cells from rejection by the host immune system, techniques to enhance cellular integration of the transplant within the host tissue, strategies for in vivo detection and monitoring of the cellular implants, and new techniques to deliver genes to cells without eliciting a host immune response. Overcoming these obstacles will be of considerable benefit, as it allows understanding, visualizing, and controlling cellular interactions at a submicron level. Nanotechnology is a multidisciplinary field that allows us to manipulate materials, tissues, cells, and DNA at the level of and within the individual cell. As such, nanotechnology may be well suited to optimize the generally encouraging results already achieved in cell transplantation (Halberstadt et al. 2006).

#### 2.4.1.4 Visualizing and Structuring Neural Prosthetic Interfaces

As we noted in the introduction to this chapter, a series of informal conversations with leading researchers in the field of neural prosthetics revealed little interest in, or engagement with, work in nanotechnology. It was with significant surprise therefore, that we discovered in the database a growing number of studies applying nanotechnology research to the design and analysis of neural prosthetics. This work can be divided into two major domains, consistent with other fields of nano-neuro research: visualization and imaging studies, and active application of nanotechnology to prosthetic design. Overall, nanotechnology research is being applied in numerous ways to the field of brain-machine interfaces and neural prosthetics in two areas. First, by enhancing both the capacity to visualize and therefore study the behavior of interfaces between biological and implant materials (Wrobel et al. 2008). Second, by enhancing the ability of researchers to create highly functional and optimized interfaces between electronic and biological systems (Sarje and Thakor 2004; Hu et al. 2006).

In the field of imaging, studies have used electron microscopy tools to explore nanoscale features of nerve growth and attachment and rates of bacterial growth on both diverse implant materials and diversely structured implant interfaces (Brors et al. 2002; Selvakumaran et al. 2002; Pawlowki et al. 2005). Other studies have explored the degradation of implant materials over time, examining rates of degradation and affect on signal transmission between nerve and implant (Mlynski et al. 2007; Trabandt et al. 2005). A final study examined the nanostructuring of the biological environment, (and potential damage to it), within which the implant is placed, to determine optimal approaches to placing the implant device (Glueckert et al. 2005).

Nanoscale techniques also offer unique capacities to create structure at the interface between biological systems and implanted devices. Three primary approaches have been adopted. First, the use of single-walled nanotubes, quantum dots, and tethered nanopolymers as extensions from conventional microelectrodes into the nerve cell to improve conductivity and signal transmission (Wickramanayake et al. 2005). Second, the patterning of implant surfaces at nanoscales to improve either biocompatibility, (e.g., by reducing microbial growth or implant surface degradation), or electrical conductivity (Pawlowki et al. 2005; Johansson et al. 2006). Third, the development of nanocoatings, again to improve various aspects of implant functioning (Turck et al. 2007). The most ambitious research efforts use arrays of nanofiber electrodes to allow simultaneous sampling either from multiple neurons or from multiple parts of the same neuron. This last could be used to improve implant conductivity and signal transmission quality (McKnight et al. 2006).

Illustrative examples of work in these areas include:

- Efforts to enhance the electrical contact between biological and electrical systems through the design of new nanoscale molecules, (e.g., polymer coatings, polymer molecules, quantum dots, and carbon nanotubes), with the goal of ultimately designing interfaces capable of high-bandwidth transmission between nerves and either biosensors or robotic devices that might be controlled by the nervous system (Widge et al. 2004).
- The use of high precision machining tools, such as focused ion beams, to modify surface patterns at nanoscales in ways that influence cell adhesion and other interface characteristics. Scales approximately tens of nanometers to 100 nm, which is the typical scale of cellular interaction in biological systems, seem to generate optimum results (Raffa et al. 2007; Johansson et al. 2006).
- The design and characterization of nanoelectronic arrays out of functionalized carbon nanotubes and quantum dot layer-by-layer assemblies that allow for facilitation of neuron growth and development and multi-site communication between

electronic systems and multi-cellular matrices (Mazzatenta et al. 2007; Greve et al. 2007).

- Visual studies of the cytotoxicity of diverse materials that might be used in implant devices to assess their biocompatibility for long-term implantation in the human body (Liopo et al. 2006; Gomez et al. 2005).
- Efforts to evaluate and optimize the nanostructural characteristics of materials for brain-machine interfaces, including pore size, material composition, geometrical features, surface chemistry, etc. (Johansson et al. 2005; Moxon et al. 2004).

#### 2.4.1.5 Improved Cancer Detection and Identification

Among their many applications in neuroscience, one interesting emerging area of research is in the use of quantum dots to detect and identify cancer cells in the brain. Research has demonstrated that brain tumor cells do take up quantum dots that can then be used to tag cancer cells using fluorescent imaging techniques. Quantum dots were functionalized to target epidermal growth factor receptors, which are believed to be involved in a number of brain cancer types. Consequently, quantum dot tagging may be usable as a tool for not only detecting cancer cells, but also for diagnosing specific cancers. Following up this work, other research has found that the fluorescent brightness varies between healthy and cancerous brain cells labeled with quantum dots, with the result that quantum dot imaging tools can potentially also be used to help make clearer differentiations between healthy and tumor tissues (Cai et al. 2006; Farias et al. 2006, 2008). Still other research has suggested quantum dot imaging can help identify radiation damage to tissues in cancer treatments, assess tissue viability, and potentially provide other valuable insights into cancer diagnosis and therapy in the brain and central nervous system (Toms et al. 2006).

Recent reviews of nanotechnology applications to brain cancer suggest the value of nanotechnology may be even bigger in the future:

> Recent developments in nanotechnology have provided researchers with new tools for cancer imaging and treatment. This technology has enabled the development of nanoscale devices that can be conjugated with several functional molecules simultaneously, including tumor-specific ligands, antibodies, anticancer drugs, and imaging probes. Since these nanodevices are 100–1,000-fold smaller than cancer cells, they can be easily transferred through leaky blood vessels and interact with targeted tumor-specific proteins both on the surface of and inside cancer cells. Therefore, their application as cancer cell-specific delivery vehicles will be a significant addition to the currently available armory for cancer therapeutics and imaging. (Wang et al. 2008)

> Nanotechnology is a multidisciplinary field, which covers a vast and diverse array of devices derived from engineering, biology, physics, and chemistry. These devices include nanovectors for the targeted delivery of anticancer drugs and imaging contrast agents. Nanowires and nanocantilever arrays are among the leading approaches under development for the early detection of precancerous and malignant lesions from biological fluids. These and other nanodevices can provide essential breakthroughs in the fight against cancer (Ferrari 2005).

## 2.5 Conclusion

The application of nanotechnology research in the field of neuroscience is growing rapidly and suggesting a wide range of applications. Like most other fields of nanotechnology research, the primary applications to neuroscience include imaging of nanoscale structures, the design of nanoscale materials, and the development of new nano-sensors and devices. Given this rapid growth of research activity, assessments of these new technologies, their potential applications, and their long-term social and ethical dimensions, is essential and timely. Many of the long-term ambitions of neuroscientists, with regard to understanding the brain and nervous system, assessing disease, creating new therapies, designing highly capable neuroprosthetics, and even manipulating and enhancing brain function through new drugs and devices, are likely to depend on nanotechnology and to be shaped by the capabilities of nanoscale science and engineering. Understanding this kind of dynamic interaction between two fields of scientific research is thus critical to assessing future technological applications and their ethics.

As nano-neuro research experiences rapid growth, there is a need to balance applied research with the various fields of study responsible for critically examining nano-neuro research and assessing how nanotechnology should be used in neuroscience, as well as creating a policy by which nanotechnology will be used. Introducing nano-neuro discussions into critical fields of study early on may help to safeguard the use of such new and emerging nanotechnologies, limiting abuse or misuse. An absence of this focus may signal a lack of foundational framework by which to systematically implement and regulate nanotechnologies in neuroscience.

Overall, the research summarized above makes something of a mockery of both the claim that the application of nanotechnology to neuroscience is speculative and the claim that ethical analysis of this field can be put off until some unspecified date in the future. The application of nanotechnology to research on the human brain is here, today, and the transition from research to clinical application is imminent if not already occurring. It is essential that ethical analyses of this research and its potential applications go forward in conjunction with broader work in nano-ethics and neuro-ethics.

## References

Barben, D. 2008. Anticipatory governance of nanotechnology: Foresight, engagement, and integration. In *The handbook of science and technology studies*, 3rd ed, ed. Edward J. Hackett, Olga Amsterdamska, Michael Lynch, and Judy Wajcman. Cambridge: MIT Press.

Biral, D., et al. 2008. Atrophy-resistant fibers in permanent peripheral denervation of human skeletal muscle. *Neurological Research* 30(2): 137–144.

Bouzigues, C., et al. 2004. Tracking of single GABA receptors in nerve growth cones using organic dyes and quantum dots. *Biophysical Journal* 86(1): 602A.

Bouzigues, C., et al. 2007. Asymmetric redistribution of GABA receptors during GABA gradient sensing by nerve growth cones analyzed by single quantum dot imaging. *Proceedings of the National Academy of Sciences* 104(27): 11251–11256.

Brors, D., et al. 2002. Interactions of spiral ganglion neuron processes with alloplastic materials in vitro. *Hearing Research* 167(1–2): 110–121.

Cai, W., et al. 2006. Peptide-labeled near-infrared quantum dots for imaging tumor vasculature in living subjects. *Nano Letters* 6(4): 669–674.

Chakrabortty, S., et al. 2000. Choroid plexus ependymal cells enhance neurite outgrowth from dorsal root ganglion nerves in vitro. *Journal of Neurocytology* 29(1): 707–717.

Chapman, S., et al. 2008. New tools for in vivo fluorescence tagging. *Current Opinion in Plant Biology* 8(6): 565–573.

Christie, J., and U. Kompella. 2008. Ophthalmic light sensitive nanocarrier systems. *Drug Discovery Today* 13(3–4): 124–134.

Cui, B., et al. 2007. One at a time, live tracking of NGF axonal transport using quantum dots. *Proceedings of the National Academy of Sciences* 104(34): 13666–13671.

Echarte, M., et al. 2007. Quantitative single particle tracking of NGF-receptor complexes: Transport is bidirectional but biased by longer retrograde run lengths. *FEBS Letters* 581(16): 2905–2913.

Farias, P., et al. 2006. Application of colloidal semiconductor quantum dots as fluorescent labels for diagnosis of brain glial cancer. In *Colloidal quantum dots for biomedical applications*. Bellingham: ISOE.

Farias, P., et al. 2008. Fluorescent II-VI semiconductor quantum dots: Potential tools for biolabeling and diagnostic. *Journal of the Brazilian Chemical Society* 19(2): 352–356.

Ferrari, M. 2005. Cancer nanotechnology: Opportunities and challenges. *Nature Reviews. Cancer* 5(3): 161–167.

Glueckert, R., et al. 2005. The human spiral ganglion: New insights into ultrastructure, survival rate, and implications for cochlear implants. *Audiology & Neuro-Otology* 10(5): 258–273.

Gomez, N., et al. 2005. Challenges in quantum dot-neuron active interfacing. *Talanta* 67(3): 462–471.

Greve, F., et al. 2007. Molecular design and characterization of the neuron-microelectrode array interface. *Biomaterials* 28(35): 5246–5258.

Halberstadt, C., et al. 2006. Combining cell therapy and nanotechnology. *Expert Opinion on Biological Therapy* 6(1): 781–791.

Howarth, M., et al. 2008. Monovalent, reduced-size quantum dots for imaging receptors on living cells. *Nature Methods* 5(5): 397–399.

Hu, Z., et al. 2006. Nanopowder molding method for creating implantable high aspect ratio electrodes on thin flexible substrates. *Biomaterials* 27(9): 2009–2017.

Ji, X., et al. 2006. An alternative approach to amyloid fibrils morphology: CdSe/ZnS quantum dots labeled beta-amyloid peptide fragments A beta (31-35), A beta (1-40), and A beta (1-42). *Colloids and Surfaces. B, Biointerfaces* 50(2): 104–111.

Johansson, F., et al. 2005. Guidance of neurons on porous, patterned silicon: Is pore size important? *Physica Status Solidi C* 9: 3258–3262.

Johansson, F., et al. 2006. Axonal outgrowth on nano-imprinted patterns. *Biomaterials* 27(8): 1251–1258.

Liang, R., et al. 2005. An oligonucleotide microarray for microRNA expression analysis based on labeling RNA with quantum dot and nanogold probe. *Nucleic Acids Research* 33(2): 8.

Liopo, A., et al. 2006. Biocompatibility of native and functionalized single-walled carbon nanotubes for neuronal interface. *Journal of Nanoscience and Nanotechnology* 6(5): 1365–1374.

Liu, K., et al. 2008. Alpha-bungarotoxin binding to target cell in a developing visual system by carboxylated nanodiamond. *Nanotechnology* 19(20).

Mazzatenta, A., et al. 2007. Interfacing neurons with carbon nanotubes: Electrical signal transfer and synaptic stimulation in cultured brain circuits. *Journal of Neuroscience* 27(26): 6931–6936.

McKnight, T., et al. 2006. Resident neuroelectrochemical interfacing using carbon nanofiber arrays. *The Journal of Physical Chemistry. B* 110(31): 15317–15327.

Mlynski, R., et al. 2007. Interaction of cochlear nucleus explants with semiconductor materials. *Laryngoscope* 177(7): 1216–1222.

Moxon, K., et al. 2004. Nanostructured surface modification of ceramic-based microelectrodes to enhance biocompatibility for a direct brain-machine interface. *IEEE Transactions on Biomedical Engineering* 51(6): 881–889.

Nazem, Amir, and G. Ali Mansoori. 2008. Nanotechnology solutions for alzheimer's disease: Advances in research tools, diagnostic methods and therapeutic agents. *Journal of Alzheimer's Disease* 13: 199–223.

O'Connell, K., et al. 2006. Kv2.1 potassium channels are retained within dynamic cell surface micro domains that are defined by a perimeter fence. *Journal of Neuroscience* 26(38): 9609–9618.

Pawlowki, K., et al. 2005. Bacterial biofilm formation on a human cochlear implant. *Otology & Neurotology* 26(5): 972–975.

Porter, A., and J. Youtie. 2009. How interdisciplinary is nanotechnology? *Journal of Nanoparticle Research* 11(5): 1023–1041.

Porter, A., J. Youtie, P. Shapira, and D. Schoeneck. 2008. Refining search terms for nanotechnology. *Journal of Nanoparticle Research* 10(5): 715–728.

Raffa, V., et al. 2007. Design criteria of neuron/electrode interface: The focused ion beam technology as an analytical method to investigate the effect of electrode surface morphology on neurocompatibility. *Biomedical Microdevices* 9(3): 371–383.

Rajan, S., and T. Vu. 2006. Quantum dots monitor TrkA receptor dynamics in the interior of neural PC12 cells. *Nano Letters* 6(9): 2049–2059.

Rochkind, S., et al. 2006. Development of a tissue-engineered composite implant for treating traumatic paraplegia in rats. *European Spine Journal* 15(2): 234–245.

Sarje, A., and N. Thakor. 2004. Neural interfacing. In *Conference Proceedings: 26th Annual International Conference of the IEEE Engineering in Medicine and Biology Society*. Piscataway: IEEE.

Sclove, R. 1995. *Democracy and technology*. New York: Guilford Press.

Selvakumaran, J., et al. 2002. Assessing biocompatibility of materials for implantable microelectrodes using cytotoxicity and protein adsorption studies. In *Proceedings of the 2nd Annual International IEEE-EMBS Special Topic Conference on Microtechnologies in Medicine and Biology*. Piscataway: IEEE.

Silva, G. 2006. Neuroscience nanotechnology: Progress, opportunities, and challenges. *Nature Reviews Neuroscience* 7: 65–74.

Tator, C. 1995. Update on the pathophysiology and pathology of acute spinal cord injury. *Brain Pathology* 5(4): 407–413.

Tomlinson, I. 2006. High affinity inhibitors of the dopamine transporter (DAT): Novel biotinylated ligands for conjugation to quantum dots. *Bioorganic and Medical Chemistry Letters* 16(17): 4664–4667.

Toms, S., et al. 2006. Neuro-oncological applications of optical spectroscopy. *Technology in Cancer Research & Treatment* 5(3): 231–238.

Trabandt, N., et al. 2005. Limitations of titanium dioxide and aluminum dioxide as ossicular replacement materials: An evaluation of the effects of porosity on ceramic prostheses. *Otology & Neurotology* 25(5): 682–693.

Tsai, C., et al. 2008. High-contrast paramagnetic fluorescent mesoporous silica nanorods as a multifunctional cell imaging probe. *Small* 4(2): 186–191.

Voss, J., D. Bauknecht, and R. Kemp. 2006. *Reflexive governance for sustainable development*. Cheltenham: Edward Elgar.

Wang, J., et al. 2005. A fluorescence microscopy study of quantum dots as fluorescent probes for brain tumor diagnosis. In *Plasmonics in biology and medicine II*. Bellingham: ISOE.

Wang, X., et al. 2008. Application of nanotechnology in cancer therapy and imaging. *CA: A Cancer Journal for Clinicians* 58(2): 97–110.

Werner, H., et al. 2007. Proteolipid protein is required for transport of sirtuin 2 into CNS myelin. *Journal of Neuroscience* 27(29): 7717–7730.

Wickramanayake, W., et al. 2005. Controlled photostimulation of neuron cells and activation of calcium ions on semiconductor quantum dot layer-by-layer assemblies. In *2005 AIChE Annual Meeting and Fall Showcase, Conference Proceedings*. New York: AIChE.

Widge, A., et al. 2004. Conductive polymer 'Molecular Wires' for neuro-robotic interfaces. In *Proceedings of the 2004 IEEE Conference on Robotics and Automation.* Piscataway: IEEE.

Wilsdon, J., and R. Willis. 2004. *See through science: Why public engagement needs to move upstream.* London: Demos.

Wrobel, G., et al. 2008. Transmission electron microscopy study of the cell-sensor interface. *Journal of the Royal Society, Interface* 5(10): 222–231.

Yamamoto, S., et al. 2007. Visualizing vitreous using quantum dots as imaging agents. *IEEE Transactions on Nanobioscience* 6(1): 94–98.

Zonneveld, L., H. Dijstelbloem, and D. Ringoir. 2008. *Reshaping the human condition: Exploring human enhancement.* The Hague: Rathenau Institute.

# Chapter 3
# Public Attitudes Towards Nanotechnology-Enabled Cognitive Enhancement in the United States

**Sean A. Hays, Clark A. Miller, and Michael D. Cobb**

## 3.1 Introduction

Anticipatory governance of emergent technologies depends on a comprehensive understanding of the values in society that shape public understanding of new and emerging technologies, as well as their response to related technologies already available within the culture (Barben et al. 2008; Guston and Sarewitz 2002). One method of contributing to the understanding of public values is to measure them directly through survey research. In this chapter, we present results from a 2008 national survey about nanotechnology and human enhancements. More specifically, the survey was designed to evaluate the public's support for potential nano-enabled cognitive enhancement technologies. To the best of our knowledge, it was the first nationally representative survey about human enhancements to be conducted in the United States. Where appropriate, we also report some preliminary findings from a follow-up survey in 2010 that supplement our analysis of the 2008 study, but we intend to report the bulk of the 2010 survey elsewhere.

In this survey we measured how much the public knows about such technologies, how they perceive them in general, which ethical paradigms they would employ in making decisions about specific technologies, and how they would react to the introduction of such technologies into the competitive structures of our democracy. We asked respondents to consider these questions with respect to both themselves and, in some instances, their children. Our results indicate to us that the average citizen approaches these technologies differently than one would predict from simple

S.A. Hays • C.A. Miller (✉)
The Center for Nanotechnology in Society, Arizona State University,
P.O. Box 875603, AZ 85287-5603, Tempe, USA
e-mail: seanahays@gmail.com; clark.miller@asu.edu

D. Cobb
Department of Political Science, North Carolina State University,
223 Caldwell, Campus Box 8102 Raleigh, NC 27695
e-mail: mike_cobb@ncsu.edu

S.A. Hays et al. (eds.), *Nanotechnology, the Brain, and the Future*,
Yearbook of Nanotechnology in Society 3, DOI 10.1007/978-94-007-1787-9_3,

cue-taking models of public opinion. They appear to apply a number of heuristics simultaneously, as well as particular and pragmatic considerations, when deciding which technologies to support and which to incorporate into their own lives.

The remainder of this chapter is organized as follows. The first section summarizes the major findings of the survey. After presenting this overview, we turn to providing details about the survey methodology and instrument. We then present our findings, starting with the presentation of basic descriptive data about knowledge and familiarity with human enhancement technologies, and continue presenting descriptive data about a range of opinions. In addition, we analyze relationships between variables where appropriate, such as the importance of prior knowledge for supporting types of enhancements. Like prior research, we examine the role of trust in institutions for supporting these technologies, and, following recent studies, we review the impact of specific technological applications on opinions rather than just analyzing opinions about nanotechnology in the abstract. Our final data come from a framing experiment designed to parse the importance of ideology and religiosity on support for human enhancement. We end with some conclusions about the likely fate of these technologies in the court of public opinion. It is clear to us that the specific type of application is the principle driver behind support for, or opposition to, human enhancement, despite ethicists' arguments that such differences lack real meaning.

## 3.2 A Brief Summary of Major Findings

Similar to findings of past surveys, the public is generally unaware of these emerging technologies (Scheufele and Lewenstein 2005). Four in ten Americans say they have heard "nothing at all" about nanotechnology, and six in ten admit to hearing nothing at all about nanotechnology used for human enhancements. Besides having only a limited awareness of these new technologies, public perceptions of how they are being used are skewed. The vast majority of respondents tend to associate nano with "machines and computers" rather than "consumer products." While the association with computers is intuitive given that nano is playing a major role in the production of computer memory and processing technologies, the failure to recognize it is also being used as an additive in an extensive array of consumer products, from sun block to tennis rackets, is alarming. Nevertheless, respondents already associated it with some of the areas with which we are concerned, such as brain research and the engineering of humans.

Despite their lack of knowledge, most respondents expressed opinions about these technologies. We find that Americans are divided in their assessments of the risks and benefits. A plurality of Americans believed the risks of nanotechnology for human enhancements will be equal to the benefits, while roughly equal but smaller percentages thought risks would outweigh benefits, or vice-versa. The 2010 survey finds more people see risks as outweighing benefits, but the comparison over time is muddled because risk perceptions were measured differently across the two surveys. When it comes to protecting the public from the risks, Americans had the greatest confidence in university scientists. Respondents had the least confidence in the business community, the mass media, and the federal government.

While it is true that some applications garnered positive support levels overall, and some demographics were likely to generally support these technologies, the overall disposition of the American public to nanotechnology and human enhancement technologies is a negative one. Some of these attitudes were malleable to contextual factors, but increased support depending on the context never turned into overall majority support for such technologies. Clearly, a principal driver of support for nano-enabled human enhancement is the nature of the application in question. Respondents supported applications with obvious therapeutic value, even when we made it apparent through the question that the application would also have enhancement benefits outside of a therapeutic setting. The added optical features possible in an implant to restore the eyesight of a blind person, for example, that could also confer some advantages over the normally sighted (better visual acuity, vision in other spectrums), did not seem to bother respondents, so long as the initial application was obviously therapeutic. Looking to the economics of these technologies, Americans expect human enhancements to be costly and have decidedly mixed reactions about how to deal with this. While most said that only the wealthiest Americans will be able to afford enhancements, a majority also reported be unworried about potentially being unable to afford them. Also, while a large majority said that that the government should guarantee equal access should these enhancements become available, most also said that individuals, not insurance, should pay for them out of pocket.

Finally, we conducted a framing experiment to identify the impact of exposure different ways of thinking about these technologies. We presented some respondents with an argument against embracing nanotechnologies framed negatively as "playing God." Other respondents heard about nanotechnology framed positively as helping improve humankind. A third group of respondents heard no arguments, while a fourth group heard both positive and negative frames. Overall, a slim majority agreed that we should "avoid playing God with new technologies" rather than "embrace new enhancement technologies to improve humankind." Yet, those in the control group were most inclined to endorse "avoid playing God." Not only did respondents' support for nano increase after hearing university scientists endorse nanotechnologies as a way to improve humankind, but support also increased after hearing religious figures framing nano negatively. These results add to our conclusion that context is important if not decisive. The specific framing in which a technology is presented can activate different aspects of the respondent's evaluation matrix, and produce a different emotional and intellectual relationship to the technology.

## 3.3 Methods and Demographics

### *3.3.1 Methods*

The 2008 phone survey consists of a sample of 556 American adults aged 18 and older that were selected through random digit dialing (RDD). It was fielded by the Survey Research Lab at the University of Wisconsin between July and October of 2008, had a response rate of 28%, and a margin of error of $^{+/-}$4.1%. In addition to measuring multiple respondent demographics, such as religiosity, political ideology, sex, race, etc., the survey included 15 core substantive questions measuring public awareness, perceptions and preferences about nano-enabled human enhancements. (A copy of the survey instrument is reproduced in Appendix 1.) Several of these questions were asked as part of a framing experiment and are discussed in more detail elsewhere because their analysis requires statistics beyond the reporting of simple frequencies as we are employing here.

Instead of a phone survey, the 2010 survey was conducted over the Internet by Knowledge Network (KN) in order to include an experiment that included a visual representation of a nano-enabled human enhancement. Despite taking place over the Internet, KN conducts representative sampling by recruiting a large nationally representative panel of potential survey respondents. In this survey, some 1,231 panelists were randomly drawn from the KN panel; 849 responded to the invitation, yielding a final stage completion rate of 69.0%. The recruitment rate for this study, reported by Knowledge Networks, was 18.9% and the profile rate was 53.6%, for a cumulative response rate of 7.0%.

### *3.3.2 Demographics*

We only discuss the 2008 sample demographics because these data are our focus. (Details can be provided upon request.) Most of the sample demographics match up well with census data, with a few exceptions. The sample was 55% female, and 82% non-Hispanic, white, while the median household income and level of educational attainment were, respectively, US$50,000–US$75,000 and "up to three years in college/technical degree". In terms of religiosity and sectarian identification, our sample divided roughly by thirds in their description of the frequency with which they seek religious guidance as low, medium, and high. Further, they identified as predominantly Protestant and Catholic. In terms of ideology, 34% of our respondents self-identified as socially liberal and 40% as socially conservative. Lastly, the average age of a respondent was 55 years old. As a result, our sample is older, more socially conservative, and it consisted of more whites than the US population as whole. While whites and non-whites tended to give nearly identical answers to our survey questions, age and ideology were often correlated with attitudes. Future studies, such as the 2010 survey, will help determine whether these slight imbalances in our demographics affect the overall distribution of opinions reported here.

## 3.4 Survey Findings – Background Questions

This section reports on what we consider background questions, including measures of knowledge and perceived importance of the topic as well as general perceptions about risks and institutions that might protect the public from them. These items help place the substantive opinions that we analyze later into context.

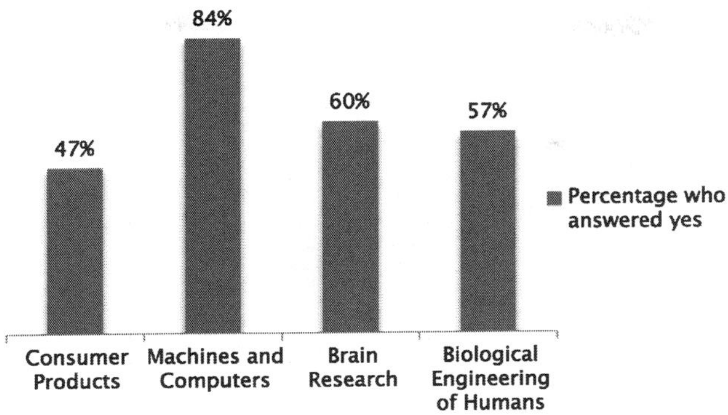


**Fig. 3.1** Do you associate nanotech with…?

### *3.4.1 Knowledge of Nanotechnology and Human Enhancement*

How much do people know about nanotechnology and human enhancement? The short answer is, not much. Respondents first self-rated their knowledge of nanotechnology in general on a scale of one to ten, with "10" standing for "very much" knowledge and "1" representing having heard nothing at all about it. Knowledge was reported, on average, as 2.9, and the plurality (38%) admitted to knowing nothing at all, or "1". Another 31% answered "2" or "3", while just 2% answered "9" or "10". Yet, nearly a third of the sample answered by picking a number from "4" through "8", suggesting a sizeable portion of the public has more than passing familiarity with nanotechnology. Turning to knowledge about nanotechnology for human enhancement specifically, awareness was even less widespread. Respondents again self-rated their knowledge on the 10-point scale, and the average score was only 2.1. Fully 61% of respondents answered that they had heard nothing at all, and another 20% answered "2" or "3".

### *3.4.2 Association of Nanotechnology with Fields of Application*

Survey respondents were asked to identify which of the four following areas of application they associated nanotechnology with: machines and computers, brain research, biological engineering, and consumer products. These data are displayed in Fig. 3.1.

Importantly, this question was only asked if respondents said they had heard something about nanotechnology. Among this sample of respondents (N = 345), the highest percentage associated nanotechnology with machines and computers (84%),

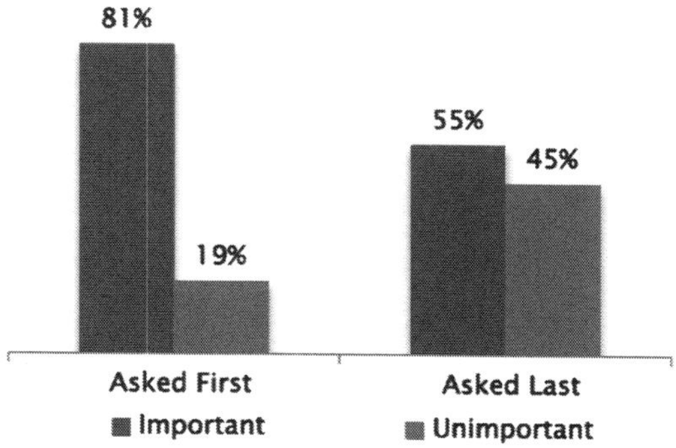


**Fig. 3.2** Estimation of human enhancement's overall importance

followed by brain research (60%) and biological engineering (57%). Consumer products was indicated as least likely (47%) to be associated with nanotechnology, which is interesting given the increasingly common presence of nanotechnology in consumer products.

The consumer products distribution is somewhat unsettling because the presence of nanotechnology is increasingly widespread in consumer products. Nano $TiO_2$ (nano titanium dioxide), for example, is widely used in cosmetics, sunscreens, and other skin products, and its long-term implications for consumer health and the environment are still not fully understood (Wiesner et al. 2006). Likewise, the use of nano-silver as an embedded antimicrobial in a variety of products is increasing in frequency, and, like nano $TiO_2$, its broader implications are relatively unknown. The fact that the public's associative awareness is misaligned with the heaviest areas of industrial development for nanotechnology is significant for our understanding of the sociotechnical contexts of both nanotechnology and human enhancement. Many in the public do not appear to be prepared to deal with these technologies in the order in which they are being introduced, or, more pointedly, the order in which they are likely to become most relevant to their individual lives.

### *3.4.3 Importance of Human Enhancement*

Respondents were asked to indicate the importance of enhancing mental, emotional, and physical capabilities to them personally. When asking this question, we conducted it as a question-order experiment to evaluate the importance of the issue to respondents. At random, half of the respondents were asked this question about importance at the beginning of the survey while the other half were asked the same exact question at the end of the survey. The results reveal a striking difference, as illustrated in Fig. 3.2.

At the beginning of the survey, a large majority of respondents indicated that this topic was important. After being exposed to the topic of human enhancement by the survey questions, however, significantly fewer respondents thought the subject was important.

The decline in importance ratings at the end of the survey has several possible interpretations. First, the decline suggests quite literally that respondents care less about it once they know more about it. If true, however, why is the case? One reason could be that the enhancement applications were initially deemed as threatening. In this scenario, importance ratings were highly correlated with anxiety about them. Importance declined at the end, then, because exposure to specific examples of potential applications made them less scary than assumed. However, the 2010 survey measured high levels of anxiety about these technologies at the end of the survey, suggesting that the decrease in importance in the 2008 survey was not driven by an increase in respondents' comfort levels as they became more knowledgeable about the types of applications we had in mind. Alternatively, the specific content in the survey questions, which are not representative of all human enhancement broadly considered, might have failed to impress respondents when compared to the hypothetical applications that they immediately conjured up in their minds at the start of the survey. Yet another possibility is that respondents deemed these applications implausible once more was known about them. In this case, respondents determined at the end of the survey that these technologies were purely speculative, and, therefore, were not likely to be relevant to them. Unfortunately, we do not have the kinds of measures on either of the two surveys to better explain this finding. Nevertheless, it also should be noted that even with the significant decrease in the percentage of respondents saying enhancement technologies were important to them, a majority of respondents always said they were important. Thus, as these technologies progress towards producing concrete applications, they have the potential to be highly salient to Americans.

### *3.4.4 Balance of Risks and Benefits*

Respondents were next asked to assess the relative balance of risks and benefits that they expect from using nanotechnology for human enhancement. We display these results in Fig. 3.3.

The plurality of respondents thought the risks and benefits would be equal to one another, while roughly equal remainders saw either the risks outweighing the benefits or vice-versa. How to interpret these responses is challenging. On the one hand, combining those seeing a balance with those seeing greater benefits suggests that most Americans do not see a net risk. On the other hand, fully one-third of Americans saw the risks as outweighing the benefits, which is a significant number. At the same time, another third thought the risks would be comparable to the benefits. While we did not do comparisons in this survey to other technologies, it seems unlikely that most technologies would be viewed as this risky. Think about it

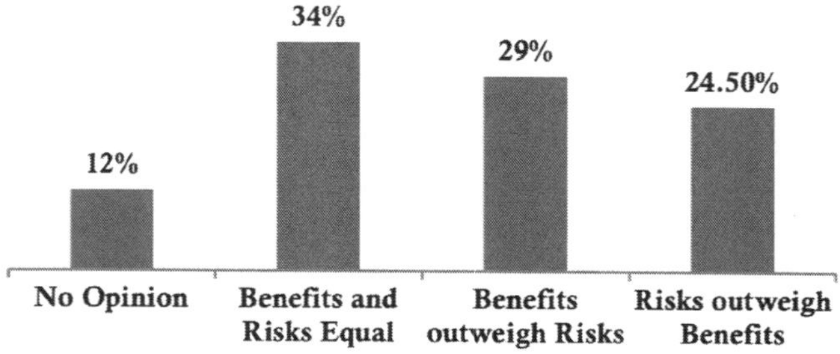


**Fig. 3.3** Evaluation of risks and benefits of nanotechnology for human enhancement

this way, if Americans thought the risks of cars were roughly equal to their benefits, would they drive them as much as they do? Digging deeper, we find that women were more likely to see risks. Among women, less than a quarter saw the application of nanotechnology to human enhancement as mostly beneficial, while 62% saw the risks as being equal or exceeding any benefits that might be gained.

Our data from 2010 support this somewhat skeptical attitude toward nanotechnologies used for human enhancement. In that study, respondents were asked to indicate how much they agreed or disagreed with the statement, "Using nanotechnology for human enhancement will be risky." Answers were recorded on a 10-point scale where "10" stands for "agree very much". The average score was over seven, indicating heightened risk perceptions. Of course, the scales are not directly comparable across surveys, and there is no reason to believe Americans became more aware about the risks of these technologies over time. One other difference across the surveys is that risks were assessed in the 2010 survey after many questions had already been asked about specific applications of the technologies, but they were assessed before these kinds of questions in the 2008 study. One tentative conclusion is that increased public awareness will be associated with increased perceptions of risks. Nevertheless, in evaluating how Americans are likely to respond to these technologies, we are beginning to see that context matters for evaluating them. Rather than making sweeping generalizations about risk and benefit, opinions appear to be contingent on the frame of reference.

### *3.4.5 Confidence in Protection from Risks*

Despite being relatively untrusting of most institutions to protect them from the risks of nanotechnology and human enhancement, nearly half reported trusting university scientists. As we show in Fig. 3.4, scientists were by far the most trusted source.

Just 3% trusted the mass media, and only 7% trusted government. Faring slightly better, 17% trusted business, 26% trusted environmental groups, and 27% trusted

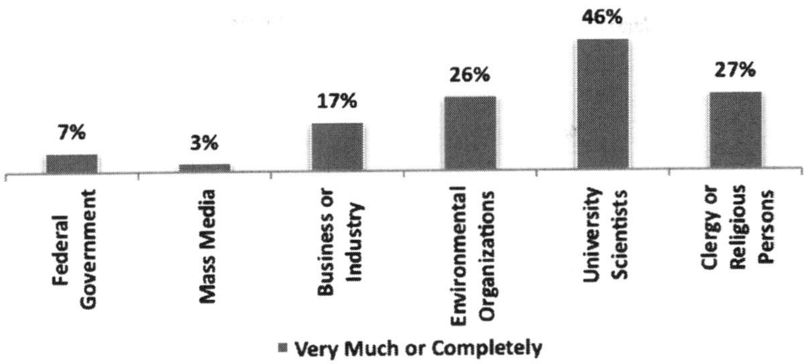


**Fig. 3.4** Public trust to protect against technological risk by institution in 2008

clergy and religious persons. The latter 2% are somewhat surprising. It is not patently obvious why religious figures would be trusted sources about nanotechnologies, (but see Scheufele et al. 2009), and mass media arguably portray environmental groups as untrustworthy. We repeated these trust questions in the 2010 survey, except for environmental groups, and the percentages were nearly identical except for university scientists. Although they remained the most trusted institution, only 28% said they very much or completely trusted university scientists in 2010. While a plurality (45%) said they "somewhat" trusted them, nearly identical to the 2008 results, distrust in scientists climbed from 10% to 25%. The decline could be due to several prominent public debates in between the two surveys in which university scientists were prominently featured, healthcare reform and climate change.

Interestingly, levels of trust across institutions were sometimes highly correlated in non-obvious ways. For example, while trust in university scientists is intuitively correlated with trust in environmental organizations at .50 ($p<.01$), it is also identically correlated at .50 ($p<.01$) with business. Otherwise, trust is only marginally correlated across most institutions, ranging from a high of .42 ($p<.01$) between government and business, and a low of .17 ($p<.01$) between clergy and religious persons and the mass media.

## 3.5 Survey Findings – Applications of Human Enhancement

The second major set of questions contained in the survey asked respondents about their views on specific applications of nanotechnology to human enhancement. We noted in the introduction that application, specifically medical applications, were a major driver of support for nano-enabled human enhancement, and in this section we will explore that finding in more depth.

### 3.5.1 Use of Nanotechnology to Enhance Visual Capabilities

We start by examining support for using nanotechnology to connect a video camera to the brain. We examined this topic experimentally, using a split ballot question. Roughly, half of the respondents received one version that described the potential benefit generically as allowing artificial eyesight. The other half heard the same technology but also described as curing blindness and possibly improving soldiers' vision on the battlefield. We designed this experiment to examine whether articulating a clear application that did, or did not, include enhancement potential would alter people's responses to the use of nanotechnology.

Surprisingly, the question wording experiment did not significantly affect responses. Support for using nanotechnology to connect a video camera to the brain was high for both the generic version (87%) and the version that discusses both curing blindness and improving battlefield vision (89%). The only meaningful difference resulted from a small number of individuals (4%) who volunteered in response to the applied version that they would support some uses but oppose others. The stipulation of nuance in response to this question by some respondents was mirrored in the surveyor notes on many questions throughout the survey, where respondents attempted to clarify some of the finer details of the technologies and scenarios before they would commit to support or opposition.

We then probed the moral acceptability of the technology. Regardless of the question wording condition, similar percentages of respondents said the application was morally acceptable (84% and 88%). Most of these people believed strongly that it was moral. One interpretation of the support for this technology is that if the primary application is therapeutic, then the technology is likely to be well received, no matter how obvious the enhancement potential may be. In considering this, it is important to remain conscious of how new drugs trend toward off-label usage. Even in a therapeutic context, if a technology has an enhancement capability, it will eventually be used that way, and this leaves an opening for advancements in enhancement technology research and development (R & D) that Americans might not otherwise approve of.

While most demographics did not predict levels of support, respondents' sex had a small impact. In the generic version of the eyesight question, women were slightly less likely than men to support this application of nanotechnology (82% vs. 88%, respectively). Women were also more likely than men (18% vs. 12%) to find the idea morally unacceptable. When both military and medical applications are mentioned, 87% of women and 91% of men supported it.

### 3.5.2 Comparison of Support for Diverse Human Enhancement Applications

Survey questions were also asked to directly compare four distinct applications of nanotechnology to human enhancement: drugs developed to prevent prisoner

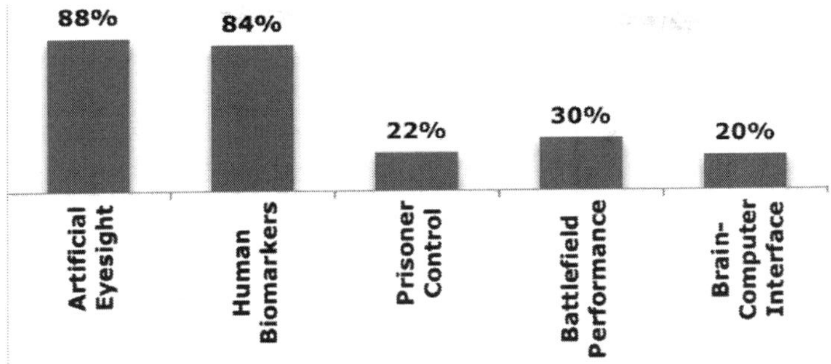


**Fig. 3.5** Support for cognitive enhancement by application

escapes; battlefield implants to improve soldier performance; medical devices for early disease detection; and implants to connect your brain to your computer. The results are presented in Fig. 3.5.

Respondents viewed these applications quite differently from one another. The only application receiving support from a majority of respondents (84%) was using biomarkers for early disease detection. Just 30% supported battlefield implants, 22% supported drugs to stop prisoner escapes, and 20% supported brain-to-computer interface technology. If we assume that respondents had a clear understanding of the kind of technology we were talking about, then the remarkably high level of support for medical devices to catch diseases early is indicative of the power of therapy to trump many other potential ethical and practical considerations. Arguably, using nanotechnology for bioassays to continuously monitor disease markers raises a large number of ethical concerns, from privacy concerns to the mechanisms for delivering and explaining results, but the potential for individual health benefits seems to override these concerns. For example, despite opposing brain-to-computer interface, it could be argued that this application raises fewer difficult political and ethical questions, and is simply a question of the personal desirability of sharing information with a computer. Yet, respondents saw it as the least favorable option.

We also find that some demographics matter for levels of support. Across these four questions, women consistently gave less support to human enhancement applications than men. These differences range from 4% on the question about the use of drugs to prevent prison escapes to 13% for the question about linking computers to brains. We see similar, and sometimes more dramatic, gender differences in the 2010 survey. For example, 73% of women oppose the use of drugs that could be used to help a person perform better during a job interview, but just 56% of men are opposed to this use of drugs. Our data indicates that gender will certainly play an important role in shaping the sociotechnical context for both nanotechnology and human enhancement, but it does not appear that it will play a decisive role given that men and women equally oppose or support these technologies overall.

On the margins, age also helps to explain opinions about human enhancement. Across all four applications, respondents over 50 were less likely to strongly support a particular human enhancement application than were respondents under 50, and they were more likely to strongly oppose rather than somewhat oppose them. Whether these differences merely represent different question answering propensities among older respondents or real differences in the intensity of opinions is unclear. Our first cut of the 2010 data suggests the age difference is replicated and larger, but also that it is more concentrated among those 18–29 compared to the rest of the population. Thus, while younger Americans are supportive of some of these technologies, we are unable to say whether this is a permanent cultural shift, or if they will age-out of this enthusiasm for enhancement technology.

Finally, we looked at the role of religiosity in mediating opinions. We measured religiosity as the degree of reported guidance provided by religion in the person's life, (originally measured on a 10-point scale, and collapsed into "low," "medium" and "high"). Prior research has identified religion as a key variable shaping views about emerging technologies like nanotechnology (Scheufele et al. 2009), but we find more equivocal evidence that religion substantively shapes opinions about four potential applications of enhancement technologies. Religious guidance is associated with, at most, a shift of about 10% support/opposition to any of these applications. For example, 22% of respondents support the use of drugs to prevent prisoner escapes regardless of their reported level of religious guidance. Yet, slight differences by religious guidance emerged on the remaining three items. Among those reporting low levels of religion as a guide, 35% support equipping soldiers with battlefield implants, while 26% feel this way if they reported a high level of guidance. Likewise, a 10% opinion gap emerges between those at the bottom and at the top of religious guidance for support for technologies to directly connect computers to brains (26% versus 16%, respectively). Finally, while respondents at each of the three levels of religious guidance support the use of drugs to detect cancerous biomarkers, those at the low end of religious guidance are more supportive of this particular application than those at the high end (91% versus 80%, respectively).

### *3.5.3 Comparison of Support for Enhancement of Child for Diverse Purposes*

To measure cultural acceptance of these technologies, we asked respondents to rate the likelihood that they would support a decision by their child to obtain an enhancement for the purposes of competing in four different activities. Although we did not measure whether the respondent indeed had a child, marital status did not affect answers to these questions, suggesting respondents treated the questions as a serious hypothetical. We view the four activities – getting a job, playing amateur sports, taking college entrance exams, and running for political office – as proxies for the four primary types of social competition within American political culture: economic, non-economic social, educational, and political. These questions help us

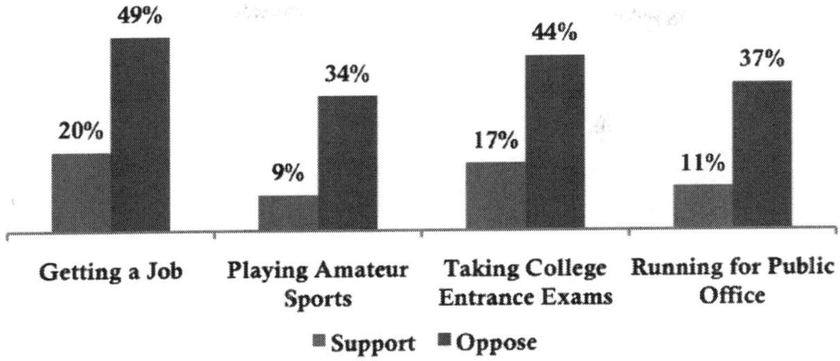


**Fig. 3.6** How likely would you be to support your child seeking enhancement for...?

probe how competitive context and the ethic of free and fair competition inherent in American political culture would affect Americans' evaluation of human enhancement. In fact, we would argue that the specific applications are less important here than the social context of the enhancement activity.

In accord with our understanding of how Americans tend to view social competition, respondents indicated opposition to the use of enhancements by their children on all four forms of competitive enterprise. This information is located in Fig. 3.6.

In each case, a majority of respondents indicated that they would be unlikely to support their child to seeking enhancements for social competition. On the other hand, in the two categories of social competition we believe are most relevant to Americans' individual self-interest – economic and educational – there are substantial numbers of people who either outright support or are neutral toward these applications. It is possible that sports was not the best proxy for non-economic social competition, as Americans tend to apply a very particular ethical system to sports, and doping, while increasingly common, is viewed with special disfavor. At the same time, responses to doping in sports may indicate that greater public exposure to and controversy over enhancement technologies may lead to greater negative polarization in the future. Enhancement for political purposes may also not have seemed as relevant or believable without specific examples, such as the use of cognitive enhancements by candidates. Results from the 2010 survey largely replicate these findings. Americans are more likely to support enhancement for the purposes of getting a job or taking a college entrance exam, and support levels vary according to factors like age and geographic region.

As with the responses to the prior question about support for particular applications of cognitive enhancement, women were less likely to support the use of some cognitive enhancement by their children. For example, there is a 12% gender gap on the question of using enhancements to get a job. Women were 6% less likely to support and 6% more likely to oppose their use for this purpose. Similarly, women were

8% less likely to support and 2% more likely to oppose enhancements for running for public office. In addition, we find that individuals reporting greater religiosity were less likely to support enhancing children. Overall, however, demographic factors like gender and religion seem to play a secondary and marginal role in determining support for these technologies.

## 3.6 Survey Findings – Values for Evaluating Human Enhancement Outcomes

The survey also asked a series of questions about the costs of these technologies and how to pay for them. Answers to these questions provide a first glimpse of the economic evaluations Americans might employ in assessing the value of enhancement technologies.

### *3.6.1 Thoughts on How to Pay for Human Enhancements*

In Fig. 3.7, we present findings about how expensive respondents thought human enhancements based on new technologies would be.

Overwhelmingly, respondents indicated that they expected enhancements based on new technologies would be quite expensive. Only 4% felt they would be affordable for most Americans. Another 62% felt they would be available to only the wealthiest Americans, while the remaining 32% felt enhancements would be quite costly for the average American. Subsequent analysis finds that these attitudes do not vary significantly by demographics like gender or age.

When asked to indicate how they thought society should decide who receives enhancements should they prove to be expensive, most (72%) believed that government should guarantee that everyone has equal access to these enhancements.

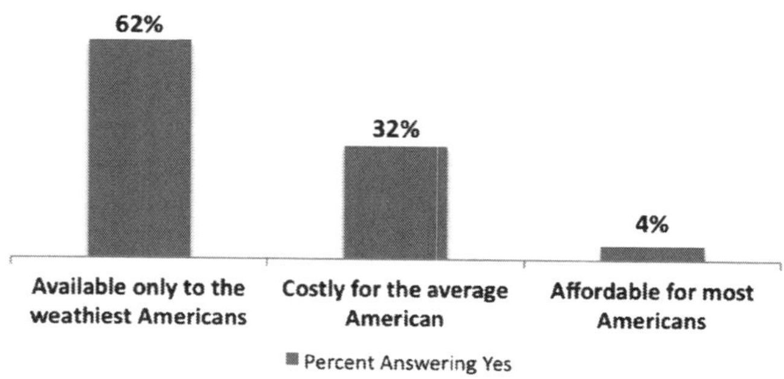


**Fig. 3.7** How costly will enhancement technologies be?

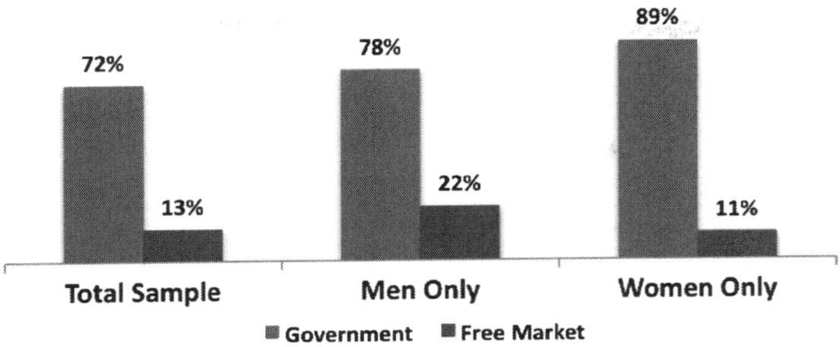


**Fig. 3.8** How should access to enhancements be guaranteed?

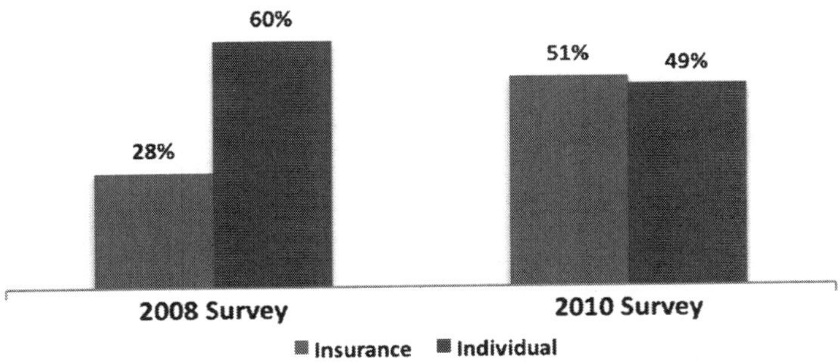


**Fig. 3.9** Who should have to pay for enhancement?

Conversely, just 13% thought we should let the free market decide who gets access to human enhancements. These results were sensitive to gender. If we discard respondents who answered "don't know," then the percentage of men willing to leave access decisions up to the market was double (22%) the percentage of women willing to do so (11%). These results are shown in Fig. 3.8.

Yet, if these technologies developed into actual enhancement products, respondents preferred individuals to have to pay out of pocket to receive them, not for medical insurance to cover them. Among those having an opinion, about seven in ten Americans said enhancements should be paid for out of pocket. This result seems to be at odds with their preference for government guaranteeing equal access since one obvious approach would be for government to require that they be covered by health insurance. Perhaps this latter finding indicates a sentiment that medical insurance is already too expensive, or perhaps it represents the difference between disliked enhancements for other purposes compared to favored therapeutic ones. Further complicating the matter, our 2010 results suggests Americans had warmed to the idea of insurance companies paying for enhancements. In the latter poll, shown in Fig. 3.9, 51% indicating that insurance should be required to cover them.

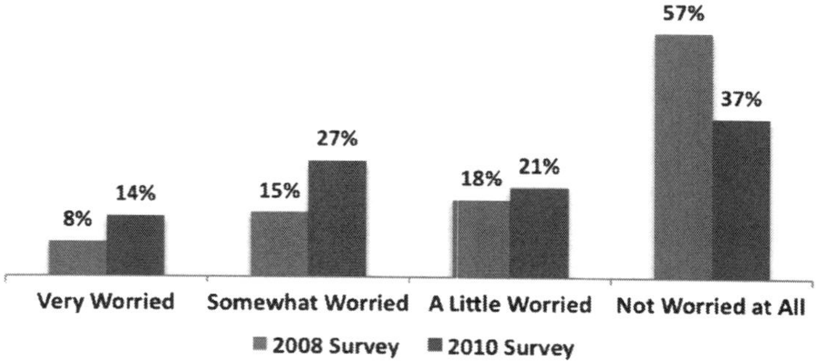


**Fig. 3.10** Are you worried about being able to afford enhancements for your family?

During this time, the health care debate was beginning, debated prominently, and passed, which might have affected Americans' opinions about what insurance should cover.

A final clue to the conflicting views comes from a question measuring how much respondents worried about being unable to afford enhancements for their family should they become available. Despite thinking they will be costly, survey respondents said they were not worried about being able to afford human enhancements. We show these data in Fig. 3.10.

Just 8% indicated feeling very worried, while 15% reported being somewhat worried. On the hand, 57% were not worried at all. Data from 2010 suggest worrying about their affordability increased, but even here 37% remained unworried at all. Like before, it is difficult to sort out the changing attitudes across the two surveys because of real world events (i.e., the health care debate) and potential question order effects across the two surveys, (i.e., "worry" was measured after exposure to different kinds of applications in 2000 versus 2010). Worry about affordability may also be impacted by two different potential reasons for that worry (or its lack): concern about its costs and relevance to the individual. In other words, individuals who are not worried about affordability may consider these technologies affordable (although this is disputed by our direct data on perceptions of cost) or because they do not expect to be buying them (which is supported by our data showing low importance for some respondents).

## 3.7 Framing Overall Support or Opposition to Human Enhancements

We also conducted a framing experiment. We explored support or opposition to enhancements when they were alternatively framed by two important discourses in society that surround new and emerging technologies: "improving humankind"

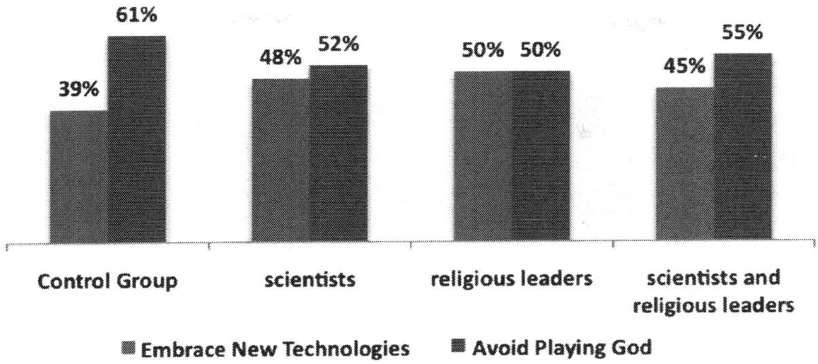


**Fig. 3.11** Technology framing experiment: variations in support when endorsed by scientists or opposed by religious leaders

versus "playing God". We also enhanced the frames by attributing specific sources to each position. Respondents heard that university scientists endorsed the "improving humankind" frame and "religious figures" said we should not "play God." The experiment proceeded by randomly assigning respondents to one of four conditions: (1) a control group unexposed to framing; (2) improving mankind frame when supported by scientists; (3) playing God frame when opposed by religious leaders; and (4) a competitive framing condition pitting scientists against religious leaders. We anticipated these frames would be effective, and that public support for embracing new technologies would rise and fall compared to the control group (no framing), depending on the direction of the frame (pro or con), but it was not obvious which frame will be more powerful, and what would occur when the two frames were presented simultaneously.

The results presented in Fig. 3.11 indicate that, surprisingly, any framing of the technology increases the opinion that we should embrace new technologies, even when the framing is opposed to them. When the question is framed as scientists believing these technologies will make us smarter, healthier, and live longer, it increases support for embracing them. However, when the question is framed so that religious figures believe we should avoid playing God with these technologies, support for embracing the technologies also goes up. Likewise, when respondents hear both ways of framing these technologies, support still increases compared to the control group.

These framing effects are unexpected to say the least. One possible explanation for why they did not work as expected is because the source trumps the frame only in the case of religious figures. We know, for example, that university scientists were the most trusted source for protecting the public against risks of technologies like these. Thus, it makes sense that support for embracing new technologies increases by 9% when respondents hear that university scientists say humankind

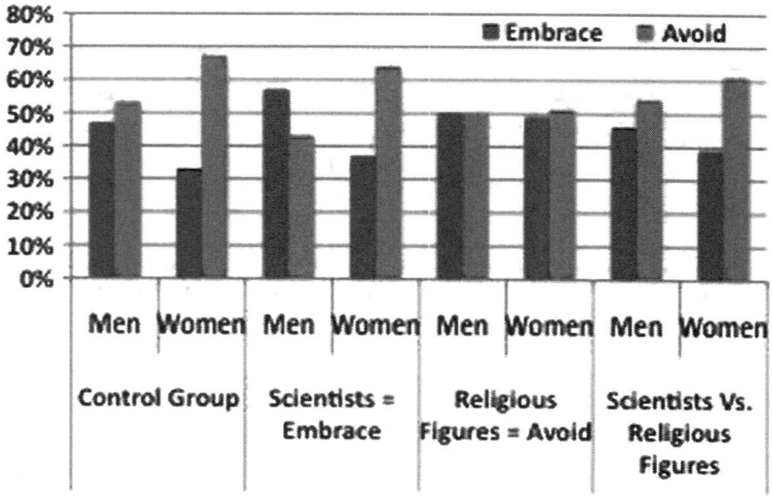


**Fig. 3.12** Gender breakdown of framing experiment

will be improved. Conversely, if religious figures are trusted less, their message about playing God might have been rejected. Yet, when we examined how trust mediated reactions to the frames (results not reported here but available upon request), this explanation did not hold up. Instead, it appears that gender mediates the reactions to the frames, and we show this in Fig. 3.12.

We find that men, but not women, were strongly influenced to embrace new technologies when university scientists endorsed them. Women, by contrast, are more opposed to these technologies in the control group, but their opposition recedes if religious figures also endorse this view. In the control group, 67% of women say we should avoid playing God, while 53% of men take this position. In the university scientists framing condition, men's initial opposition turns into 57% support while women's opposition only declines by 3% (from 67% to 64%). Conversely, women, but not men, respond quite negatively to religious figures telling them that we should avoid playing God with technologies. Compared to the control group, women's opposition drops from 67% to 50%, but men's declines just 3% (from 53% to 50%). Although the pattern is consistent, it is still not obvious why gender matters. We find, for example, no equivalent backlash among women in the two-sided framing condition. Furthermore, there is no gender gap in respondents' trust of religious or university sources. Moreover, women are much more likely than men to report religion is an important guide in their daily lives (43% versus 24% rate it as "high"). Future studies, though, should take care to examine for gender differences in responses to different kinds of messages about these emerging technologies when they are attributed to different sources, particularly religious ones.

## 3.8 Conclusion

It should be obvious by now that the opinion data we have gathered presents a rather complicated portrait of Americans' views about nanotechnology. Thus, many of the observations we make about American attitudes toward nanotechnology, and nano-enabled human enhancement technology, are suggestive and preliminary rather than definitive. Yet, some of these data are robust and sufficiently consistent that we can reach several important conclusions.

First, Americans are largely unaware of these technologies. There is little evidence that knowledge about them is increasing over time, (although see Corley and Scheufele 2010). This fact matters because production of various kinds of applications could outpace public acceptance of them (Barben 2010). Second, we can also say that Americans perceive such technologies as at least somewhat risky. In some sense, this is the case with all new technologies, and it partly is a function of their "newness." Yet, both nanotechnology and human enhancement technology represent a significant advance beyond the technology with which Americans are already familiar, and, as always, it is a mistake to try to draw linear historical analogies in an attempt to predict how people will respond to technological innovation. In this sense, the trepidation with which American's view these new technologies is important in understanding how the various institutional actors involved in the innovation communities producing them should proceed (Walker et al. 1999). How should industry communicate risks and benefits to the public or how should policymakers prepare themselves to respond to the shifts in social and political dynamics that will accompany these technologies?

The third and most obvious result is that there is substantial value in the "therapy" label. Devices described, or perceived, as being therapeutic garner public acceptance even while significant opposition exists to the general concept of nano-enabled human enhancement technology. The power of the "therapy" label could be both a blessing and a curse. On the one hand, it indicates that there is some flexibility in the amount of development Americans are likely to tolerate in terms of human enhancement technology. If the development and introduction of these technologies is properly managed, it is possible to avoid much of the opposition that met genetically modified foods. On the other hand, the "therapy" label can also act as a device to prevent critical public evaluation of technologies with obvious enhancement capabilities. If we take psycho pharmaceuticals as a model, their therapeutic label has, largely, served to mask the fact that they are being prescribed off-label for enhancement purposes in ever-greater numbers. Despite the attempt by Greely et al. (2008) to stir the pot in their *Nature* commentary, there has not been a substantial national discussion of the role of these drugs in lives of college, and even high school, students, and their impact on the competitive structures embedded within the American political culture. The implications for American meritocracy, if such a thing truly exists, are substantial, (see, e.g., Carson 2007), and, yet, in part because of their therapeutic label, there has not been a sustained attempt to evaluate the role of drugs like *Adderall* and *Provigil* in skewing competitive outcomes at the college level, and beyond.

Finally, we can also say that the economic dimension to human enhancement technologies, nano-enabled and otherwise, will likely be problematic. Americans believe that these devices are going to be costly, and their answers to other questions suggest they will demand that either the government or insurance companies guarantee access to them. Thus, these technologies have the potential to place significant additional strain on the American healthcare system, and, possibly, on federal entitlement outlays. At the same time, most people expressed surprisingly little worry about their cost to their own families. It is important going forward to reconcile these seemingly different accounts. If, for example, the discrepancy between predicted costliness and the lack of worrying about their affordability stems from a lack of interest in the actual products or the perception that these things are fantastical, then they are not prepared to start thinking in advance about the economic changes they are likely to produce, or the political actions that will be necessary to cope with these changes. If, however, the lack of worry stems from the belief that government will eventually guarantee access, then that places added pressure on the government to intervene against the outcomes of market forces that are already being determined. Clearly, future research is needed to further plumb the process of opinion formation about these technologies so that these kinds of data can be of greater value to policymakers, university scientists, and the business community.

## Appendix 1

Q1. Recently there has been some talk about new technologies usually referred to as nanotechnology, or nanotech for short. How much have you heard or seen about this topic? Using a 10 point scale with 1 being nothing at all and 10 being very much, which number between 1 and 10 would best represent how much you have heard, read or seen about nanotechnology?

Q2. Now I would like to know which of the following areas, if any, you associate with nanotechnology and its potential uses. Do you associate nanotech with: (a) consumer products, (b) machines and computers, (c) brain research, and (d) biological engineering?

Q3. For the following questions, we would like you to think about nanotechnology as new technologies that allow scientists to manipulate materials at the level of tiny molecules. Recently there has been some talk about using nanotechnology to enhance human mental, emotional, or physical abilities. Using a 10 point scale with 1 being nothing at all and 10 being very much, which number between 1 and 10 would best represent how much you have heard, read or seen about using nanotechnology to enhance human mental, emotional, or physical abilities?

Q4. How important to you is the issue of enhancing human mental, emotional, and physical abilities? Would you say very important, somewhat important, somewhat unimportant, or not at all important?

Q5. What do you think about the risks and benefits of using nanotechnology for human enhancement? Do you think the risks of using nanotechnology for human enhancement will outweigh its benefits, the risks and benefits will be about equal, or will the benefits outweigh its risks?

Q6. How much confidence do you have in (a) the federal government or (b) the business industry protecting the public from significant risks associated with nanotechnology? Would you say you have no confidence, very little confidence, some confidence, quite a bit of confidence, or complete confidence?

Q7. Some people trust the following people and institutions to balance the risks and benefits of human enhancement, while other people do not trust these institutions to balance the risks and benefits. I will read a list of these people and institutions, and for each one, I would like you to tell me how much YOU trust them to balance the risks and benefits of human enhancement. First, the federal government. Would you say you trust the federal government not at all, not very much, somewhat, very much, or completely? (a) The federal government; (b) the mass media; (c) business or industry scientists; (d) environmental organizations; (e) university scientists; (f) clergy or religious persons.

Q8a. One possible use of nanotechnology is to connect a video camera to the human brain to allow artificial eyesight. Would you support or oppose this use of nanotechnology? Do you feel that this use of nanotechnology is morally acceptable or morally unacceptable? How strongly do you feel this way? Strongly or not very strongly?

Q8b. One possible use of nanotechnology is to connect a video camera to the human brain to allow artificial eyesight. This might allow doctors to cure blindness, or soldiers to improve their vision on the battlefield. Would you support or oppose this use of nanotechnology? Do you feel that this use of nanotechnology is morally acceptable or morally unacceptable? How strongly do you feel this way? Strongly or not very strongly?

Q9a. The next questions are about how much you support or oppose certain kinds of human enhancements that might 1 day be possible. First, the administration of drugs to prisoners to prevent prison escapes. Would you say you strongly support, somewhat support, somewhat oppose, or strongly oppose this practice, or are you neutral?

Q9b. Next, the use of battlefield computer implants, to help soldiers perform better.

Q9c. The use of medical devices to detect changes in human biomarkers such as blood pressure or protein levels to catch diseases before they become dangerous.

Q9d. The use of implants to transmit computer information to the brain while asleep, or to plug in to virtual realities on the Internet?

Q10a. Getting a job? Would you be very likely, somewhat likely, somewhat unlikely, or very unlikely to allow your child to seek enhancements in order to get a job, or do you not have an opinion?

Q10b. Competing in amateur team sports? Would you be very likely, somewhat likely, somewhat unlikely, or very unlikely to allow your child to seek enhancements in order to compete in amateur team sports, or do you not have an opinion?

Q10c. Taking college entrance exams? Would you be very likely, somewhat likely, somewhat unlikely, or very unlikely to allow your child to seek enhancements in order to take college entrance exams, or do you not have an opinion?

Q10d. Running for public office? Would you be very likely, somewhat likely, somewhat unlikely, or very unlikely to allow your child to seek enhancements in order to run for public office, or do you not have an opinion?

Q11. Do you think that when human enhancements based on new technologies are brought to the market that they will be affordable for most Americans, quite costly for the average American, or available to only the wealthiest Americans?

Q12. If it is very expensive to obtain human enhancements, how should we decide who receives them? Do you think a person's wealth should determine access to them, or should the government guarantee everyone has equal access to these enhancements?

Q13. Once they become available, should medical insurance be required to cover most kinds of human enhancement, or should people that want enhancements have to pay out of their own pocket?

Q14. How worried are you that you and your family will not be able to afford drugs and other treatments for human enhancements once they become available? Are you very worried, somewhat worried, only worried a little, or not worried at all?

Q15. Now we have some questions about some of your general views about American society and how it works. I will be reading some statements. After hearing each statement, please tell me if you strongly agree, somewhat agree, neither agree nor disagree, somewhat disagree, or strongly disagree?

a. The government interferes far too much in our everyday lives.
b. It is not the government's business to try to protect people from themselves.
c. People should be able to rely on the government for help when they need it.
d. Our society would be better off if the distribution of wealth was more equal.
e. We have gone too far in pushing equal rights in this country.
f. Discrimination against minorities is still a very serious problem in our society.

Q16. Next, I would like to ask about your feelings about these new enhancement technologies.

a. Do you think we should embrace new enhancement technologies to improve humankind or should we avoid playing God with new enhancement technologies?
b. According to university scientists, nanotechnologies used for human enhancements will make us smarter, healthier, and live longer lives. What do you think?

Should we embrace new enhancement technologies to improve humankind or should we avoid playing God with new enhancement technologies?

c. According to religious figures, nanotechnologies used for human enhancements are like playing God and should be avoided. What do you think? Should we embrace new enhancement technologies to improve humankind or should we avoid playing God with new enhancement technologies?

d. According to university scientists, nanotechnologies used for human enhancements will make us smarter, healthier, and live longer lives. According to religious figures, however, nanotechnologies used for human enhancements are like playing God and should be avoided. What do you think? Should we embrace new enhancement technologies to improve humankind or should we avoid playing God with new enhancement technologies?

## References

Barben, Daniel. 2010. Analyzing acceptance politics: Towards an epistemological shift in the public understanding of science. *Public Understanding of Science* 19(3): 274–292.

Barben, Daniel, et al. 2008. Anticipatory governance of nanotechnology: Foresight, engagement, and integration. In *The handbook of science and technology studies*, ed. E. Hackett et al. Cambridge: MIT Press.

Carson, John. 2007. *The measure of merit: Talents, intelligence, and inequality in the French and American Republics, 1750–1940*. Princeton: Princeton University Press.

Corley, Elizabeth, and Dietram Scheufele. 2010. Outreach gone wrong? When we talk nano to the public, we are leaving key audiences behind. *Scientist* 24(1): 22.

Greely, Hank et al. 2008. Reprinted as Ch. 14 in this volume.

Guston, David, and Daniel Sarewitz. 2002. Real-time technology assessment. *Technology in Society* 24(1–2): 93–109.

Scheufele, Dietram, and Bruce Lewenstein. 2005. The public and nanotechnology: How citizens make sense of emerging technologies. *Journal of Nanoparticle Research* 7(6): 659–667.

Scheufele, Dietram, et al. 2009. Religious beliefs and public attitudes toward nanotechnology in Europe and the United States. *Nature Nanotechnology* 4: 91–94.

Walker, Gordon, et al. 1999. Risk communication, public participation, and the Seveso II Directive. *Journal of Hazardous Materials* 65(1–2): 179–190.

Wiesner, Mark, et al. 2006. Assessing the risks of manufactured nanomaterials. *Environmental Science and Technology* 40(14): 4336–4345.

# Chapter 4
# U.S. News Coverage of Neuroscience Nanotechnology: How U.S. Newspapers Have Covered Neuroscience Nanotechnology During the Last Decade

**Doo-Hun Choi, Anthony Dudo, and Dietram A. Scheufele**

## 4.1 Introduction

Nanotechnology has become one of the fastest-growing emerging technologies in the United States. President Bill Clinton established the National Nanotechnology Initiative (NNI) to increase federal investment in nanotechnology research and development in 2001. In 2003, the U S Congress enacted the twenty-first Century Nanotechnology Research and Development Act to evaluate and promote federal nanotechnology research, development, and other activities. This research and development in nanotechnology has led to the advancement of nanotechnology applications and products. For example, consumers can find over 1,000 nanotechnology applications on the market, ranging from tennis rackets to skin care products (Scheufele and Dudo 2010). The social penetration of nanotechnology has raised concerns as an important social issue associated with its potential benefits and risks (Scheufele and Lewenstein 2005). News media have been covering nanotechnology issues for decades, helping to shape how the public understand and perceive the new technology, and whether they support the technology or not (Scheufele and Lewenstein 2005; Weaver et al. 2009).

The development of nanotechnology has great potential to produce a wide range of applications in areas such as agriculture, food, medicine, health, the environment, energy, and information technology (Roco and Bainbridge 2001). In particular, neuroscience nanotechnology, one of the most important applications in nanotechnology, is designed to resolve many of the challenges and issues of neuroscience and medicine (Silva 2006a). Neuroscience nanotechnology explores how the nervous system operates, how the technology contributes to medical care and prevention,

D.-H. Choi (✉) • A. Dudo • D.A. Scheufele
Life Sciences Communication, College of Agricultural & Life Sciences, University of Wisconsin-Madison Hiram Smith Hall, 1545 Observatory Drive, WI 53706-1215, Madison

S.A. Hays et al. (eds.), *Nanotechnology, the Brain, and the Future*, Yearbook of Nanotechnology in Society 3, DOI 10.1007/978-94-007-1787-9_4,

and how brain-machine interfaces help us to understand brain function, addressing both positive and negative aspects (Parpura 2008; Silva 2006a). From a positive perspective, neuroscience nanotechnology promises better health care for humans. It plays an important role in diagnosing and treating neural (brain) diseases such as Parkinson's, epilepsy, and depression (Keefer et al. 2008), delivering drugs and small molecules across the blood-brain barrier (Silva 2004), preventing retinal disorders that cause blindness (Silva 2006b), and promoting neuron cell growth (Kuzma 2007). However, the increasing use of nanotechnology in neuroscience has raised concerns about the potential risks associated with nanoparticles, nanomaterials, and carbon nanotubes. Recent research has suggested that some nanomaterials have negative effects on brain, skin, and lung tissues and accumulate in the human body (Barnard 2009; Hoet et al. 2004). For example, it is possible that a drug delivery system mediated by nanoparticles across the blood-brain barrier amplifies neurotoxicity or moves into unintended passages (Hoet et al. 2004; Kuzma 2007).

## 4.2 Public Perceptions of Nanotechnology

Public opinion plays an important role in determining the continuing development of nanotechnology and science, more broadly (Mooney 2010). Much of the research examining public attitudes toward nanotechnology has explored the public familiarity with the emerging technology and how the public perceives the risks and benefits it poses (Currall 2009). Recent research has shown that the public is largely uninformed about nanotechnology (Cobb and Macoubrie 2004; Scheufele and Lewenstein 2005), and that levels of actual knowledge about the technology have remained low since 2004 (Scheufele et al. 2009). This lack of knowledge, however, has not stopped the public from forming attitudes about the technology. Studies have shown that the American public perceives that the benefits of nanotechnology outweigh the risks and that overall opinions toward the emerging technology are positive (Bainbridge 2002; Cobb and Macoubrie 2004).

How do citizens form their opinions toward nanotechnology with little or no knowledge? Instead of direct experiences with nanotechnology, it is likely that individuals rely on cognitive shortcuts or heuristics, such as ideological orientations, religious beliefs, or cues from the mass media, in order to form their attitudes about the scientific issues (Scheufele and Lewenstein 2005; Scheufele et al. 2009). In particular, media portrayals of the issue play a key role as a heuristic in forming public attitudes toward nanotechnology (Scheufele 2006). As an important source of information about scientific issues in general, and nanotechnology in particular, news media often present messages in certain ways that can influence audience interpretation (Scheufele and Lewenstein 2005; Scheufele 2006). For example, news media have often emphasized the benefits and positive aspects of agricultural biotechnology through their common focus on scientific developments and economic progress (Brossard and Nisbet 2006; Nisbet and Lewenstein 2002). Moreover, recent research has shown that the positive framing of nanotechnology in news media coverage contributes to promoting positive perceptions among news audiences (Scheufele and Lewenstein 2005). In short, for novel issues like neuroscience nanotechnology, news media play a significant role in raising public awareness of the issue and its potential risks and benefits.

## 4.3 Media Coverage of Nanotechnology

Numerous studies have explored patterns in the media coverage of nanotechnology, particularly focusing on the number of news stories, the overall tone, and the thematic characteristics (e.g., Dudo et al. 2009; Friedman and Egolf 2005; Gaskell et al. 2005; Stephens 2005; Weaver et al. 2009). These analyses of media coverage have provided a relatively comprehensive understanding of how the mass media have portrayed nanotechnology.

Despite the relatively low level of public awareness about the technology, there has been steadily increasing coverage of nanotechnology over time, except for a small decline after 2006 (Dudo et al. 2009). A number of significant key events in the nanotechnology realm, including the discovery of nanotubes in 1993 and the establishment of the NNI proposed by President Bill Clinton in 2001, have led to an increase in media coverage about the emerging technology (Dudo et al. 2009). Since 2006, however, the amount of news articles on nanotechnology has shown a modest downward trend because newsworthy moments related to nanotechnology have been rare over the past few years (Dudo et al. 2009).

Much of the content analyses also investigated the overall tone and the thematic emphases presented in the articles. Previous studies have found that news stories about nanotechnology have tended to, overall, carry a positive tone and emphasize the beneficial aspects of nanotechnology (Dudo et al. 2009; Gaskell et al. 2005; Stephens 2005). Even though the labels are different across studies, the thematic patterns of media coverage have also steadily emphasized "progress" (Weaver et al. 2009) and "business" (Dudo et al. 2009) in nanotechnology. It is possible that many news stories have largely covered the development of the engineering and commercial aspects of nanotechnology (Dudo et al. 2009). However, some thematic shifts in coverage might be occurring, as recent studies have noted increase in the presence of coverage, discussing "regulations," "health," and "potential risks" (Dudo et al. 2009; Weaver et al. 2009). It is likely that news stories have stressed the advancement of health-related nanotechnology applications, the concerns about negative effects on health, and a growing voice for the regulation of nanotechnology over the past few years.

Building on this body of content analytic research, we explore how media portray a specific application of nanotechnology: neuroscience nanotechnology. This study tracks the evolution of American newspaper coverage of neuroscience nanotechnology, identifying the descriptive and thematic emphases that have characterized this coverage over time.

## 4.4 Research Questions

Our examination of neuroscience and nanotechnology newspaper coverage focuses on the volume of coverage, authorship patterns, and thematic emphases. Our first question examines the number of newspaper articles about neuroscience nanotechnology. Since our sample covers a 30-year period, from 1980 to 2009, we are able to explore this question over almost the whole life span of nanotechnology.

RQ1: How has the number of neuroscience and nanotechnology news stories changed over time?

Given the scientific nature of neuroscience nanotechnology, it is possible that the mass media consider the issue as being specialized and allocate its coverage to science journalists and experts. This may result in an "inner club" of a select few journalists who dominate coverage of the topic (Dunwoody 1980). It is also possible given the drastic transformations occurring within the newspaper industry (e.g., downsizing journalists, eliminating science sections, etc.) that unspecialized journalists will be authoring a large amount of neuroscience nanotechnology coverage. With our second research question, we examine the concentration of the number of journalists writing about this topic.

RQ2: Which journalists are authoring newspaper articles about neuroscience nanotechnology?

Identifying the presence of thematic emphases, helps illuminate dominant media narratives in media coverage. We examine our sample for the presence of two broad themes: "benefits" and "risks". It is likely that news media often use the benefits and risks themes to transform the complicated issues into a simplified format. It can influence news audiences, perceiving their opinions particularly in the benefit and the risk aspects toward neuroscience nanotechnology. The benefits theme includes positive and advantageous perspectives, while the risks theme reflects negative and harmful perspectives of this new application of nanotechnology.

RQ3: How does the presence of the risks and benefits vary in newspaper coverage of neuroscience nanotechnology?

## 4.5 Methods

Our analysis was conducted on a representative sample of U.S. newspaper coverage of nanotechnology. We developed a complex Boolean search term and used the LexisNexis database to identify nanotechnology-related news articles published in the 21 different newspapers between January 1, 1980 and December 31, 2009 (See Appendix 1 for more details about the sample). The search term enabled us to maximize our ability to identify articles about nanotechnology, while simultaneously limiting the number of false positives (e.g., stories about Apple's iPod nano product) (See Appendix 2 for the Boolean search term). The final sample included 1,965 news stories about nanotechnology. We created a sub-sample of news stories about neuroscience applications of nanotechnology from the main sample that were published between 1980 and 2009. The sub-sample contained keyword stems representing neuroscience nanotechnology: brain, neur, nerv, cerebr, and mri. The extraction process yielded 345 news articles about neuroscience nanotechnology. Neuroscience nanotechnology is closely related to the issue of health nanotechnology, so to compare patterns of the number of news stories about the two issues, we created another sub-sample of news articles about health nanotechnology from the main sample. The health sub-sample included the keywords health, medicine, cancer, toxin, and asbestos, and resulted in 727 news stories about health nanotechnology. The two sub-samples, neuroscience nanotechnology and health nanotechnology, were not mutually exclusive (i.e., if an article included a keyword [or keywords] from both lists, it was included in both the neuroscience and health sample).

We examined descriptive features, such as the number of published news stories per year and patterns of authorship, and the presence of generalized risk and benefit language in the news coverage about neuroscience nanotechnology. The presence of generalized risk and benefit language was measured by counting mentions of specific sets of root words. The risk theme included the following eight root words: risk, hazard, danger, threat, harm, exposure, peril, and adverse. The benefit theme was composed of the following eight root words: benefit, breakthrough, promise, advantage, revolution, innovation, discovery, and useful. Where appropriate, multiple forms of each root word were also counted. To best address our research question dealing with the risks and benefits themes, we controlled for the number of newspaper stories published each year so as to provide a more valid assessment of how often each theme appeared over time. We did not assess inter-coder reliability because our use of computer-based analysis renders perfect reliability.

## 4.6 Results

### *4.6.1 Amount of Coverage*

Figure 4.1 describes the amount of US newspaper coverage about nanotechnology, and the relative proportions of that coverage that were about neuroscience nanotechnology and health nanotechnology, during the last 30 years (RQ1). It shows that the general tendency toward coverage of nanotechnology has been consistently increasing during the period with the exception of the last 3 years. Coverage of neuroscience nanotechnology and health nanotechnology have each reflected this pattern, but on a smaller scale. News stories about nanotechnology started to appear in the mid- to late-1980s, while coverage about neuroscience and health, nanotechnology did not

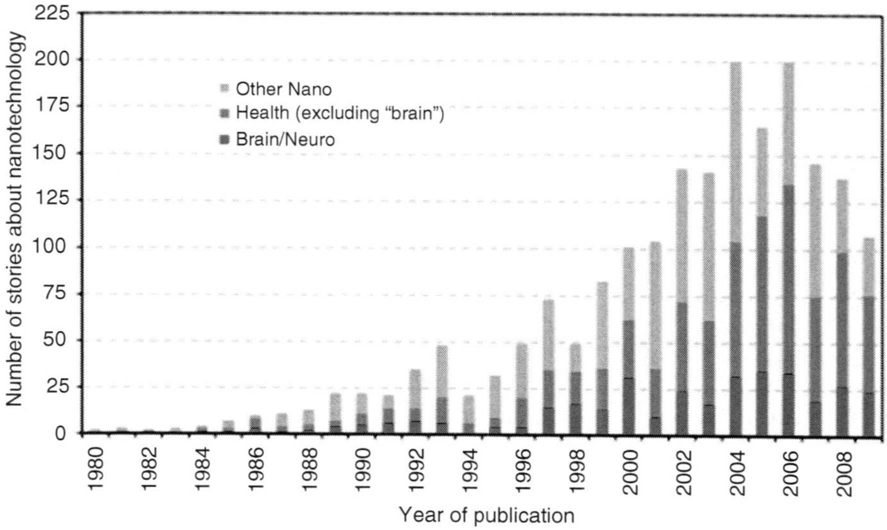


**Fig. 4.1** Coverage of nanotechnology, neuroscience, and health nanotechnology in U.S. newspapers

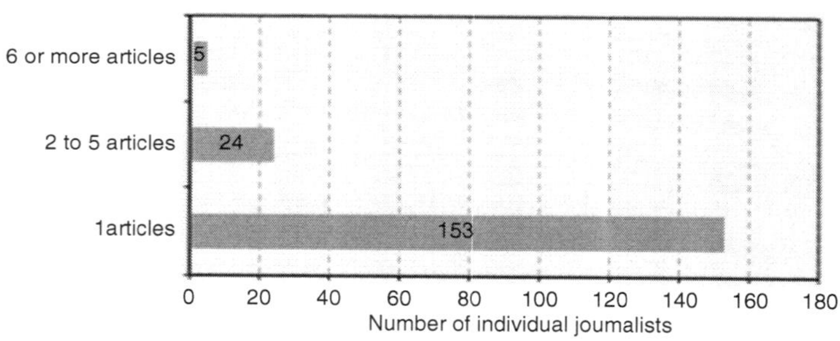


**Fig. 4.2** Authorship frequency of newspaper articles about neuroscience nanotechnology

begin to emerge until the late 1990s. In particular, neuroscience nanotechnology-related news stories experienced peak periods in 2004 ($n=32$), 2005 ($n=35$), and 2006 ($n=34$), and, like nanotechnology coverage, declined in the following years. News coverage of health nanotechnology also reached its zenith in 2004 ($n=72$), 2005 ($n=83$), and 2006 ($n=101$), and decreased in the ensuing years. As shown below, the authorship and conceptual themes (risks and benefits) of neuroscience nanotechnology (RQ2 and RQ3) were analyzed from the beginning of 1997–2009 because 1997 was the first year that more than ten stories about the neuroscience application were printed in the newspapers included in our sample.

**Table 4.1** Journalists in our sample who have wrote more than six stories about neuroscience nanotechnology[a]

| Journalist: | Newspaper affiliation: | Number of articles: |
|---|---|---|
| Barnaby J. Feder[b] | New York Times | 12 |
| Rick Weiss[b] | Washington Post | 10 |
| Henry Fountain | New York Times | 8 |
| Eric Berger | The Houston Chronicle | 8 |
| Kenneth Chang | New York Times | 6 |

[a] This table reflects only the newspapers that were included in our sample and subsequent analyses, and is not an exhaustive list of all U.S. journalists who have written extensively about nanotechnology. This table is based on an analysis of news coverage about neuroscience nanotechnology published between 1997 and 2009

[b] These journalists are no longer working in the print newspaper industry

### 4.6.2 Authorship

RQ2 asks about authorship trends in US newspaper coverage of neuroscience nanotechnology. Figure 4.2 indicates 208 different journalists wrote stories about this issue, and that the majority of these journalists authored only one article about neuroscience nanotechnology. Only six journalists penned six or more stories, and a mere 31 journalists wrote between two and five stories about neuroscience nanotechnology. Table 4.1 provides details about the most productive group of neuroscience nanotechnology journalists, including their names, newspaper affiliations, and the number of stories they wrote. All of them are affiliated with large-circulation newspapers, and two of these five journalists, Barnaby J. Feder of *The New York Times* and Rick Weiss of the *Washington Post*, no longer work in the newspaper business.

### 4.6.3 Themes

RQ3 asks how often the two themes, risks and benefits, have appeared in US newspaper stories about neuroscience nanotechnology. Figure 4.3 shows the average number of mentions per story for the two themes between 1997 and 2009. Overall, benefits have been mentioned more (2.38 mentions per story) than risks (1.82 mentions per story) over this time. With the exception of 2001, benefits were mentioned more often than risks from 1997 through 2002. Since 2003, however, the mentions of risks and benefits in coverage of neuroscience nanotechnology have been much more similar. In fact, the risks have been mentioned more than benefits in 2003, 2005, and 2007. These trends suggest that mentions of risks are becoming increasingly prevalent in journalistic accounts of this issue.

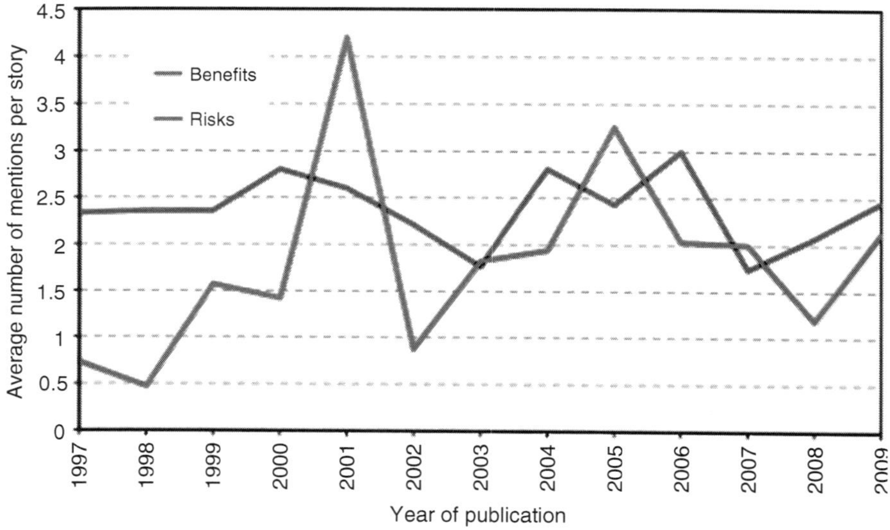


**Fig. 4.3** Presence of conceptual themes, benefits and risks, in newspaper stories about neuroscience nanotechnology

### *4.6.4 Discussion*

Public acceptance will largely determine the future development of nanotechnology, including its application to neuroscience. Mass media have been shown to play a significant role in shaping public perception about nanotechnology (Scheufele and Lewenstein 2005; Brossard et al. 2008). Seeking to understand the role of media in the formation of public perceptions of neuroscience nanotechnology, this study tracked the evolution of US newspaper coverage of this issue, exploring the amount of coverage, authorship patterns, and thematic emphases that have characterized this reporting. To our knowledge, this is the first study to empirically examine journalistic coverage of neuroscience nanotechnology.

Before discussing the implications of our findings, we should mention some limitations. First, although our sampling technique provided a substantial number of newspaper articles ($N = 1{,}965$), a hybrid of LexisNexis and other online sampling methods, such as Google News, would likely have produced an even larger sample (Weaver et al. 2009). Second, our computer-aided content analysis method maximized the reliability of our analyses, but limits our ability to capture more latent aspects of news coverage (Riffe et al. 2005). We encourage researchers to build on our study using other methods that can explore latent meanings in neuroscience and nanotechnology media coverage. Third, news audiences are increasingly moving to media outlets other than newspapers to find information about science. For example, approximately 40% and 20% of all Americans are acquiring science and technology news from television and the Internet, respectively (Horrigan 2006). Additional research to explore media portrayals about neuroscience nanotechnology should include television and Internet coverage.

Despite these limitations, we found a continuous increase in coverage of neuroscience nanotechnology over time, with a slightly decreasing trend after 2006. In particular, since 1997, the emergence of neuroscience and nanotechnology news stories has also coincided with that of general nanotechnology and health nanotechnology coverage. Since the late 1990s, neuroscience has been interested in applying nanotechnology to health issues, particularly to healing brain and nerve diseases and improving brain function. Those scientific developments might be enough to attract moderate attention from journalists. However, despite the continued growth of neuroscience nanotechnology research and development, there could be at least two possible explanations for the recent decline in press coverage. First, it is possible that the number of newsworthy events that would attract media attention, such as the Nobel Prize for the discovery of buckyballs in 1997, have been rare in recent years. Another possible explanation for the recent dip in neuroscience nanotechnology is the shifts underway within journalism. The number of science writers and reporters has declined in recent years (Mooney 2008), and mass media, including newspapers, are also assigning only a small percentage of their news hole to science and technology issues compared to other topics (The Pew Research Center's Project for Excellence in Journalism 2008). These changes are likely contributors to the decreasing volume of neuroscience nanotechnology coverage.

Our authorship analysis in this study found that a small number of journalists have covered a large proportion of news stories about neuroscience nanotechnology. As Fig. 4.2 shows, more than 80% of the journalists in the sample wrote only one article, while only six individual journalists authored six or more stories and 31 journalists penned between two and five stories, accounting for less than 20% of the total coverage of neuroscience nanotechnology. This pattern has an important implication. While many scientific issues are commonly covered by a group of specialist journalists on science and technology, in the case of neuroscience nanotechnology, this expert group is small. It is likely those journalists whose major area of specialty is not in nanotechnology, or even science journalism, are writing the majority of neuroscience nanotechnology articles. Whether or not the coverage generated by non-specialist journalists is inferior to the coverage generated by specialist is an empirical question, but it is likely that this coverage may differ in numerous ways, some of which may have ramifications for how readers come to perceive neuroscience nanotechnology.

These changes will have an effect on the news stories of neuroscience nanotechnology in terms of how frequently and how well it is covered. Taken together, the current media transformation and audience shifts portend future coverage of neuroscience nanotechnology that is likely to become more generic and event-driven.

Our content analysis also suggests that the disparity between mentions of risks and benefits in coverage of neuroscience nanotechnology has reduced in the years since this coverage first started appearing in 1997. Unlikely early coverage, which mentioned benefits more often, mentions of risks and benefits have been more equitable since 2003. This result is consistent with patterns unearthed in research examining media coverage of nanotechnology in general (Dudo et al. 2009) and suggests that the issue cycle for neuroscience nanotechnology, while still young, is moving into a stage that more often emphasizes the potential risks it poses for society (Downs 1972). Researchers should continue monitoring media coverage of this emergent application of nanotechnology, recognizing that media portrayals will contribute to public perceptions of this, yet, novel issue.

## Appendices

### *Appendix 1 Newspaper Sources with Content About Nanotechnology Included in the Sample*

| Medium | Large circulation[a] | Medium circulation | Small circulation |
|---|---|---|---|
| Newspaper[b] | The New York Times | The Plain Dealer (Cleveland) | The Augusta |
| | Washington Post | Milwaukee Journal | Chronicle |
| | Houston Chronicle | Sentinel | Santa Fe New Mexican |
| | The Boston Globe | The Seattle Times | Bangor Daily News |
| | The Atlanta Journal-Constitution | St. Louis Post-Dispatch | Lewiston Morning Tribune |
| | USA Today | St. Petersburg Times | The Herald (Rock Hill) |
| | Star Tribune (Minneapolis) | The Sacramento Bee | Star-News |
| | | Pittsburgh Post-Gazette | Wyoming Tribune-Eagle |

[a]Large circulation was defined as >500,000
Medium circulation was defined as 100,000–499,999
Small circulation was defined as <99,999
[b]In 2009, we were not longer able to access The Boston Globe, The Sacramento Bee, and the Seattle Times via LexisNexis. We replaced them with The Philadelphia Inquirer, The Oregonian, and The Baltimore Sun

### *Appendix 2 Boolean Term Used to Search the LexisNexis Academic Database for News Coverage of Nanotechnology*

Atleast3(nanotech!) OR nanosci! OR nanoscal! OR nanocrystal* OR nanotube*OR nanomat! OR (nanometer* NOT W/15 light or laser or wavelength or UV) OR nano-dot* OR nanomed! OR nanopart! OR nanowir! OR nanoeng! OR nanocomp! OR nanoelectric! OR nanoelectronic! OR nanobot* OR nanomachine* OR fullerene* OR buckminsterfullerene* OR fullerite* OR buckyball* OR buckypaper* OR buckytube* OR molecular assembl! OR molecular manufactur! OR micromachine* OR quantum dot* OR quantum wire* OR quantum well* OR sub micron OR (individual atom* w/5 manipulate or move or build) OR (scanning w/3 microscope*) OR (tunneling w/3 microscope*) AND NOT nanosecond* AND NOT apple AND NOT iPod AND NOT mp3 AND NOT digest AND NOT news w/2 brief* AND NOT business w/2 brief* AND NOT news summary

## References


Bainbridge, W.S. 2002. Public attitudes toward nanotechnology. *Journal of Nanoparticle Research* 4(6): 561–570.

Barnard, A.S. 2009. How can ab initio simulations address risks in nanotech? *Nature Nanotechnology* 4(6): 332–335.

Brossard, D., and M.C. Nisbet. 2006. Deference to scientific authority among a low information public: Understanding U.S. opinion on agricultural biotechnology. *International Journal of Public Opinion Research* 19: 24–52.

Brossard, D., D.A. Scheufele, E. Kim, and B.V. Lewenstein. 2008. Religiosity as a perceptual filter: Examining processes of opinion formation about nanotechnology. *Public Understanding of Science* 18(5): 546–558.

Cobb, M., and J. Macoubrie. 2004. Public perceptions about nanotechnology: Risks, benefits, and trust. *Journal of Nanoparticle Research* 6: 395–405.

Currall, S.C. 2009. Nanotechnology and society: New insights into public perceptions. *Nature Nanotechnology* 4: 79–80.

Downs, A. 1972. Up and down with ecology: The issue-attention cycle. *The Public Interest* 28: 38–50.

Dudo, A., S. Dunwoody, and D.A. Scheufele. 2009. The emergence of nano news: Tracking thematic trends and changes in media coverage of nanotechnology. Paper presented at the annual convention of the Association for Education in Journalism and Mass Communication, Boston, MA.

Dunwoody, S. 1980. The science writing inner club: A communications link between science and the lay public. *Science, Technology, and Human Values* 5: 14–22.

Friedman, S.M., and B.P. Egolf. 2005. Nanotechnology risks and the media. *IEEE Technology and Society Magazine* 24(4): 5–11.

Gaskell, G., T. Ten Eyck, J. Jackson, and G. Veltri. 2005. Imagining nanotechnology: Cultural support for technological innovation in Europe and the United States. *Public Understanding of Science* 14(1): 81–90.

Hoet, P.H.M., I. Brüske-Hohlfeld, and O.V. Salata. 2004. Nanoparticles-known and unknown health risks. *Journal of Nanobiotechnology* 2(1): 12. doi:10.1186/1477-3155-2-12.

Horrigan, J. 2006. The internet as a resource for news and information about science. Pew Internet and American Life Project. http://www.pewinternet.org/Reports/2006/The-Internet-as-a-Resource-for-News-andInformation-about-Science.aspx. Accessed 21 June 2010.

Keefer, E.W., B.R. Botterman, M.I. Romero, A.F. Rossi, and G.W. Gross. 2008. Carbon nanotube coating improves neuronal recordings. *Nature Nanotechnology* 3(7): 434–439.

Kuzma, J. 2007. Moving forward responsibly: Oversight for the nanotechnology-biology interface. *Journal of Nanoparticle Research* 9(1): 165–182.

Mooney, C. 2008. The science writer's lament. Scienceprogress.com. http://www.scienceprogress.org/2008/10/the-science-writers-lament/. Accessed 24 June 2010.

Mooney, C. 2010. *Do scientists understand the public?* Cambridge: American Academies of Arts and Sciences.

Nisbet, M.C., and B.V. Lewenstein. 2002. Biotechnology and the American public: The policy process and the elite press, 1970 to 1999. *Science Communication* 23: 359–384.

Parpura, V. 2008. Instrumentation: Carbon nanotubes on the brain. *Nature Nanotechnology* 3(7): 384–385.

Riffe, D., S. Lacy, and F.G. Fico. 2005. *Analyzing media messages: Using quantitative content analysis in research*, 2nd ed. Mahwah: Lawrence Erlbaum Associates.
Roco, M.C., and W.S. Bainbridge (eds.). 2001. *Societal implications of nanoscience and nanotechnology*. Boston: Kluwer Academic Publishers.
Scheufele, D.A. 2006. Messages and heuristics: How audiences form attitudes about emerging technologies. In *Engaging science: Thoughts, deeds, analysis and action*, ed. J. Turney, 20–5. London: The Wellcome Trust.
Scheufele, D.A., and A. Dudo. 2010. Emerging agendas at the intersection of political and science communication: The case of nanotechnology. In *Communication yearbook 34*, ed. C.T. Salmon, 143–167. New York: Routledge.
Scheufele, D.A., and B.V. Lewenstein. 2005. The public and nanotechnology: How citizens make sense of emerging technologies. *Journal of Nanoparticle Research* 7(6): 659–667.
Scheufele, D.A., E.A. Corley, T.-J. Shih, K.E. Dalrymple, and S.S. Ho. 2009. Religious beliefs and public attitudes to nanotechnology in Europe and the US. *Nature Nanotechnology* 4(2): 91–94.
Silva, G.A. 2004. Introduction to nanotechnology and its applications to medicine. *Surgical Neurology* 61(3): 216–220.
Silva, G.A. 2006a. Neuroscience nanotechnology: Progress, challenges, and opportunities. *Nature Reviews Neuroscience* 7: 65–74.
Silva, G.A. 2006b. Nanomedicine: Seeing the benefits of ceria. *Nature Nanotechnology* 1(2): 92–94.
Stephens, L.F. 2005. News narratives about nano SandT in major US and non-US newspapers. *Science Communication* 27(2): 175–199.
The Pew Research Center's Project for Excellence in Journalism. 2008. The state of the news media: An annual report on American journalism. http://www.stateofthenewsmedia.com/2008/index.php. Accessed 24 June 2010.
Weaver, D.A., E. Lively, and B. Bimber. 2009. Searching for a frame: News media tell the story of technological progress, risk, and regulation. *Science Communication* 31(2): 139–166.

# Chapter 5
# Nanotechnology, the Brain, and the Future: Ethical Considerations

**Valerye Milleson**

## 5.1 Introduction

The task of this chapter is to present and briefly discuss the main ethical implications of current and prospective neural applications of nanotechnology. While the task seems simple, undertaking it is by no means so. In part, this is because there is very little literature specifically focused on the intersection of nano- and neurotechnology. At the time of this chapter's completion, there appear to be approximately a dozen articles that provide substantial ethical consideration specific to nanotechnology and the brain. While this provides a substantial base from which to evaluate the ethical considerations, the dearth of original research material on nanotechnology and neuroscience makes difficult an original thoroughgoing analysis of the ethical implications of nano and neuro. I have drawn from the larger literatures on "neuroethics" and "nanoethics" to supplement the extant articles on the ethics of nanotechnology and neuroscience.

In particular, I have developed a grid depicting, based on the results of my research, the key ethical issues related to nanotechnology. I group these issues according to the broad categories that I believe best encompass them – risk, humanness, and justice. I also propose subcategories within each category that, taken together, cover all of the major ethical issues within the debate. I then provide a brief description of each of these ethical subcategories with an explanation of how they relate to nanotechnology.

After presenting the ethical issues as they relate to nanotechnology generally, I offer an account, based on the grid, of how they apply specifically to the intersection between nanotechnology and neuroscience. These results are partly based on an extrapolation from prior analysis of the ethical issues specific to neuroscience itself, and partly based on what ethical issues appear most likely given current nano-neural

V. Milleson (✉)
The Center for Nanotechnology in Society, Arizona State University,
P.O. Box 875603, AZ 85287-5603, Tempe, USA
e-mail: Valerye.milleson@asu.edu

S.A. Hays et al. (eds.), *Nanotechnology, the Brain, and the Future*,
Yearbook of Nanotechnology in Society 3, DOI 10.1007/978-94-007-1787-9_5,

applications and research, which, while few and far between, do provide some measure of foresight into what is likely to develop at this scientific and technological intersection. I will then proceed with a brief discussion of those ethical issues of the nano-neuro convergence that are highlighted in this yearbook – risk, enhancement, and machine-brain interfacing. Since these issues are presented more thoroughly over the course of the rest of this volume, I will conclude by offering only some general remarks about what was discovered about the current debate over the course of the preceding analysis.

## 5.2 Methods

Options for determining the ethical issues relevant to nanotechnology include real-time technology assessment (Guston and Sarewitz 2002) and a network approach (van de Poel 2008; Zwart et al. 2006), both of which appear to be quite promising in directing where ethical analysis in nanotechnology ought to go and determining relevant ethical distinctions between various stakeholders. However, the most frequently applied method in the recent literature on ethics and nanotechnology is essentially some version of what Bruce et al. have termed the "stepwise methodological approach" (Bruce et al. 2006, 405), which basically searches current discussion in the literature and on the web in an attempt to identify relevant items connecting "nanotechnology" and "ethics". Followers of this approach generally then evaluate the revealed material to see if and where any ethical issues are raised that are either new topics for ethical reasoning or specific to nanotechnology. This would include issues that presumably cannot be resolved via already-given normative frameworks. Finally, they try to identify potential gaps within the current discussions on nanotechnology and ethics (Bruce et al. 2006, 405).

However, this method is not without its downsides. For one thing, it is not particularly systematic; it will almost certainly leave things out. Likewise, the piecemeal plan of scanning through literature, searching out relevant articles references, and the references of the references, etc. becomes an exceedingly tedious task and in no way assures that all of the relevant literature is covered. This method does not, on the face of it, offer any way to distinguish between what *should* be considered as an ethical issue as opposed to what *is* being considered as an ethical issue. Bruce *et al.* offer the new/specific criterion, but it is not obvious that that is a good criterion to be applying. Ibo van de Poel points out, quite rightly I believe, that this probably places far too much emphasis on the newness of an ethical issue. This may be a good criterion for determining whether nanoethics warrants a distinct field from the rest of applied ethics, but does actually very little to capture relevant ethical notions generally (van de Poel 2008, 28). As a related problem, the stepwise method runs the risk of leaving out ethically relevant issues while at the same time focusing too much on issues that may not be nearly as important normatively – but are nonetheless widely discussed.

These downsides notwithstanding, the stepwise method is still quite valuable at times, and is essentially the method employed in this chapter. Moreover, since the goal herein is largely to provide a survey of the current literature as it pertains to the ethics of nanotechnology, it really is the best-suited method available. However, in an attempt to overcome the aforementioned downsides, various precautionary measures were taken in the research for this chapter. First, the range of material included in the survey was greatly expanded upon. In addition to the standard literature searches and academic web searches, this paper also utilized the Georgia Tech nanotechnology database, which proposes to include all of the published scientific literature related to nanotechnology (for details about the database, see Nulle et al., this volume). Two separate searches were initiated using the database: one specific to ethics and one specific to the convergence between nanotechnology and the brain. These two smaller databases were then both searched again for relevant material.

The second precautionary measure taken in the research for this chapter was to include an expanded word search.[1] The words included in this search were revised over several runs through different search engines[2] in an effort to hone in on the most relevant ethical issues (or broad issue categories) present in the current discussion. While it is nearly impossible to capture every relevant paper or article in even a compilation of several searches, certain general patterns made themselves repeatedly apparent, and these patterns form the basis of the categorization of ethical issues seen later in the chapter. That these patterns arose in multiple searches provides evidence that the grid presented in the next section is a fairly accurate reflection of the main issues that make up the topic, nonetheless.

## 5.3 The Issues

As Mihail Roco points out, "[The social and ethical] implications of nanotechnology apply in a variety of areas, including technological, economic, environmental, health, and educational, ethical, moral, philosophical." (Roco 2003, 185) For some, "[t] he ethical issues fall into the areas of equity, privacy, security, environment, and metaphysical questions concerning human-machine interaction" (Mnyusiwalla et al. 2003, R11). Others consider primarily privacy and control, longevity, and runaway nanobots (Moor and Weckert 2004), or issues of safety and environment, nanomedicine, individual privacy, and justice (Litton 2007) or even a much wider range of ethical issues, including environmental issues, workforce issues, privacy issues, national and international political issues, intellectual property issues, and human enhancement (Lewenstein 2005). Some try to capture all of the ethical issues into three main groups: safety, social justice, and the transformation of humanity

[1] For example, an early draft of the search terms utilized, in addition to nano*, included the following: risk, risk+environment, health, security, war, terror*, nanobot, grey goo (or gray goo), privacy, privacy+screen*, surveillance, equity, equity+national, equity+global, cost, distribut*, equality, welfare, welfare+human, welfare+social, welfare+political, regulat*, regulat*+legal, regulat*+political, human, human+identity, human+enhance*, human+artificial, transhumanism, disability, justice, nonmaleficence, beneficence, autonomy, and neuro*.

[2] Academic Search Premier, The Philosopher's Index, Web of Science, and Google Scholar.

(Keiper 2007). Others try to catalogue the ethical issues in four main areas: risk assessment; questions of human and personal identity; the possibility of human and personal enhancement; and the justice of the distribution of risks and benefits (Lenk and Biller-Adorno 2007, 175). While these are not necessarily bad categorizations, in the following grid you will see that I have my own preferences.

The grid itself was primarily generated in response to the recurrent patterns apparent in the database searches I undertook; whether explicitly mentioned or not, there were certain connections and broad generalizations that were being repeatedly made. Additionally, this table was guided in part by the suggestions of Jason Scott Robert, whose previous table of "[s] tock ethical, societal, and policy issues" (Robert 2008, 227–8) was part of what *Nanoethics* reviewer Travis Rieder called "one of the best essays I have come across in the nanoethics literature" (Rieder 2008, 330). It is most similar to Lenk and Biller-Adorno (2007), at least as far as the general categories – risk, humanness, and justice – are concerned (Table 5.1).

### 5.3.1 Risk

In most of the nanotechnology discussion, risk is seen as the primary ethical issue (e.g. Commission 2008; Grunwald 2005; Levi-Faur and Comaneshter 2007; Litton 2007). Within this broad category of risk, the risk to human health was often of greater concern than other risks (Bawa and Johnson 2008; Litton 2007; Mody 2008; Wilson 2006), although environmental risk was also mentioned frequently (Litton 2007; Roco 2003). For the former, there is often mention of the "asbestos" analogy with respect to nanotechnology (Wilson 2006), as well as concern about the potential for nanoscale particles to harm human health by crossing the blood-brain barrier (Bawa and Johnson 2008). In terms of environmental risk, the concern is primarily what will happen to the natural environment if there is a widespread release of manufactured nanoscale particles into it (Roco 2003). These two risks generally dominate the ethical discussion, and the general sentiment is well expressed by Litton: "The most pressing concern is whether nanomaterials pose dangers to health and the environment.... Our primary ethical challenge will be gauging the adequacy of the toxicology studies and the developed standards for introducing nanomaterials into common usage" (Litton 2007, 23, 25).

Another frequently discussed risk – although it was often set apart as a separate category – is nanotechnology's potential risk to privacy. Given that nanoscale components may allow for smaller, more powerful surveillance and screening devices – with potential forms of "smart dust", global positioning systems, and improved information databases (e.g. Weckert 2007) – there is a potentially heightened threat of these new technologies to personal privacy and autonomy (Bond 2003; Mnyusiwalla et al. 2003; van den Hoven 2008; Weckert 2007).

Discussed on a much smaller scale, but still present in the literature, are security risks and risks of a catastrophic nature. For security, some fear that nano-engineered chemicals or drugs will enable improved bioterrorism. Another, albeit extended,

**Table 5.1** Kinds of Ethical Issues in Nanotechnology

| Key ethical issues | General description | Applications to nanotechnology |
|---|---|---|
| *Risk* | | |
| Health risk | Risk of possible toxicity to human health resulting from exposure: medical, environmental or workplace; through inhalation, ingestion, dermal contact, therapeutic/ diagnostic intervention or contact in utero | Given their size, nanoscale components will presumably be able to enter the human body in many ways with so far relatively unknown consequences, potentially putting human health at risk |
| Environmental risk | Risk of possible toxicity to or destruction of the environment | The consequences of introducing nanotechnologies into the environment are relatively unknown and potentially risky |
| Privacy risk | Risk that the development and implementation of a technology will lead to diminished privacy and personal autonomy | Nanoscale components may allow for smaller, more powerful surveillance and screening devices |
| Security risk | Risk that the development and implementation of a technology will lead to increased or improved terrorism or warfare | Nanoscale drugs and weaponry may allow for increases and improvements in terrorism. The use of nanoscale psychopharmacology and implants may be used to create better soldiers and, ergo, better warfare |
| Catastrophe risk | Risk that the development and implementation of a technology will lead to catastrophic consequences | Nanotechnology may lead to irreversible damage to human health, the environment, etc |
| *Humanness* | | |
| Identity | Potential that the application of a technology will challenge certain core issues of identity, personal, societal, and species-wide | The increased role and integration of nanotechnology into our lives is thought to be challenging to various aspects of human identity, both at an individual level and at a wider scale |
| Human enhancement | Potential that a technology will be used to enhance humans physically and/or cognitively to the point of becoming "trans"-human | Nanotechnology is thought to be a key component of the new era of human enhancement (e.g. NBIC) |
| Human-machine interfacing | Potential that a technology will be used to interface and integrate "machine" components to the point of becoming "trans"-human | Nanotechnology may enable improved interfacing and integration of mechanical components with the human body |

(continued)

**Table 5.1** (continued)

| Key ethical issues | General description | Applications to nanotechnology |
|---|---|---|
| "Trans"-human consequences | Potential that a technology will enable "trans"-human, resulting in destabilization or destruction of the human race, human family, human society, human-as-made-in-God's-image, etc | Insofar as nanotechnology better enables transhumanism, it better enables the relevant transhuman consequences |
| *Justice* | | |
| Funding and regulatory justice | Problems of fairness resulting from who has to fund the research, development, implementation, etc. of the technology and who gets to regulate it | As with any emerging and potentially lucrative technology, the question of fairness in funding and regulating nanotechnology is of great importance |
| Distributive justice | Problems of fairness resulting from who has access to the technology, how it is distributed, and the social and political consequences thereof | Given the many promised benefits of nanotechnology, there are a number of concerns about who will have access to it and whether or not it will result in a "nanodivide" |
| Global justice | Problems of fairness as they exist on a global scale | Since the wealthier countries stand a better chance of developing and implementing nanotechnologies, there is a risk of a nanodivide on a global scale |

consequence is the fear that the use of nanoscale psychopharmacology and implants will be used to create better, "super" soldiers, and this in turn will lead to either increased or more efficient warfare (Shipbaugh 2006). With respect to potential catastrophe, there is the fear that nanotechnology may lead to irreversible damage to either human health or the environment. This is essentially what the fear of nanobots run amok – the "grey goo" scenario – is about. (Drexler 1986)

## *5.3.2 Humanness*

According to some, the increased role and integration of nanotechnology into our lives will likely be challenging to various aspects of human identity. For example, according to Lenk and Biller-Adorno, "a biotechnology that comes closer and closer to the natural functions and mechanisms of the human body also raises doubts about what are the genuinely "human" qualities of the body. Similar to applications such as xenotransplantation, i.e. the transfer of non-human organs to patients, the invention of nanotechnology poses the question of human identity as well as the distinction of body and machine" (Lenk and Biller-Adorno 2007, 178). Moreover, increased use of nanotechnology poses the potential for transgressions or commodification of humanity (Laurent and Petit 2005; Langer 2008) and the "artifactualization of human beings" and "anthropomorphization of technology" could lead to a breakdown of the human/machine, organic/inorganic distinctions (Gordijn 2006, 730; see also Naufel, this volume).

The challenge to human identity especially appears in the literature with respect to the convergence of nano-bio-info-technologies. For example:

> Widespread use of NBIC enabled enhancement technologies will result in an increasingly close association of the body with technology, which could trigger a negative change in attitudes towards the body…If a human being were regularly subjected to new NBIC-enabled technological interventions to improve his sensory, motor, or cognitive abilities, it could become increasingly difficult to answer [questions about his physical or moral capacity] (Gordijn 2006, 729–730).

Even if nanotechnology is not necessarily incorporated into the human body, some argue that it will necessarily alter human self-conception because of its narrative existence (Berne 2004; Mordini 2007).

Insofar as it *is* incorporated into the human body, the key ethical issues being discussed are its potential enhancement or human-machine interfacing uses. As Grunwald rightly points out, "Nanotechnology, in combination with biotechnology and medicine, opens perspectives for fundamentally altering and rebuilding the human body" (Grunwald 2005, 197). Whether via improved drug delivery, cognitive or physical enhancement, or human-machine interfacing, nanotechnology – individually or part of NBIC – holds the potential to radically alter human nature, and this potential brings up several ethical issues (e.g. Bond 2003; Jotterand 2008).

For some, this potential for altering human nature with nanotechnology brings about a worry about large-scale species and societal consequences; there is a fear that this technology will result in the destabilization or destruction of the human race, human family, human society, human-as-made-in-God's-image, and so on. Some argue against nanotechnology on essentially theological grounds (Foltz and Foltz 2005; Langer 2008); others maintain that the incorporation of nanotechnology will require radical reformation of human relationships and human societies, especially the family unit, and with it a radical re-conception of "religion" (Bainbridge 2007).

### 5.3.3 *Justice*

Despite the obvious fact that, as Weckert notes, "Undoubtedly there will be winners and losers from nanotechnology" (Weckert 2007, 60) the area of ethical analysis least prominent in the nanotechnology discussion is the issue of justice. The matter of funding and regulation is discussed quite a bit, but, generally, as a matter of policy formation, not with a mind to the underlying ethical issues. There is some discussion with respect to distributive justice (c.f. Sparrow 2007), focusing on the matter of a potential "nano-divide" between those with access to the technology and those without it. There is also some discussion with the matter of justice, as it exists on a global scale (Barker et al. 2008; Roco 2003; Schummer 2008). However, mostly the matter of justice remains largely unexplored.[3] Many of the most issues most salient to the category of justice are related to the convergence of nanotechnology, neuroscience, and other emergent technologies for the purpose of enhancement.

Current research trends in attempting to utilize nanoscale science and engineering for various brain applications – such as drug delivery, cancer targeting, potential machine interfacing, and so on – raise ethical issues arising from the convergence between nano and the brain. While the ethical implications of various types of neuroscience technologies – particularly psychosurgery, psychopharmacology, deep-brain stimulation, neuroimaging and neuroimplantation – are already in widespread discussion, the answers to these implications are by no means clear. Even if they were, what is still in need of further discussion is where and how these technologies, nano and neuro, intersect and what the ethical consequences of those intersections will be. (For discussion, see Naufel, this volume.) In the following sections, I will delve deeper into the three categories of ethical concern, as seen through the lenses of specific applications of nanotechnology and neuroscience.

As neuroethicist Walter Glannon explains, "…the ability to map, intervene in, and alter the neural correlates of the mind raises important ethical questions. Indeed, these questions are arguably weightier and more momentous than any other set of questions in any other area of bioethics. This is because techniques that target the brain can reveal and modify the source of the mind and affect personal identity, the will, and other aspects of our selves" (Glannon 2006a, b, 38). Fellow neuroethicists Martha Farah and Paul Root Wolpe second this position: "The brain is the organ of the mind and consciousness, the locus of our sense of selfhood. Interventions in the brain therefore have different ethical implications than interventions in other organs" (Farah and Wolpe 2004, 36). Therefore, given these heightened ethical issues related to neuroscience, we can expect the convergence between neuroscience and nanotechnology to have potentially even greater ethical weight.

## 5.4 Some Current Research with Nano-Neuro Connections

While there are a wide variety of issues being discussed with respect to "nanomedicine" (Jain 2005; Moghimi 2005), the key areas where this intersects with the brain involve primarily diagnostics, drug delivery, therapy targeting, and machine interface. According to some, diagnostics will be the primary brain use of nanomedicine. For example, according to Sanjeeb Sahoo, "One of the first applications of nanomedicine will be improved fluorescent markers for diagnostic and screening purposes" (Sahoo 2005, 1049), including quantum dots for cancer detection, as well as "lab-on-a-chip"

[3] Volume 2 of the *Yearbook of Nanotechnology in Society* focused on issues of equity and responsibility in the development of nanoscale science and engineering, and was in preparation at the same time as this chapter.

in diagnostics (Moghimi 2005; Bawa and Johnson 2008). According to others, the ability to send targeted genes (Mastrobattista et al. 2006) or drug delivery (Bawa and Johnson 2008; Greenfield 2005) will be the key application. With respect to human-brain interfacing, the research is currently a bit more hypothetical (Adamson and Wilkinson 2005; Farah et al. 2004; Keiper 2006; Khushf 2007a, b; Berger et al. 2008, this volume), but the general notion is that nanoscale science and engineering will enable smaller and more labile devices to create better connections between brain and machine.

Underlying this primacy of nanotechnology usage in these three areas is the benefits resulting from the radical reduction in scale. With respect to diagnostics, size reduction of biomedical lab tests has several advantages: Not only does it lead to a marked reduction of the sample volume needed for testing, but it also results in a marked reduction of (potentially expensive) reagents such as monoclonal antibodies. Finally, yet importantly, it may lead to a significant reduction in the time required. (Sahoo 2005, 1049). Similarly, for drug delivery and gene therapy, since, as neuroscientist Susan Greenfield illustrates, "The particular difficulty for administering protein drugs for brain diseases is the lack of access because of the blood-brain barrier" (Greenfield 2005, 34–5). Thus, the inclusion of nanotechnology – which itself promises to be able to cross the blood-brain barrier – into drug delivery for brain diseases would be most advantageous (Bawa and Johnson 2008). For human-brain interfacing, there is already some evidence that nanoparticle coatings can improve biotic acceptance of abiotic units in other areas of the body (Adamson and Wilkinson 2005); there is then promise that the use of nanoparticle coating will make the already existing field of brain-machine interface more practical.

## 5.5 The Issues Revisited

In looking at the implications for the brain and the current research with nano-neuro connections, we can see to some extent how the convergence between nanotechnology and neuroscience will affect our chart of associated ethical issues. These extrapolations on the ethical issues resulting from the intersection between nanotechnology and neuroscience hint at the following, modified version of the previous grid (Table 5.2).

## 5.6 The Future

So, what are the upshots of the foregoing analysis? The first is that, while discussion has already been present for some time on the ethical issues of both nanotechnology and of neuroscience (or neurotechnology), it has been heretofore greatly neglected the ethical issues that exist specifically where these two technologies (or sciences) converge. In spite of this, however, both of these separate discussions often center on the issues of risk (including health, environmental, privacy, security and catastrophe),

**Table 5.2** Reinterpretation of Kinds of Ethical Issues in Nanotechnology

| Key ethical issues | General description | Applications to nanotechnology | Applications to the brain |
|---|---|---|---|
| *Risk* | | | |
| Health risk | Risk of possible toxicity to human health resulting from exposure: medical, environmental or workplace; through inhalation, ingestion, dermal contact, therapeutic/diagnostic intervention or contact in utero | Given their size, nanoscale components will presumably be able to enter the human body in many ways with so far relatively unknown consequences, potentially putting human health at risk | Nanoscale components will also presumably be able to cross the blood-brain barrier, making almost every implementation of nanotechnology a potential health risk to the human brain |
| Environmental risk | Risk of possible toxicity to or destruction of the environment | The consequences of introducing nanotechnologies into the environment are relatively unknown and potentially risky | Nanotechnology initially implemented in the brain may have environmental consequences through human elimination |
| Privacy risk | Risk that the development and implementation of a technology will lead to diminished privacy and personal autonomy | Nanoscale components may allow for smaller, more powerful surveillance and screening devices | Enhanced neuroimaging may allow for increased access to personal information. Nanoscale or nano-enabled neural implants may allow for improved scanning and tracking, perhaps even mind control |
| Security risk | Risk that the development and implementation of a technology will lead to increased or improved terrorism or warfare | Nanoscale drugs and weaponry may allow for increases and improvements in terrorism. The use of nanoscale psychopharmacology and implants may be used to create better soldiers and, ergo, better warfare | Nanoscale biochemicals may be able to reach and affect a greater number of people, with more likely irreversible consequences, thus enabling improved biochemical terrorism. The neural uses of nanotechnology in creating super soldiers may lead to better warfare |
| Catastrophe risk | Risk that the development and implementation of a technology will lead to catastrophic consequences | Nanotechnology may lead to irreversible damage to human health, the environment, etc | Neural applications of nanotechnology may lead to widespread, irreversible damage to human brains |
| *Humanness* | | | |
| Identity | Potential that the application of a technology will challenge certain core issues of identity, personal, societal, and species-wide | The increased role and integration of nanotechnology into our lives is thought to be challenging to various aspects of human identity, both at an individual level and at a wider scale | Given the pivotal role of the brain in human identity, and nanotechnological implementation holds the potential of radically altering that identity |

| | | | |
|---|---|---|---|
| Human enhancement | Potential that a technology will be used to enhance humans physically and/or cognitively to the point of becoming “trans”-human | Nanotechnology is thought to be a key component of the new era of human enhancement (e.g. NBIC) | Neural applications of nanotechnology – in particular psychopharmaceuticals and neuroimplantation – may enable more fundamental alteration and enhancement of human beings |
| Human-machine interfacing | Potential that a technology will be used to interface and integrate “machine” components to the point of becoming “trans”-human | Nanotechnology may enable improved interfacing and integration of mechanical components with the human body | Nanotechnology may better enable improved brain biotic/abiotic hybridization and integration |
| “Trans”-human consequences | Potential that a technology will enable “trans”-human, resulting in destabilization or destruction of the human race, human family, human society, human-as-made-in-God’s-image, etc | Insofar as nanotechnology better enables transhumanism, it better enables the relevant transhuman consequences | Insofar as neural applications of nanotechnology – in particular fundamental human alteration via psychopharmaceuticals, neuroimplantation and human-machine interfacing – better enable transhumanism, they better enable the transhuman consequences |
| *Justice* | | | |
| Funding and regulatory justice | Problems of fairness resulting from who has to fund the research, development, implementation, etc. of the technology and who gets to regulate it | As with any emerging and potentially lucrative technology, the question of fairness in funding and regulating nanotechnology is of great importance | The problem of whether or not to fund R&D into certain neural applications of nanotechnology – particularly enhancement applications, and even therapeutic ones – may be of great importance |
| Distributive justice | Problems of fairness resulting from who has access to the technology, how it is distributed, and the social and political consequences thereof | Given the many promised benefits of nanotechnology, there are a number of concerns about who will have access to it and whether or not it will result in a “nanodivide” | Particularly with nanotechnologically-enabled neural enhancements, there is a worry about who will have access to them and what social stigmatization might occur for those who do not |
| Global justice | Problems of fairness as they exist on a global scale | Since the wealthier countries stand a better chance of developing and implementing nanotechnologies, there is a risk of a nanodivide on a global scale | Insofar as access to nanotechnologically-enabled neural enhancements is divided along a global line, there will be problems of global justice |

(continued)

**Table 5.2** (continued)

| Key ethical issues | General description | Applications to nanotechnology | Applications to the brain |
|---|---|---|---|
| Human enhancement | Potential that a technology will be used to enhance humans physically and/or cognitively to the point of becoming "trans"-human | Nanotechnology is thought to be a key component of the new era of human enhancement (e.g. NBIC) | Neural applications of nanotechnology – in particular psychopharmaceuticals and neuroimplantation – may enable more fundamental alteration and enhancement of human beings |
| Human-machine interfacing | Potential that a technology will be used to interface and integrate "machine" components to the point of becoming "trans"-human | Nanotechnology may enable improved interfacing and integration of mechanical components with the human body | Nanotechnology may better enable improved brain biotic/abiotic hybridization and integration |
| "Trans"-human consequences | Potential that a technology will enable "trans"-human, resulting in destabilization or destruction of the human race, human family, human society, human-as-made-in-God's-image, etc | Insofar as nanotechnology better enables transhumanism, it better enables the relevant transhuman consequences | Insofar as neural applications of nanotechnology – in particular fundamental human alteration via psychopharmaceuticals, neuroimplantation and human-machine interfacing – better enable transhumanism, they better enable the transhuman consequences |

humanness (including identity, human enhancement, human-machine interfacing and "trans"-human consequences) and justice (including funding and regulatory justice, distributive justice, and global justice). Most illuminating is that where they specifically intersect, the ethical issues surrounding nano- and neurotechnologies are often new or have heightened importance, particularly with regard to the issues of risk, enhancement and human-machine interaction.

Another upshot of this survey is that the issues of risk and, to a lesser extent, human enhancement, dominate much of the ethical discussion. In his 2007, web search-engine comparison Gregor Wolbring found that "the nanotechnology discourse is highly myopic towards covering medical health and environmental safety issues, and mostly ignores social safety, social health and social justice issues" (Wolbring 2007, 15). As seen in this chapter – as well as in this volume – this absence of social ethical discussion remains. Yet, these sorts of issues appear to be of primary importance to many citizens (National Citizens Forum, this volume; Nanotechnology Survey, this volume). According to Wolbring, this indicates a lack of diversity in the stakeholders controlling the ethical debate (Wolbring 2007, 24). In addition to promoting further discussion of the ethical issues residing at the convergence of nanotechnology and neuroscience, one hope of this *Yearbook* and this chapter is to help fill that gap, thereby guiding the ethical debate at this new technological frontier along a more inclusive – and more just – orientation.

## References


Ackerman, Sandra J. 2006. *Hard science, hard choices: Facts, ethics, and policies guiding brain science today*. Washington, DC: Dana Press.

Adamson, George, and J. Malcolm Wilkinson. 2005. Nanotechnology: What it is and how it can be applied to healthcare. *Asia Pacific Biotech News* 9(20): 1078–1082.

Allhoff, Fritz. 2008. On the autonomy and justification of nanoethics. In *Nanotechnology & society: Current and emerging ethical issues*, ed. Fritz Allhoff and Patrick Lin, 3–38. New York: Springer Science and Business Media.

Allhoff, Fritz, and Patrick Lin. 2008. *Nanotechnology and society: Current and emerging ethical issues*. New York: Springer Science+Business Media, B. V.

Allhoff, Fritz, Patrick Lin, James Moor, and John Weckert (eds.). 2007. *Nanoethics: The ethical and social implications of nanotechnology*. Hoboken: Wiley.

Alpert, Sheri. 2008. Neuroethics and nanoethics: Do we risk ethical myopia? *Neuroethics* 1: 55–68.

Bailey, Ronald. 2005. *Liberation biology: The scientific and moral case for the biotech revolution*. Amherst: Prometheus Books.

Bainbridge, William Sims. 2007. Converging technologies and human destiny. *The Journal of Medicine and Philosophy* 32: 197–216.

Barker, Todd F., Leili Fatehi, Michael T. Lesnick, Timothy J. Mealey, and Rex R. Raimond. 2008. Nanotechnology and the poor: Opportunities and risks for developing countries. In *Nanotechnology & society: Current and emerging ethical issues*, ed. Fritz Allhoff and Patrick Lin, 243–263. New York: Springer Science and Business Media.

Bawa, Raj, and Summer Johnson. 2008. Emerging issues in nanomedicine and ethics. In *Nanotechnology & society: Current and emerging ethical issues*, ed. Fritz Allhoff and Patrick Lin, 207–223. New York: Springer Science and Business Media.

Bennett-Woods, Deb. 2008. *Nanotechnology: Ethics and society*. Boca Raton: CRC Press, of Taylor & Francis Group, LLC.

Berger, Francois, Sjef Gevers, and Ludwig Siep. 2008. Ethical, legal and social aspects of brain-implants using nano-scale materials and techniques. *NanoEthics* 2: 241–249.

Berne, R.W. 2004. Towards the conscientious development of ethical nanotechnology. *Science and Engineering Ethics* 10: 627–638.

Berne, Rosalyn W. 2006. *Nanotalk: Conversations with scientists and engineers about ethics, meaning, and belief in the development of nanotechnology*. Mahwah: Lawrence Erlbaum Associates.

Bond, Phillip J. 2003. Preparing the path for nanotechnology. In *Nanotechnology: Societal implications – Maximizing benefits for humanity: Report of the national nanotechnology initiative workshop*, December 2–3, 2003, ed. Mihail C. Roco and Williams Sims Bainbridge, 16–21. Arlington: National Science Foundation.

Bruce, Donald. 2007. *Human enhancement? Ethical reflections of emerging nanobio-technologies: Report on an expert working group on converging technologies for human functional enhancement*. NanoBio-RAISE EC FP6 Science and Society Co-ordination Action.

Bruce, Harold, Ernst Holger, Grunwald Armin, Grunwald Werner, Hogmann Heinrich, Krug Harald, Janich Peter, Mayor Marcel, Rathgeber Wolfgang, Schmid Günter, Simon Ulrich, Vogel Viola, and Danil Wyrwa. 2006. *Nanotechnology: Assessment and perspectives*. Berlin: Springer.

Canli, Turhan, Susan Brandon, William Casebeer, Philip J. Crowley, Don DuRousseau, Hendry T. Greely, and Alvaro Pascual-Leone. 2007. Neuroethics and national security. *The American Journal of Bioethics* 7(5): 3–13.

Chase, Victor D. 2006. *Shattered nerves: How science is solving modern medicine's most perplexing problem*. Baltimore: The Johns Hopkins University Press.

Chatterjee, A. 2006. The promise and predicament of cosmetic neurology. *Journal of Medical Ethics* 32: 110–113.

Clark, Andy. 2007. Re-inventing ourselves: The plasticity of embodiment, sensing, and mind. *The Journal of Medicine and Philosophy* 32: 263–282.

Commision de l'Ethique de la Science et de la Technologie. 2008. Ethics, risk, and nanotechnology: Responsible approaches to dealing with risk. In *Nanotechnology & society: Current and emerging ethical issues*, ed. F. Allhoff and P. Lin, 75–89. New York: Springer Science and Business Media.

Dana Foundation. 2002. *Neuroethics: Mapping the field*. Conference, San Francisco, CA, 13–14 May 2002.

DeGrazia, David. 2005. Enhancement technologies and human identity. *The Journal of Medicine and Philosophy* 30: 261–283.

Drexler, K.Eric. 1986. *Engines of creation: The coming era of nanotechnology*. New York: Anchor Books.

Drexler, K.Eric, Chris Peterson, and Gayle Pergamit. 1991. *Unbounding the future: The nanotechnology revolution*. New York: William Morrow and Company, Inc.

Dupuy, Jean-Pierre. 2007. Some pitfalls in the philosophical foundations of nanoethics. *The Journal of Medicine and Philosophy* 32: 237–261.

European Group on Ethics. 2008. Ethical aspects of nanomedicine: A condensed version of the EGE Opinion 21. In *Nanotechnology & society: Current and emerging ethical issues*, ed. Fritz Allhoff and Patrick Lin, 187–206. New York: Springer Science and Business Media.

Farah, Martha J., and Andrea S. Heberlein. 2007. Personhood and neuroscience: Naturalizing or nihilating. *The American Journal of Bioethics* 7(1): 37–48.

Farah, Martha J., and Paul Root Wolpe. 2004. Monitoring and manipulating brain function: New neuroscience technologies and their ethical implications. *The Hastings Center Report* 34(3): 35–45.

Farah, Martha J., Judy Illes, Robert Cook-Deegan, Howard Gardner, Eric Kandel, Patricia King, Eric Parens, Barbara Sahakian, and Paul Root Wolpe. 2004. Neurocognitive enhancement: What can we do and what should we do? *Nature Reviews* 5: 421–425.

Foltz, Franz A., and Frederick A. Foltz. 2005. The societal and ethical implications of nanotechnology: A Christian response. *Journal of Technology Studies* 32: 104–114.

Freitas, R.A.J. 1998. Exploratory design in medical nanotechnology: A mechanical artificial red cell, artificial cells, blood substitutes, and biotechnology [formerly artificial cells, blood substitutes, and immobile]. *Biotech* 26(4): 411–430.

Garcia, Tamara, and Ronald Sandler. 2008. Enhancing justice? *NanoEthics* 2: 277–287.

Garreau, Joel. 2005. *Radical evolution: The promise and peril of enhancing our minds, our bodies – And what it means to be human*. New York: Broadway Books.

Gillett, Grant. 2006. Cyborgs and moral identity. *Journal of Medical Ethics* 32: 79–83.

Glannon, Walter. 2006a. *Bioethics and the brain*. New York: Oxford University Press.

Glannon, Walter. 2006b. Neuroethics. *Bioethics* 20(1): 37–52.

Godman, Marion. 2008. But is it unique to nanotechnology? *Science and Engineering Ethics* 14: 391–403.

Gordijn, Bert. 2005. Nanoethics: From utopian dreams and apocalyptic nightmares towards a more balanced view. *Science and Engineering Ethics* 11: 521–533.

Gordijn, Bert. 2006. Converging NBIC technologies for improving human performance: A critical assessment of the novelty and the prospects of the project. *The Journal of Law, Medicine & Ethics* 34: 726–732.

Greenfield, Susan A. 2005. Biotechnology, the brain and the future. *Trends in Biotechnology* 23(1): 34–41.

Grunwald, Armin. 2000. Against over-estimating the role of ethics in technology development. *Science and Engineering Ethics* 6: 181–196.

Grunwald, Armin. 2005. Nanotechnology – A new field of ethical inquiry? *Science and Engineering Ethics* 11: 187–201.

Guston, D.H., and D. Sarewitz. 2002. Real-time technology assessment. *Technology in Society* 24: 93–109.

Hansson, Sven Ove. 2004. Great uncertainty about small things. *Technè* 8(2): 26–35.

Hansson, Sven Ove. 2005. Implant ethics. *Journal of Medical Ethics* 31: 519–525.

Hardman, Ron. 2006. A toxicological review of quantum dots: Toxicity depends on physicochemical and environmental factors. *Environmental Health Perspectives* 114: 165–172.

Hassoun, Nicole. 2008. Nanotechnology, enhancement, and human nature. *NanoEthics* 2: 289–304.

Haugeland, J. 1998. Mind embodied and embedded. In *Having thought*, ed. J. Haugeland, 207–237. Cambridge: MIT Press.

Hodge, Graeme, Bowman Diana, and Ludlow Karinne. 2007. Introduction: Big questions for small technologies. In *New global frontiers in regulation: The age of nanotechnology*, ed. Graeme Hodge, Bowman Diana, and Ludlow Karinne, 3–26. Northampton: Edward Elgar Publishing, Inc.

Holm, Sören. 2005. *Does nanotechnology require a new 'nanoethics'?* London: Cardiff Centre for Ethics, Law & Society. August 2005.

Illes, Judy. 2005. *Neuroethics: Defining the issues in theory, practice and policy*. New York: Oxford University Press.

Jain, Kewal K. 2005. The role of nanobiotechnology in drug discovery. *Drug Discovery Today* 10(21): 1435–1442.

Johnson, Deborah G. 2007. Ethics and technology 'in the making': An essay on the challenge of nanoethics. *NanoEthics* 1: 21–30.

Jotterand, Fabrice. 2008. Beyond therapy and enhancement: The alteration of human nature. *NanoEthics* 2: 15–23.

Kaiser, Mario. 2006. Drawing boundaries of nanoscience – Rationalizing the concerns? *The Journal of Law, Medicine & Ethics* 34: 667–674.

Keiper, Adam. 2006. The age of neuroelectronics. *The New Atlantis* 11: 4–41.

Keiper, Adam. 2007. Nanoethics as a discipline? *The New Atlantis* 16: 55–67.

Khushf, George. 2007a. The ethics of NBIC convergence. *The Journal of Medicine and Philosophy* 32: 185–196.

Khushf, George. 2007b. Open questions in the ethics of convergence. *The Journal of Medicine and Philosophy* 32: 299–310.
Langer, Richard. 2008. Humans, commodities, and humans-in-a-sense. *Philosophia Christi* 10(1): 101–118.
Laurent, Louis, and Jean-Claude Petit. 2005. Nanoscience and its convergence with other technologies. *HYLE – International Journal for Philosophy of Chemistry* 11: 45–76.
Lenk, Christian, and Nikola Biller-Adorno. 2007. Nanomedicine – Emerging or re-emerging ethical issues? A discussion of four ethical themes. *Medicine, Health Care and Philosophy* 10: 173–184.
Levi-Faur, David, and Hanna Comaneshter. 2007. The risks of regulation and the regulation of risks: The governance of nanotechnology. In *New global frontiers in regulation: The age of nanotechnology*, ed. Hodge Graeme, Bowman Diana, and Ludlow Karinne, 149–165. Northampton: Edward Elgar Publishing, Inc.
Levy, Neil. 2007. *Neuroethics: Challenges for the 21st century*. New York: Cambridge University Press.
Lewenstein, Bruce V. 2005. What counts as a 'social and ethical issue' in nanotechnology? *HYLE – International Journal for Philosophy of Chemistry* 11: 5–18.
Lin, Patrick, and Fritz Allhoff. 2007. Nanoscience and nanoethics: Defining the discipline. In *Nanoethics: The ethical and social implications of nanotechnology*, ed. Fritz Allhoff, Patrick Lin, James Moor, and John Weckert, 3–16. Hoboken: Wiley.
Litton, Paul. 2007. "Nanoethics"? What's new? *The Hastings Center Report* 37: 22–25.
Mastrobattista, E., A.E.M. van der Aa Marieke, Wim E. Hennink, and Daan J.A. Crommelin. 2006. Artificial viruses: a nanotechnological approach to gene delivery. *Nature Reviews. Drug Discovery* 5: 115–121.
McGee, Ellen M., and Gerald Q. Maguire Jr. 2007. Becoming Borg to become immortal: Regulating brain implant technologies. *Cambridge Quarterly of Healthcare Ethics* 16: 291–302.
Meaney, Mark E. 2006. Lessons from the sustainability movement: Toward an integrative decision-making framework for nanotechnology. *The Journal of Law, Medicine & Ethics* 34: 682–688.
Mnyusiwalla, Anisa, Abdallah S. Daar, and Peter A. Singer. 2003. Mind the gap: Science and ethics in nanotechnology. *Nanotechnology* 14: R9–R13.
Mody, Cyrus C.M. 2008. The larger world of nano. *Physics Today* 61: 38–44.
Moghimi, S.Moein. 2005. Nanomedicine: Prospective diagnostic and therapeutic potentials. *Asia Pacific Biotech News* 9(20): 1072–1077.
Moor, James, and John Weckert. 2004. Nanoethics: Assessing the nanoscale from an ethical point of view. In *Discovering the nanoscale*, ed. D. Baird, A. Nordmann, and J. Schummer, 301–310. Amsterdam: Ios Press.
Mordini, Emilio. 2007. Nanotechnology, society and collective imaginary: setting the research agenda. In *New global frontiers in regulation: The age of nanotechnology*, ed. Graeme Hodge, Diana Bowman, and Karinne Ludlow, 29–48. Northampton: Edward Elgar Publishing, Inc.
Nicolelis, Miguel. 2002. Human-machine interaction: Potential impact of nanotechnology on design of neuroprosthetic devices aimed at restoring or augmenting human performance. In *Converging technologies for improving human performance: Nanotechnology, biotechnology, information technology and cognitive science*, ed. M. Roco and W.S. Bainbridge, 223–227. Arlington: National Science Foundation and Department of Commerce.
Nordmann, Alfred. 2007. If and then: A critique of speculative nanoethics. *NanoEthics* 1: 31–46.
Parens, Erik, and Josaphine Johnston. 2007. Does it make sense to speak of neuroethics? *European Molecular Biology Organization* 8: S61–S64.
Parr, Douglas. 2005. Will nanotechnology make the world a better place? *Trends in Biotechnology* 23(8): 395–398.
Patel, Geeta M., Gayatri C. Patel, Ritesh B. Patel, Jayvadan K. Patel, and Madhabhai Patel. 2006. Nanorobot: A versatile tool in nanomedicine. *Journal of Drug Targeting* 14(2): 63–67.

Pidgeon, Nick, and Tee Rogers-Hayden. 2007. Opening up nanotechnology dialogue with the public: Risk communication or 'upstream engagement'? *Health, Risk and Society* 9(2): 191–210.

Pontius, A.A. 1973. Neuro-ethics of 'walking' in the newborn. *Perception and Motor Skills* 37: 235–245.

Priestly, Brian, and Andrew Harford. 2007. The human health risk assessment (HRA) of nanomaterials. In *New global frontiers in regulation: The age of nanotechnology*, ed. Graeme Hodge, Diana Bowman, and Karinne Ludlow, 134–148. Northampton: Edward Elgar Publishing, Inc.

Rajzi, Alex. 2008. One danger of biomedical enhancements. *Bioethics* 22(6): 328–336.

Rees, Dai, and Steven Rose (eds.). 2008. *The new brain sciences: Perils and prospects*. New York: Cambridge University Press.

Rieder, Travis N. 2008. Fritz Allhoff and Patrick Lin (Eds.), Nanotechnology and society: Current and emerging issues. Nanoethics 2: 329–331.

Robert, Jason S. 2008. Nanoscience, nanoscientists, and controversy. In *Nanotechnology & society: Current and emerging ethical issues*, ed. Fritz Allhoff and Patrick Lin, 225–239. New York: Springer Science and Business Media.

Roco, Mihail C. 2003. Broader societal issues of nanotechnology. *Journal of Nanoparticle Research* 5: 181–189.

Roco, Mihail C., and W.S. Bainbridge (eds.). 2003. *Converging technologies for improving human performance: Nanotechnology, biotechnology, information technology and cognitive science*. Dordrecht/Boston: Kluwer Academic.

Rose, Steven. 2006. *The future of the brain: The promise and perils of tomorrow's neuroscience*. New York: Oxford University Press.

Sahoo, Sanjeeb K. 2005. Applications of nanomedicine. *Asia Pacific Biotech News* 9: 1048–1050.

Sandberg, Anders, and Nick Bostrom. 2006. Converging cognitive enhancements. *Annals of the New York Academy of Sciences* 1093: 201–227.

Schummer, Joachim. 2008. Cultural diversity in nanotechnology ethics. In *Nanotechnology & society: Current and emerging ethical issues*, ed. Fritz Allhoff and Patrick Lin, 265–280. New York: Springer Science and Business Media.

Shipbaugh, Calvin. 2006. Offense-defense aspects of nanotechnologies: A forecast of potential military applications. *The Journal of Law, Medicine & Ethics* 34: 741–747.

Sparrow, Rob. 2007. Negotiating the nanodivide. In *New global frontiers in regulation: The age of nanotechnology*, ed. Graeme Hodge, Diana Bowman, and Karinne Ludlow, 87–107. Northampton: Edward Elgar Publishing, Inc.

Stock, Gregory. 1993. *Metaman: The merging of humans and machines into a global superorganism*. Toronto: Doubleday Canada Limited.

Turk, Volder, and Christa Liedtke. 2007. Invisible but tangible? Societal aspects and their consideration in the advancement of a new technology. In *New global frontiers in regulation: The age of nanotechnology*, ed. Graeme Hodge, Diana Bowman, and Karinne Ludlow, 67–86. Northampton: Edward Elgar Publishing, Inc.

van de Poel, Ibo. 2008. How should we do nanoethics? A network approach for discerning ethical issues in nanotechnology. *NanoEthics* 2: 25–38.

van den Hoven, Jeroen. 2008. The tangled web of tiny things: Privacy implications of nano-electronics. In *Nanotechnology & society: Current and emerging ethical issues*, ed. Fritz Allhoff and Patrick Lin, 147–162. New York: Springer Science and Business Media.

van den Hoven, Jeroen, and Peter E. Vermaas. 2007. Nano-technology and privacy: On continuous surveillance outside the panopticon. *The Journal of Medicine and Philosophy* 32: 283–297.

Weckert, John. 2007. An approach to nanoethics. In *New global frontiers in regulation: The age of nanotechnology*, ed. Graeme Hodge, Dianna Bowman, and Karinne Ludlow, 49–66. Northampton: Edward Elgar Publishing, Inc.

Wilson, Robin Fretwell. 2006. Nanotechnology: The challenge of regulating known unknowns. *The Journal of Law, Medicine & Ethics* 34: 704–713.

Wolbring, Gregor. 2007. Nano-engagement: Some critical issues. *Journal of Health and Development* 3: 9–29.
Wolbring, Gregor. 2008. Why NBIC? Why human performance enhancement? *Innovation – The European Journal of Social Science Research* 21(2): 25–40.
Wolpe, Paul. 2002. Neurotechnology, cyborgs, and the sense of self. In *Neuroethics: Mapping the field*, May 13–14, 2002. San Francisco, CA: The Dana Foundation.
Zebrowski, Robin L. 2006. Altering the body: Nanotechnology and human nature. *International Journal of Applied Philosophy* 20(2): 229–246.
Zwart, Sjoerd D., Ibo van de Poel, Harald van Mil, and Michael Brumsen. 2006. A network approach for distinguishing ethical issues in research and development. *Science and Engineering Ethics* 12: 663–684.

# Chapter 6
# A New Model for Public Engagement: The Dialogue on Nanotechnology and Religion

**Richard Milford and Jameson M. Wetmore**

## 6.1 Introduction

Public engagement and deliberation about new and emerging technologies and their potential future trajectories and meanings in society occupies an important place in the idea of anticipatory governance. However, what should public engagement and deliberation look like in anticipatory governance? Here we offer a few ideas that took form in a Dialogue on Nanotechnology and Religion conducted by the Center for Nanotechnology in Society at Arizona State University (CNS-ASU) in 2008.

Anticipatory governance imagines not only a thorough discussion of the uncertainties that accompany new technologies but also "the cultivation of a societal capacity for foresight" with regards to the various futures that nanotechnology could make possible (Barben et al. 2008, 991). Desirable outcomes of anticipatory governance include a public that is not only well informed about nanotechnology's technical specifics but also able to provide valuable guidance and input into future nanotechnology research, development, and policy. This latter outcome requires, in turn, a broad and deep sense of how new nanotechnology applications intersect with the values, goals, concerns, wishes, and identities of a wide range of individuals and communities in society. Thus, methods of upstream public engagement must be employed that give diverse groups meaningful voice and opportunities to explore potential nanotechnological futures and contribute to debates about how to pursue nanotechnology innovations.

Unfortunately, many current approaches to public engagement in nanotechnology do not achieve this ambitious goal. First, they typically engage the public as a monolithic whole, without significant variation or, at least, in terms of variation that can be adequately dealt with by choosing an appropriately representative sample. They thus routinely fail to engage the deep sensibilities of diverse social groups

R. Milford • J.M. Wetmore (✉)
The Center for Nanotechnology in Society, Arizona State University,
P.O. Box 875603, AZ 85287-5603, Tempe, USA
e-mail: wetmore@asu.edu

S.A. Hays et al. (eds.), *Nanotechnology, the Brain, and the Future*,
Yearbook of Nanotechnology in Society 3, DOI 10.1007/978-94-007-1787-9_6,

(see Wolbring 2007). Second, they traditionally engage scientists and engineers solely as experts, rather than as members of the public whose views and perspectives could valuably enhance ethical and normative deliberation (Berne 2006). Third, they often focus in a general way on nanotechnology rather than on more specific applications that are likely to raise the most focused and acute ethical challenges. Therefore, nanotechnology public engagements tend to generate generic, bland results that wash out the varieties and depths of human experience while focusing solely on the most visible and widely discussed ethical domains – such as environmental and health risks. These issues will be discussed further in the next section.

In this chapter, we present the results of an experiment in public engagement designed to try to address these three design challenges. The Dialogue on Nanotechnology and Religion sponsored in 2008 by CNS-ASU sought to cultivate conversation within a narrow segment of the public chosen for its interest in engaging nanotechnology under the thematic umbrella of religion. In this way, the Dialogue sought to pay heed to one part of the broad spectrum of different worldviews and perceptions that inform and will continue to color the public's views about new and emerging technologies. It thus sought to foster discourse among only some of the many publics that together comprise the public as a whole. Additionally, it sought to supplement prior nanotechnology engagements with a more in-depth focus on the thoughts, ideas, and ethics of specific communities who are able to draw upon their own unique values, experiences, identities, and vocabularies to grapple with the challenges of nanotechnology.

The chapter begins with a more expansive discussion of the shortcomings of current approaches to public engagement in nanotechnology, followed by a detailed discussion of the design elements we adopted for the Dialogue. The latter illustrates how such activities can be designed to be as participant-centered as possible, creating an environment that fosters rather than restricts the issues that can potentially be discussed. Finally, we analyze the Dialogue itself, showing how its focus on participant's religious identities helped create a potential for similar future dialogues to effectively contribute to anticipatory governance.

## 6.2 Critiques of Current Approaches to Nanotechnology Engagement

The need to improve strategies for public engagement in nanotechnology has become a common theme in recent writing. Overviews of public engagement in nanotechnology by think tanks in the UK, such as Demos and Involve, highlight the desire to improve and enhance public engagement techniques (Gavelin and Wilson 2007; Stilgoe 2007). Both organizations have carried out a variety of nano-engagement activities in recent years that have sought to deepen the participant orientation of public engagement, such that the participants are able to have a greater role in determining the issues that are discussed. One example is *Nanotechnology and Development*, a dialogue designed to function as an outlet for the needs of a Zimbabwean town (Gavelin and Wilson 2007). This project's organizers took pains, for example, to not mention nanotechnology until local participants themselves brought it up in the context of the community's challenges. NanoJury, another project that took place at the University of Newcastle in 2005, also allowed local participants to use the first few dialogue sessions to talk about an issue of their own choosing before beginning a conversation explicitly focused on nanotechnology.

TA-Swiss' "Nanotechnology, Health and the Environment" is another example of an activity in which organizers strove not to impose a specific agenda on participants in an attempt to allow them to express more fully their own "expectations, fears, and hopes" (Burri 2007). As a result, organizers learned that participants in general shared an overall positive attitude towards nanotechnology and optimism regarding its potential benefits for medicine, the working environment, and the quality of everyday life. Such findings lend further weight to the argument that open-ended forums for public deliberation can yield information that could do much to contribute to anticipatory governance.

Although these examples of innovation in public engagement are in-line with the goal of enhancing the degree to which such activities focus on issues and concerns of the participants, they are not representative of the overall nano-engagement enterprise. Public engagement, in general, remains too focused on problems identified first by experts and is insufficiently attentive to issues of interest to citizens in their day-to-day lives or of interest because of citizens' unique ethical sensibilities. Wolbring (2007) points out in a review of online keyword searches based on nanotechnology-related public engagement that nano-engagement exercises have failed to take into account the needs of a variety of social groups, including women, the disabled, indigenous communities, and the poor. Instead, they have stressed rather generic issues in health, safety, toxicology, and trade that are already circulating in the media. Thus, most engagements have failed to offer a unique perspective emanating from the public itself, in large part because they have lacked explorations of the needs of particular communities to whom these issues do not have the highest priority.

An overview of presentations given at a National Nanotechnology Initiative conference in May 2006 provides insight into how leading scholars of the public participation effort are shaping the discussion on how to make future activities more effective. It also brings to light conflicting perspectives regarding the extent to which the public's creative input should be actively sought after in these activities. One presentation identified the need to recognize the diverse groups which collectively make up the public, yet ultimately recommended the use of surveys focusing on "risk perception" to gain an understanding of their concerns regarding nanotechnology (Harthorn 2006) rather than relying on what the participants themselves bring to such discussions. Another pointed out flaws in the information deficit model in which scientifically uninformed citizens become enlightened using various educational methods. However, the presenter then argued for the use of pre- and post-discussion surveys to assess the effects of information learned during a public participation activity as if this were the most important form of output to be gained (Hudson 2006). Such methods are problematic for achieving the sort of authentic public input which anticipatory governance aims at using to potentially steer nanotechnology research and development.

While the use of surveys is effective for measuring the effects that these activities have on participants' views and knowledge relative to nanotechnology, such endeavors are ultimately unable to facilitate an understanding of the relevant values and concerns that participants bring to public engagement activities. Of course, these methods are useful for enabling public engagement scholars to demonstrate the effectiveness of nano-engagement in assessing the public's overall understanding of nanotechnology and the effects of educational efforts towards increasing this understanding. However, if public participation is to indeed become a tool of anticipatory governance, it will be necessary to move beyond activities that simply assess public knowledge towards ones which afford public participants with real opportunities to have creative input into the nanotechnology discourse. Genuine public-oriented anticipatory governance will require that these avenues are created, but such input will only be possible if public participation is designed to value the interests of participants over their knowledge regarding issues that are pre-determined by expert, non-public perspectives.

Other presentations at the May 2006 conference more clearly identified the need to allow public concerns to frame and drive these discussions. One of these summarized a National Humane Genome Research Institute's effort, entitled "The Community Genetics Forum", to both engage public participants to gain an understanding of their questions and concerns and to "aggressively engage" specific niche audiences, including underrepresented minorities and special populations, with whom to establish sustainable relationships (Bonham 2006). Another identified the importance of focusing on the participant's own values first, using these to frame the resulting questions that are asked of the participants (Sarno 2006). Clearly, the idea participant-driven engagement is also gaining traction within this community of scholars.

The overall thought expressed at this conference indicates that the community of scholars interested in the questions posed by public engagement appear to agree that the input of the public needs to be a top priority to ensure nanotechnology's eventual success. There also seemed to be consensus that the one-way information deficit model of engagement is no longer sufficient for developing a meaningful relationship between science and the public. However, they have yet to arrive at a consensus regarding the role of public participants in shaping the questions asked in these activities and the issues that are focused on. Those who advocate mechanisms such as anticipatory governance for integrating public input into research and development (R&D) should ensure that participant-driven models are given ample voice in this debate.

A second concern regarding typical approaches to public engagement has emerged surrounding the role of scientists in these exercises. In nearly all public engagements about nanotechnology to date, scientists have appeared at most as members of "expert panels" who can answer technical questions that arise in the course of discussions among the public participants in these dialogues. This was the case in NanoJury, for example, in which scientists were referred to as "witnesses" for the "jury" of participants to use as sources of information (Gavelin and Wilson 2007). While it certainly makes sense to provide non-scientist dialogue participants with the information needed to help them think through the issues, it is not clear why scientists themselves have not been taken more seriously as a source of ethical insight and why they should not be included in these conversations.

The reason to exclude scientists as participants in nano-engagements is often that the purpose of such engagements is to bring perspectives from non-scientists into nanotechnology innovation and policy. Ostensibly, this is to protect engagements from being dominated by scientists and engineers who may have little grasp of the public's needs and perspectives. In addition, there are concerns that introducing technical experts into public discussions about nanotechnology may bias the direction they take if non-scientist voices are cowed by scientists' technical authority. There may also be good reason to believe that introducing scientists into these discussions will present a considerable conflict of interest, particularly if they are personally involved in the development of the technologies in question.

The work of Rosalyn Berne suggests otherwise, however. In *Nanotalk*, she demonstrates through a series of interviews with scientists and engineers, whose expertise is often assumed to be limited to technical matters, that they may in fact have much to offer such discussions. Concerns that scientists speaking in the public arena would hijack the conversation are mitigated by her discovery that they often feel unqualified to speak out as individuals about issues of ethics and policy. Berne attributes this to both the culture of their work environments, where such concerns are often viewed as irrelevant to their expected roles. She also implicates mainstream views of scientists as "amoral and taciturn about seeking to understand and solve problems through technology…as people who have difficulty connecting to humanity because of their obsessive interest in the mundane elements of the material world of observation, processes, gadgets, and devices" (Berne 2006). It appears from Berne's interviews that scientists often feel discouraged from taking part in discussions of ethics because they have been hedged in by mainstream assumptions about their detachment from everyday human life and the supposed limits of their expertise.

More importantly, leaving out scientists prevents them from being included as concerned citizens in public engagement. This excludes their voices, ethical perspectives, and knowledge of ethical dilemmas, prevents them from participating in deliberative activities, and continues to foster sharp divides and mistrust between citizens and the public. Berne concludes: "the narratives of individual research scientists and engineers should also be included, not solely as the voices of professional experts, but as interested citizens with a story to tell. Successful public trust and understanding warrants the inclusion of individual laboratory researchers as persons, who might be willing to contribute their own stories to the wider public discourse" (Berne 2006).

Finally, analyses of the public engagement efforts carried out by Demos and Involve also highlighted a third design problem for past nanotechnology engagements. Both NanoJury and Small Talk, another nano-engagement activity that took place in the UK between 2004 and 2006, found that participants were alienated by nanotechnology's amorphous, undefined nature. These organizations concluded after running several workshops that a broad focus on nanotechnology could overwhelm participants while making it unclear where the discussion was headed. The participants' uncertainty about exactly which technologies were being discussed frequently brought the conversation to an awkward halt. That this occurred in Small

Talk is particularly striking when one considers that all participants in that particular project were self-selected for their already substantial personal interest in issues of science and technology (Gavelin and Wilson 2007).

## 6.3 Designing the Dialogue on Nanotechnology and Religion

In February 2008, the Center for Nanotechnology in Society at Arizona State University facilitated an exercise in public engagement that sought to address the three critiques of previous public engagements of nanotechnology described in the previous section. This experiment, the Dialogue on Nanotechnology and Religion: (1) specifically engaged individuals with religious viewpoints; (2) included scientists as well as non-scientists among its eight participants; and (3) focused on the ethical implications of brain-machine interfaces, one of the many applications which nanotechnology is poised to dramatically augment. Participants in the Dialogue came from a variety of religious backgrounds, including members who identified as Catholic, Protestant, Latter-Day Saint, Buddhist, and non-religious.

### *6.3.1 Why Religion?*

The design of the Dialogue sought, first and foremost, to address the critique that the vast majority of public engagement activities have failed to take into account the identities – ethnic, geographic, socioeconomic, familial, and religious – of their participants. The Dialogue sought to reverse this problem by making it clear through its title, the materials provided to participants, and the questions asked by facilitators that the religious identities of the participants would be valued and actively probed throughout the 2-day discussion. These considerations reflect an important relationship between how public engagement activities choose to engage their participants and the type of ethical discourse that they ultimately foster. Members of the public are comfortable using personal examples and beliefs to inform ethical reflections only under conditions in which dialogue facilitators make it clear that their various sources of identity are not only welcomed in the nanotechnology discourse but valued as repositories of ethical reflection. By encouraging participants to explicitly bring religious perspectives to the table, the Dialogue sought to complement existing models for dialogue, which focus primarily on the public as a singular, monolithic entity, by using methods that focus more on specific identity groups rather than samples which supposedly represent the population at large. In so doing, it sought to address Wolbring's critique of the non-inclusiveness of past public engagements. It also sought to enhance the capacity of public dialogues on nanotechnology to engage important and deeply held repertoires of ethical reflection and deliberation in contemporary societies, by encouraging participants to raise issues of ethical considerations and aspirations that often go unheard when individuals are merely identified as members of a generic public.

The Dialogue's focus on religion also reflected an interest among religious thinkers in the ethics of nanotechnology and their articulated desire to engage with scientists and the public to learn about and deliberate its future possibilities. In the *Journal of Lutheran Ethics*, several authors have expressed a desire for religious communities to engage with scientists and the public in extended ethical reflection about nanotechnology (Taylor 2004; Pearson 2006). Pearson calls on his readers to "foster continued public discourse", and Taylor encourages the formation of close personal relationships with scientists to better understand the desires and motivations informing their research. These perspectives suggest a confluence of interest between religious ethicists and those who are hopeful about democratizing nanotechnology's development in fostering ethical deliberation and dialogue.

Examples from the writings of several theologians help to highlight this growing interest in nanotechnology within the religious community and to demonstrate the diversity of viewpoints that are emerging in response. Lutheran author Noreen Herzfeld articulates both the concern and aspiration that characterize much of the nanotechnology discourse, yet does so in terms that go beyond conventional risk/benefit analysis. While posing challenges to a future scenario in which nanotechnology is potentially used to extend human life indefinitely, she also uses Christian insights to articulate a view of its possible role in bringing humanity closer to God. Herzfeld likens the Christian aspiration to "grow in [God's] image and likeness and ... ultimately be something else" (Herzfeld 2006) to nanotechnology's proposed role in changing the human body and the world around us for the better. Such thought demonstrates how, within Christian thought, issues of technology can be connected to questions of human identity and meaning. Herzfeld also compares the hopes invested in nanotechnology to the similar aspirations which framed the pursuit of medieval alchemy, as both have been connected to hopes of "unprecedented wealth, restoration of the body, and life everlasting" (Herzfeld 2006). It also indicates how awareness of one's own narratives, (i.e. the Christian story as understood by Lutherans), makes one able to openly discuss the narratives of others. This is a crucial element of ethical discussion, which has only begun to be tapped in the nanotechnology discourse. Personal identities, if encouraged in public engagement, provide participants with a foundation on which to build, evaluate, and deliberate narratives of technological futures.

Christian ethicists Ebbesen and Anderson also demonstrate how aspirations concerning nanotechnology can be shaped by religious beliefs. They argue that nanotechnology should be used for the purposes of "neighbor love," helping people in need, in lieu of "transgressing" the limits that "define humanity" (Ebbesen and Anderson 2006). Once again, in their work, it is clear that the ability to engage Christian ethical vocabularies. These vocabularies include: ethical traditions and debates; stories and narratives; registers of meaning; and widely recognized conventions for applying and appropriating values in public discourse. They allow for a deep expression of both hopes for a better future as well as reservations about the outcomes of certain interactions between human beings and technology. This indicates that religious voices may thus be able to contribute to a discussion of not only the problems that could arise in the intersections between human nature and technology, but also to our understanding of the various ways that technological applications may be used to contribute to the general welfare of those in need and to the position of the needy in potential technological futures.

What one can take from these writings is that the ability to articulate one's ethical positions using language and ethical traditions that openly draw on religious values fosters more varied discussion of ethical concerns than what has been seen thus far in typical public engagements about nanotechnology. So far, however, it is easier for such voices to be heard in journals of religious ethics that are quite removed from the majority of discussions about nanotechnology. Indeed, such deliberations reflect the views of but a few individual religious thinkers. The idea of anticipatory governance encourages more to be done to bring such points of view into mainstream discourses about nanotechnology, not only from religious perspectives but also from the perspectives of the wide variety of diverse communities that exist within society.

### 6.3.2 Including Scientists as Ethical Thinkers

Scientists are both citizens and, often, members of religious communities, and they share their own capacities for ethical reflection and dialogue. The Dialogue on Nanotechnology and Religion therefore decided to actively seek the participation of scientists and engineers in the conversation. In doing so, the Dialogue drew heavily on the findings of Rosalyn Berne, whose interviews of numerous scientists and engineers involved in fashioning nanotechnology's future investigated the ethical issues they wrestled with in their everyday work (Berne 2006). "Citizen-scientists," by virtue of the fact that they are both aware of the specific technical details relevant to the conversation and have their own opinions when it comes to ethics, are poised to bring new insights to the table. Their inclusion in a dialogue on nanotechnology and religion helped enhance the diversity of religious perspectives involved in the dialogue, providing religious scientists an outlet for ethical reflections drawn from religious perspectives that they may feel unable to discuss in their professional lives.

Recognizing the potential risks of this choice, the Dialogue moderators worked to ensure that the conversations that developed adequately reflected the voices of all those who participated. Non-scientist participants were encouraged to voice any views or questions that they have in the face of the new technical information that participating scientists may introduce, while the ethical voices of the scientists were also explicitly prompted. Likewise, moderators made sure that scientists did not seek to exclude others from the conversations while also working to ensure that participants without technical expertise were able to participate freely. Admittedly, these tasks were not easy. Yet putting in the extra resources to prepare dialogue facilitators in this way arguably made for conversations that were more satisfying to all parties involved and, ultimately, produced more interesting and useful output.

### 6.3.3 *Specific Technological Applications*

The Dialogue on Nanotechnology and Religion opted to focus primarily on brain-machine interfaces. This "niche" of potential nanotechnology applications is a good focal point for discussion because of its relatively high visibility in the media. Cyberkinetic's BrainGate TM system, for example, has gained significant attention from a clinical trial in which a paralyzed man equipped with the device used it to achieve an increased measure of control over his environment (Duncan 2005). This includes a newfound ability to move the cursor of a computer interface using only cognition. Such striking technological achievements are poised to take hold of the public imagination and provide the material for stimulating and focused discussions, especially when considering the popularity of stories like Michael Crichton's *Prey* (2002). Crichton's story is based on technological outcomes – like the advent of dangerous nanobots – which are much less feasible. Having dialogues centered on technologies early in their development ensures not only that conversations are relevant to real-world contexts of anticipatory governance but also that participants become better informed about feasible technological outcomes.

## 6.4 The Dialogue on Nanotechnology and Religion

To begin the discussion, moderators asked participants three overarching questions:

1. Is nanotechnology different from past technologies in terms of the ethical issues to which it gives rise?
2. What is the role of non-experts in making decisions about nanotechnology's development and outcomes?
3. What is the role of religion in addressing nanotechnology and making these decisions?

Over the course of two 3-h sessions, participants discussed a diverse set of concerns revolving around the complex intersections of technology, society, and ethics. Religious concepts were discussed often as participants related them to issues ranging from the distribution of technologies to human enhancement. Here we offer three brief illustrations.

### 6.4.1 *Technology and the Limits of Human Nature*

Of all the topics discussed, the most prevalent were those related to ethics and human nature. Rationing and equity, the difference between enhancement and therapy and the ethical issues which result from this distinction, the limits of human nature, and the relationship between suffering and technology were notable examples. It is perhaps unsurprising given the Dialogue's focus on religious perspectives that participants frequently chose to discuss issues related to the potential limits of human nature and whether or not they should be overstepped. While being initially articulated as a religious concern, all participants agreed that this was an important issue regardless of one's religious orientation. A physicist used his religious beliefs to articulate the notion that human beings are created in the image of God, and later on, other participants further explored this idea. Nanotechnology's supposed potential to give humans control over their bodies and the world around them can be seen as analogous with the theological notion of human beings as not only created in "God's image", but also as "imagers of God", indicating that religious beliefs can be used to articulate a vision in which human beings are "co-creators" of reality. This component of the discussion highlights the different ways in which religious ideas were used to both analyze and support a range of potential technological futures.

From the discussion on human limitations emerged concerns over the use of brain-machine interfaces to create feelings and emotions such as those that have been associated with "religious experiences." One participant, a philosopher and chemist, reflected on how a "brain chip" could be used to create feelings of love and compassion, wondering whether religious communities would perceive this favorably. Another participant who was a member of the Latter-Day Saints (LDS) community responded to this possibility with reservations that were grounded in LDS ideas about human nature. He distinguished between emotion's role in private or religious experiences and in the formation of human community, and mentioned the importance of the fact that these emotions have been fine-tuned over the course of evolutionary history and should thus not be tampered with. He connected this with his community's belief that human beings have opposing poles of temptation, which can only be navigated between using our capacities for free-will and decision-making. Other participants agreed that manipulating people's emotional states could indeed detract from these fundamental human qualities and have negative effects on communal life.

Of course, our participants did not always agree with one another. The reality of human suffering in a world of science and technology was discussed frequently, and this brought out some of the group's clearest differences in worldview. Several participants discussed how in their respective religious communities, suffering was at times understood as a source of personal growth. A former neurosurgeon responded to this by drawing upon his experience as a medical practitioner to argue that technologies with the potential to alleviate suffering usually deserve strong support. Reflecting on the abundance of disease and death in human life, he expressed doubt that religious perspectives could sufficiently address their existence and justify ways in which suffering could lead to positive outcomes. However, the physicist was quick to point out that the reflections of his own Lutheran community noted that religious perspectives have taken into account a variety of other social effects that technology can bring about. Thus, religious thinking about technology is not limited to grappling with their implications for the amelioration of suffering. In the Lutheran tradition, he pointed out, ethicists have demonstrated how the use of technology can create new inequalities, injustices, and technological divides. This illustrates how many religious thinkers have grappled with the relationship between technology and suffering, leading to views which are more nuanced than simple pro- or anti-technology stances.

These two examples help to highlight how the emphasis on religion influenced the course of the Dialogue's discussion. The discussions of brain-machine interfaces and suffering reveal an important characteristic of the conversations which took place within the Dialogue: namely, that one person's expression of particular religious beliefs, even if not shared by other participants, could nevertheless contribute to the articulation of views that others could engage with in discussion. The discussion on human suffering additionally illustrates how the articulation of a religious perspective can encourage other participants, even if they are not religious themselves, to give voice to their own beliefs. Our focus on our participant's religious beliefs – or lack thereof – ultimately created a backdrop for discussion in which all participants felt free to draw from their own personal experiences, beliefs, and ethical reflections. One thus wonders what sorts of discussion might emerge in technology-centric dialogues that focus on other aspects of identity such as race, gender, ability, family, or nationality.

### 6.4.2 The Power of Narrative

There are other reasons for believing that our emphasis on participant's religious identities, coupled with a specific technological application, allowed them to contribute more fully to the conversation. Our group agreed that, while there are times when rational, rules-based decision-making will be necessary to manage technology, the upstream engagement of technological decision-making could benefit immensely from the construction of narratives grounded more fully in the creativity and imagination of its participants. Such stories, they agreed, have been very useful in envisioning the possibilities and issues associated with future technologies. Although these narratives are not always grounded in real-world feasibility, they have often been much more influential than straightforward argumentation based on specific risks and benefits. Our neuroscientist participant described how, within the community of engineers, it is not those who are most technically talented who most often secure funding for their research. Rather, it is those who communicate better, telling a more convincing story, who end up being most successful in influencing how others conceive of the outcomes of a particular technological application.

Religious narratives are similarly influential. The stories told by both the earlier-mentioned Lutheran theologians and the Dialogue's participants were both given weight using symbols, imagery, and metaphors that accompany religious worldviews. By drawing on our participants' religious identities, we were able to encourage the telling of these stories in our Dialogue. In the discussion about the use of brain interface technology to manipulate people's emotional states, for example, the use of religious vocabulary and metaphor allowed for the envisioning of a particular future grounded in certain views about freedom and human nature. Insofar as such stories continue to shape the discourse on nanotechnology and illustrate the public's perception of the futures that it could lead to, they must be continually sought after in the attempt to better understand the attitudes that shape responses to nanotechnology.

### 6.4.3 *Reversibility*

Another topic from the Dialogue that deserves discussion is how nanotechnology, overall, and brain-machine interfaces specifically bring about issues of reversibility. Our neuroscientist discussed the difficulty and risk associated with reversing today's brain-implant technologies, and this reminded the physicist of the Drexlerian scenario in which nanomachines released into the environment could not be brought back or turned off in the case of malfunction. This led him to argue for the importance of anticipating futures in which technologies unintentionally become irreversible in their application and effects, for dramatic consequences could result if such technologies were developed without proper forethought. This aspect of the conversation was important because it demonstrated how the technical backgrounds of scientists and engineers can contribute to their awareness of specific ethical concerns of which non-experts may not be aware. This outcome of the Dialogue highlights the potential power of including scientists in these types of conversations who can contribute substantially by bringing additional ethical issues to the table.

Ultimately, our participants decided to formulate a list summarizing the ethical issues that arose throughout the course of the Dialogue as output. These include the concerns described above, and included thoughts on Luddism, the difference between moral and rational decision-making, and the rationing of technologies. This was the simplest form of output to produce given the limited time we had available. We hope that these considerations will be used in future engagement efforts and are curious to see what additional issues they may bring to mind for other participants and the discourses that could result.

## 6.5 Conclusion

Overall, it is apparent from an analysis of the Dialogue on Nanotechnology and Religion that a specific focus on religious aspects of personal identity can result in different sorts of conversations about nanotechnology, taking them beyond the discussions of risk that have become the norm for nano-engagement efforts and exposing previously unvoiced hopes and concerns about future technologies. That the Dialogue was ultimately able to foster discussions that engaged complex ethical perspectives and elicited the expression of deep personal beliefs as they related to technology strongly indicates that those who wish to improve ethical discussions within nanotechnology-related public engagement efforts should consider adopting similar strategies.

Of course, the Dialogue was limited in many ways. The views that were expressed by our participants were in no way representative of any particular single community, religious or otherwise, much less of "religious people" as a whole. Our relatively small number of participants and the short amount of time spent in discussion also contrast with other engagement activities which have been much larger in scope and longer in duration such as Small Talk, NanoJury, Citizen Science (Gavelin and Wilson 2007), Nanodialogues (Stilgoe 2007), and the National Citizens' Technology Forum (Center for Nanotechnology in Society 2008). Insofar as it is important to derive broad-level understandings of the views of specific communities, facilitators will need to equip public engagement activities to generate thorough discussions that are representative of these communities.

However, while learning about how communities or social groups think overall about technological issues is important, this need not be public engagement's only purpose, particularly if it is to be informed by a philosophy of anticipatory governance. As long as we remain interested in fostering different types of discussions that can deepen our level of ethical reflection about emerging technologies, activities on the scale of the Dialogue on Nanotechnology and Religion can be of much use. A focus on the specific and personal, rather than the broad and impersonal, characteristics of the participants in attendance can do much to illuminate the intricacies of thought on certain issues. Conversely, broad surveys of particular communities or social groups can potentially miss the rich amount of ethical deliberation that often occurs at the interpersonal level.

A recent study by Scheufele and Brossard (2008) indicates that for many Americans, religious background is an effective predictor of how they will respond to nanotechnology. This work argues that religion can function as a "filter" which affects how believers decide whether or not new technologies are morally acceptable. Scheufele and Brossard conclude that Americans' religious backgrounds have much to do with why many citizens are already taking issue with future nanotechnology applications. However, the findings of the Dialogue on Nanotechnology and Religion indicate that one's religious affiliation can also have a more subtle impact on how technological issues are thought about.

As our discussion indicated, religious ideas were more often employed to think about the ills *and* the benefits that could accrue from potential nanotechnology applications. In fact, at no point in the Dialogue were religious justifications given for the outright rejection of a technology. When religious narratives were employed to be critical of particular technologies, as seen with our LDS participant's response to the hypothetical emotion-manipulating brain chip, this did not lead him to reject the technology, but rather to draw from the vocabulary of his religious identity to ask particular ethical questions. The same is evident from the authors voiced in the *Journal of Lutheran Ethics*, who use their religious backgrounds to encourage fellow believers to engage in deeper reflection about future technologies. In such cases, it is clear that religion acts not as a cognitive filter which leads believers to make immediate judgments as to the moral acceptability of these new technologies, but rather as a template from which to derive novel questions and ways of thinking about the issues they might bring about. It is this sort of thinking that nanotechnology-based public engagement activities can and should do a better job of illuminating.

This can happen by focusing not only on the religious aspects of the public identity but also on other components such as geographic, socio-economic, and cultural backgrounds as well. Thus, we again see how participant-centered nano-engagement such as that attempted by the Dialogue on Nanotechnology and Religion could do much to enrich the current discourse about nanotechnology and ethics.

Combining such insights with what public engagement scholars have articulated regarding our need to actively include different social groups and foster different kinds of ethical discussions, we propose that the participant-centered model for public engagement applied here be put to use in future efforts to better understand the public's views about nanotechnology. In their articulation of anticipatory governance, Barben et al. remind us that the success of efforts to use democratic processes to collectively guide the direction of technology innovation requires the cultivation of society-wide reflection (Barben et al. 2008). Discussions of human nature and ethics, while not always leading directly to concrete solutions that can be applied to the development of technological applications, are extremely valuable. They contribute to a better understanding of the attitudes that inform the views of the public, and they may ultimately produce insights that could help to guide technology towards a desirable future. Advocates of anticipatory governance should thus consider such identity-centered models for engagement with the public's various communities as a valuable tool to further the goal of fostering an enlightened public and the democratizing of science and technology.

## References

Barben, D., E. Fisher, S. Selin, and D.H. Guston. 2008. Anticipatory governance of nanotechnology: Foresight, engagement, and integration. In *The handbook of science and technology studies*, 3rd ed, ed. E.J. Hackett. Cambridge: MIT Press.

Berne, R. 2006. *Nanotalk: Conversations with scientists and engineers about ethics, meaning, and belief in the development of nanotechnology*. Mahwah: Lawrence Erlbaum Associates.

Bonham, V. 2006. Community genetics forum: A model for engaging the 'public'. Public participation in nanotechnology workshop: An initial dialogue. National Nanotechnology Initiative. http://www.nano.gov/html/meetings/p2/uploads/15Bonham.pdf. Accessed 2 Sept 2009.

Burri, R.V. 2007. Deliberating risks under uncertainty: Experience, trust, and attitudes in a Swiss nanotechnology stakeholder discussion group. *NanoEthics* 1: 143–154.

Crichton, M. 2002. *Prey*. New York: Harper Collins Publishers.

Duncan, D. 2005. Implanting hope. Technology Review. Posted March, 2005. http://www.technologyreview.com/Biotech/14220/page1/. Accessed 10 Sept 2008.

Ebbesen, M., and S. Anderson. 2006. Nanoethics: General principles and Christian discourse. *Journal of Lutheran Ethics* 6(2).

Gavelin, K., and Wilson, R. 2007. Democratic technologies? The final report of the Nanotechnology Engagement Group (NEG). Involve. www.involve.org.uk. Accessed 10 Sept 2008.

Harthorn, B. 2006. How do we identify the publics to be engaged in nanotechnology? Public participation in nanotechnology workshop: An initial dialogue. National Nanotechnology Initiative. http://www.nano.gov/html/meetings/p2/uploads/07Harthorn.pdf. Accessed 2 Sept 2009.

Herzfeld, N. 2006. The alchemy of nanotechnology. *Journal of Lutheran Ethics* 6(2).

Hudson, K. 2006. The genetic town hall: Making every voice count. Public participation in nanotechnology workshop: An initial dialogue. National Nanotechnology Initiative. http://www.nano.gov/html/meetings/p2/uploads/13Hudson.pdf. Accessed September 2, 2009.

Pearson, T.D. 2006. The ethics of nanotechnology: A Lutheran reflection. *Journal of Lutheran Ethics* 6(2).

Sarno, D. 2006. Best practices in public participation. Public participation in nanotechnology workshop: An initial dialogue. National Nanotechnology Initiative. http://www.nano.gov/html/meetings/p2/uploads/06Sarno.pdf. Accessed 2 Sept 2009.

Scheufele, D.A., and D. Brossard. 2008. Nanotechnology as a moral issue? Religion and science in the US. *AAAS Professional Ethics Report* 21(1): 1–3.

Stilgoe, J. 2007. Nanodialogues: Experiments in public engagement with science. Demos. http://www.demos.co.uk/files/Nanodialogues%20-%20%20web.pdf?1240939425. Accessed 28 May 2009.

Taylor, P. 2004. Going all the way? Cybernetics and nanotechnology. *Nucleus,* pp 12–19, Apr 2004.

The Center for Nanotechnology in Society at ASU. 2008. National citizen's technology forum. http://cns.asu.edu/nctf/. Accessed 21 July 2009.

Wolbring, G. 2007. Nano-engagement: For whom? By whom? With whom? For what? What risk? Medical health? Environmental? Social? *Journal of Health and Development (India)* 25.

# Part II
# Brain Repair and Brain-Machine Implants

# Chapter 7
# The Age of Neuroelectronics

**Adam Keiper***

Every so often, when some new scientific paper is published or new experiment revealed, the press pronounces the creation of the first bionic man—part human, part machine. Science fiction, they say, has become scientific reality; the age of cyborgs is finally here.

Many of these stories are gross exaggerations. But something more is also afoot: There is legitimate scientific interest in the possibility of connecting brains and computers—from producing robotic limbs controlled directly by brain activity to altering memory and mood with implanted electrodes to the far-out prospect of becoming immortal by "uploading" our minds into machines. This area of inquiry has seen remarkable advances in recent years, many of them aimed at helping the severely disabled to replace lost functions. Yet public understanding of this research is shaped by sensationalistic and misleading coverage in the press; it is colored by decades of fantastical science fiction portrayals; and it is distorted by the utopian hopes of a small but vocal band of enthusiasts who desire to eliminate the boundaries between brains and machines as part of a larger "transhumanist" project. It is also an area of inquiry with a scientific past that reaches further back in history than we usually remember. To see the future of neuroelectronics, it makes sense to reconsider how the modern scientific understanding of the mind emerged.

---

*Adam Keiper is the editor of *The New Atlantis* and a fellow at the Ethics and Public Policy Center in Washington, DC. He can be reached at akeiper@thenewatlantis.com.

A. Keiper
The New Atlantis, 1730 M Street N.W., Suite 910, DC 20036, Washington
e-mail: akeiper@thenewatlantis.com

S.A. Hays et al. (eds.), *Nanotechnology, the Brain, and the Future*,
Yearbook of Nanotechnology in Society 3, DOI 10.1007/978-94-007-1787-9_7,

## 7.1 The Body Electric

The brain has been clearly understood to be the seat of the mind for less than four centuries. A number of anatomists, philosophers, and physicians had, since the days of the ancient Greeks, concluded that the soul was resident in the head. Pride of place was often given to the ventricles, empty spaces in the brain that were thought to be home to our intelligent and immaterial spirits. Others, however, followed Aristotle in believing that the brain was just an organ for cooling the body. The clues that suggested its true function—like the brain's proximity to most of the sensory organs, and the great safety of its bony encasement—were noticed but explained away. This is an understandable mistake. After all, how could that custard-like unmoving mass possibly house something as sublime and complex as the human mind? Likelier candidates were to be found in the heart or in the body's swirling, circulating humors.

The modern understanding of the brain as the mind's home originated with a number of seventeenth-century philosophers and scientists. Among the most important was the Englishman Thomas Willis, an early member of the Royal Society, an accomplished physician, and a keen medical observer. Willis and his colleagues carefully dissected countless human brains, gingerly scooped from the skulls of executed criminals and deceased patients. He described his anatomical findings in several books, most notably *The Anatomy of the Brain and Nerves*, which included lovely and meticulous drawings by Christopher Wren (Zimmer 2004). Willis described in detail the structure of the brain and the body's system of nerves. He assigned the nerves a critical new role in the control of the body, and considered their study worthy of a new word, *neurology*. Carl Zimmer—whose enjoyable book *Soul Made Flesh* tells the story of Willis and the age of intellectual ferment and social turmoil in which he lived—details Willis's understanding of the body's nerves:

> When Willis and his friends looked at them through a microscope they saw solid cords with small pores, like sugar cane. Reaching back to his earliest days of alchemy, Willis found a new way to account for how this sort of nerve could make a body move. He envisioned a nervous juice flowing through the nerves and animal spirits riding it like ripples of light. The spirits did not move muscles by brute force but rather carried commands from the brain to the muscles, which responded with a miniscule explosion. Each explosion, Willis imagined, made a muscle inflate (Zimmer 2004).

We know today that Willis was not far off the mark—although instead of a nervous juice, we now know that nerves transmit electrical signals. This was a discovery long in coming, even though electricity had been used in medicine off and on for millennia. In olden days, it was often obtained by rubbing stones like amber; the Romans made medical use of electric eels. Inventions in the eighteenth century made it much easier to store electric charges, and the medical use of electricity became commonplace. Ben Franklin treated patients with shocks. So did Jean-Paul Marat. So, too, did John Wesley, the Methodist; he eventually opened three clinics for electrical treatment in London. The rapid rise and broad acceptance of electrotherapy in the eighteenth century, as chronicled in Timothy Kneeland and Carol Warren's book *Pushbutton Psychiatry*, is astonishing; it was used in treating a wide range of mental and physical ailments (Kneeland and Warren 2002). Despite the exposure of several notorious quacks, electrotherapy only became more popular in the nineteenth century, reaching its zenith in the decades just before and after 1900. Today's lastingly controversial practice of electroshock therapy (also called electroconvulsive therapy, because of the seizures it induces) can be considered a latter-day descendant of the old electrotherapy.

The scientific study of electricity and the nervous system progressed in tandem with the electrotherapy craze. A few researchers early in the eighteenth century suggested that nerves might transport electricity produced in the brain, but this was all speculation until the 1770s when Luigi Galvani noticed the twitching that occurred when dead frog legs were touched by two different metals. By the 1840s, scientists had used sensitive instruments to measure the tiny currents of nerves and muscles, and in 1850, the great German physicist and physiologist Hermann von Helmholtz succeeded in measuring the speed at which electrical impulses traversed the nervous system. The impulses traveled much slower than anyone expected. This was because the electrical signals weren't transmitted with lightning speed like signals on a copper wire; instead, the impulses were propagated by a slower biochemical process discovered later.

Scientists were also coming to understand more fully the functional structure of the brain. The scrutiny of patients with sick or injured brains—like the famous case of Phineas Gage, the railroad foreman whose personality changed radically in 1848 after a spike accidentally blew through his head—suggested to anatomists that skills and behaviors could be linked to specific brain locations. These clinical discoveries were complemented by laboratory research. At the beginning of the nineteenth century, Galvani's nephew, Giovanni Aldini, showed that electrical shocks to the brains of dead animals—and later dead criminals and patients—could produce twitches in several parts of their bodies. Decades later, other researchers continued this work more systematically, electrically shocking the brains of live animals to figure out which body parts were controlled by which spots on the brain.

By the 1890s, scientists had also worked out the cellular structure of the nervous system, using a staining technique that made it easier to see the fine details of the brain, the spinal cord, and the nerves. The individual nerve cells, called neurons, branch out to make connections with a great many other neurons. There are tens of billions of neurons in the adult human brain, meaning that there are perhaps a hundred trillion synapses where neurons can transmit electrical signals to one another.

The twentieth century brought great advances in psychopharmacology, and also a bewildering assortment of imaging technologies—X-rays, CT, PET, SPECT, MEG, MRI, and fMRI—that have made it possible to observe the living brain. Just as aerial or satellite photos of your hometown can convey a richer sense of ground reality than the battered old atlas in the trunk of your car, today's imaging technologies give a breadth of real-time information unavailable to the neural cartographers of a century ago. The latest research on brain-machine interfaces relies upon these new imaging technologies, but at its core is the basic knowledge about the nervous system—its electrical signals, its localized nature, and its cellular structure—already discovered by the turn of the last century. The only thing missing was a way of getting useful information directly out of the brain.

## 7.2 Brain Waves and Background Noise

In the 1870s, Richard Caton, a British physiologist, began a series of experiments intended to measure the electrical output of the brains of living animals. He surgically exposed the brains of rabbits, dogs, and monkeys, and then used wires to connect their brains to an instrument that measured current. "The electrical currents of the gray matter appear to have a relation to its function," he wrote in 1875, noting that different actions—chewing, blinking, or just looking at food—were each accompanied by electrical activity (Caton 1875). This was the first evidence that the brain's functions could be tapped into directly, without having to be expressed in sounds, gestures, or any of the other usual ways.

Several years passed before others replicated Caton's work (in some cases, without awareness of his precedence), but even then, almost no one took notice. There was no easy way to keep records of the constant changes in their measurements of animal brain activity, so these early experimenters had to draw pictures of the activity their instruments measured. Only by 1913 did anyone manage to make the first crude photographic records of brain electrical measurements.

It wasn't until the 1920s that a researcher—German psychiatrist Hans Berger—first measured and recorded the electrical activity of human brains. As a young man, Berger had experienced an odd coincidence that led him to believe in telepathy. This influenced his decision to study the connection between mind and matter, and led him to research "psychic energy" (Millett 2001). He spent decades trying to measure the few quantifiable brain processes involving energy—the flow of blood, the transfer of heat, and electrical activity—and attempting to link those physical processes to mental work. Electrical measurement was of special interest to Berger, and whenever he could get away from his family, his patients, and his many administrative obligations, he would sequester himself in a laboratory from which he barred colleagues and visitors.

The great difficulty facing Berger was to isolate the brain's activity amidst the electrical cacophony of the body and through the thick obstruction of the skull using instruments that were barely sensitive enough for the task. His first successful measurements were on patients with fractures or other skull injuries that left spots with less bone in the way. (The recently concluded war had something to do with the availability of such patients.) Slowly improving his instrumentation through years of frustrating trial and error, by 1929 Berger was finally reliably producing records of the brain activity of subjects with intact skulls, including his son and himself. He coined the word *electroencephalogram* for his technique, and published more than a dozen papers on the subject.

Berger's electroencephalograms (EEGs) represented the brain's electrical activity as complicated lines on a graph, and he tried to discriminate between the various underlying patterns that made up the whole. He believed that certain recurring wave patterns with discernible shapes—which he called alpha waves, beta waves, and so forth—could be linked to specific mental states or activities. A few years passed before other researchers took notice of Berger's work; when they finally did, in the mid-1930s, there was rapid progress in picking apart the patterns of the EEG.

One early breakthrough was the use of the EEG to locate lesions on the brain. Another was the discovery of a particular wave pattern—an unmistakable repeating "spike-and-dome"—connected to epilepsy. This pattern was so pronounced that the United States Army Air Corps began using EEGs during World War II to screen out pilots who might have seizures. There was even some discussion about the possible use of EEG as a eugenic tool—akin to the way genetic counseling is sometimes used today. "Couples who believe in eugenics may yet exchange brain-wave records and consult an authority on heredity before they marry," said one 1941 *New York Times* article. "A man and a woman who may be outwardly free from epilepsy but whose brain waves are of the wrong shape and too fast are sure to have epileptic children" (Kaempffert 1941).

That term—"brain waves"—actually antedates the EEG by several decades. It was used as early as the 1860s to describe a "hypothetical telepathic vibration," according to the *Oxford English Dictionary*. As the public slowly came to learn about the wavy lines of the EEG, the term donned a more respectable scientific mantle, and newspaper articles during the 1940s sometimes spoke of the EEG as a "brain-wave writer" (See, for example, *The New York Times* 1944.). Perhaps Berger, whose initial impetus for neurological research was his own interest in telepathy, would have been amused by the terminological transition from fancy to fact.

What that etymological shift somewhat obscures, though, is that EEG is most assuredly *not* mind-reading. The waves of the EEG do not actually represent thoughts; they represent a sort of jumbled total of many different activities of many different neurons. Beyond that, there remains a great deal of mystery to the EEG. As James Madison University professor Joseph H. Spear recently pointed out in the journal *Perspectives in Science*, there remains a "fundamental uncertainty" in EEG research: "No one is quite certain as to what the EEG actually measures" (Spear 2004). This mystery can be depressing for EEG researchers. Spear quotes a 1993 EEG textbook that laments the "malaise" and "signs of pessimism, fatigue, and resignation" that electroencephalographers evince because of the slow theoretical progress in their field.

Here's one way to think about the great challenge facing these researchers: Imagine that your next-door neighbor is having a big dinner party with some foreign friends who speak a language you don't know. Your neighbor's windows are closed, his curtains are drawn shut, and his stereo is blasting loud music. You aren't invited, but you want to know what his guests are talking about. Well, by listening intently from outside a window to laughter and lulls and cadences, you can probably figure out whether the conversation is friendly or angry, whether the partygoers are bored or excited, maybe even whether they are talking about sports or food or the weather. You would try hard to ignore the sounds of the stereo. Maybe you would call your neighbor on the telephone and ask him to tell you what his foreign friends are talking about, although there is no guarantee that his description will be accurate. You might even try to affect the conversation—maybe by flashing a bright light at the window—just to see how the partygoers react.

EEG research is somewhat similar. Researchers try to look for patterns, provoke responses, and tune out background noise. Since the 1960s, they have been aided in their work by computers, which use increasingly sophisticated "signal processing" techniques to filter out the din of the party so that an occasional whisper can be heard. By exposing a patient to the same stimulus again and again, investigators can watch for repeating reactions. Using these methods, researchers have been able to go beyond the old system of alpha, beta, and delta waves to pick out more subtle spikes, dips, and bumps on the EEG that can be linked to action, reaction, or expectation.

Since our brains produce these tiny signals without conscious command, it is not surprising that there is interest in exploiting some of these signals to "read the mind" in the same way that pulse, galvanic skin response, and other indicators are used in lie detectors. In fact, a neuroscientist named Lawrence A. Farwell has gotten a great deal of press in the last few years for marketing tests that rely heavily on an EEG wave called P300. The P300 wave, which has been researched for decades, has been called the "Aha!" wave because it occurs a fraction of a second after the brain is exposed to an unexpected event—but before a conscious response can be formulated and expressed. Farwell, who spent 2 years working with the CIA on forensic psychophysiology, has founded a company called Brain Fingerprinting Laboratories. The company's website boasts of helping to free an innocent man serving a life sentence for murder in an Iowa prison: the P300 test "showed that the record stored in [the convict's] brain did not match the crime scene and did match his alibi"(Brain Fingerprinting Testing Ruled Admissible in Court 2011). The district court admitted the P300 test as evidence. When a key accuser was "confronted with the brain-fingerprinting evidence," he changed his story, and the convict was freed after two decades in prison. The use of the P300 test was not without controversy, however: Among the witnesses testifying against its admissibility in court was Emanuel E. Donchin, the preeminent P300 researcher, and a former teacher of and collaborator with Farwell. Donchin has repeatedly said that much more research and development needs to be done before the technique can be used and marketed responsibly.

The criticism hasn't slowed Farwell, however. His company's website describes how the P300 wave helped put a guilty man behind bars for a long-unsolved murder; it tells of Farwell's fruitless eleventh-hour efforts to save a murderer from execution because the P300 test supposedly indicated his innocence; and it discusses how the P300 test might be used to diagnose Alzheimer's disease and to "identify trained terrorists" (Counterterrorism Applications 2011). While there is little reason to believe the P300 test will be so used—after all, the traditional means of determining dementia or identifying terrorists seem simpler—it is conceivable that the P300 test or something similar will someday become more refined and more widely accepted, replacing older, and notoriously unreliable, lie-detection technology.

The P300 test relies on one of several electrical signals that the conscious mind generally cannot control. Yet one of the major applications of EEG has been to exert more conscious control upon the unconscious body. "Biofeedback" is the name of a controversial set of treatments generally classified alongside acupuncture, chiropractic, meditation, and other "alternative therapies" that millions of people swear by even though the medical establishment frowns its disapproval. Biofeedback treatments that use EEG are sometimes called "neurofeedback," and they generally work something like this: A patient wears electrodes that connect to an EEG and is given some kind of representation of the results in real-time. This is the feedback, which can be a tone, an image on a screen, a paper printout, or something similar. The patient then tries to change the feedback (or maintain it, depending on the purpose of the therapy) by thinking a certain way: clearing his mind, or concentrating very hard, or imagining a particular activity.

This may sound absurd—and indeed, much of the literature about neurofeedback is quite kooky, rife as it is with mystical mumbo-jumbo—but evidently, enough people are interested to sustain a small neurofeedback industry. Steven Johnson hilariously described visits to several neurofeedback companies in his 2004 book *Mind Wide Open* (Johnson 2004). First, he meets with representatives from The Attention Builders, a company whose Attention Trainer headset and software is intended for children with attention-deficit disorder. The company has "concocted a series of video games that reward high-attention states and discourage more distracted ones," Johnson writes. "Start zoning out while connected to the Attention Trainer software, and you'll see it reflected on the screen within a split second" (Johnson 2004). Then he visits Braincare, a neurofeedback practice in New York, and uses its similar system to control an onscreen spaceship—"and once again I find that I can control the objects on the screen with ease" (Johnson 2004). Next Johnson visits a California-based practice run by the Othmers, a couple who first encountered neurofeedback in 1985 when they were looking for some way to help their neurologically-impaired son control his behavior. Soon, the entire Othmer family was using neurofeedback therapy—mother for her hypoglycemia, brother for his hyperactivity, and father for a head injury. So convinced were the Othmers of the efficacy of neurofeedback that they made a career of it. Johnson describes his experience with the Othmers' system for training patients to control their mental "mode":

> Othmer suggests that we start with a more active, alert state. She hits a few buttons, and the session begins. I stare at the Pac-Man and wait a few seconds. Nothing happens. I try altering my mental state, but mostly I feel as though I'm altering my facial expression to convey a sense of active alertness, as though I'm sitting in the front row of a college lecture preening for the professor. After a few seconds, the Pac-Man moves a few inches forward, and the machine emits a couple of beeps. I don't really feel any different, but I remember Othmer's mantra—"be pleased that it's beeping"—and so I try to shut down the part of my brain that's focused on its own activity, and sure enough the beeping starts up again. The Pac-Man embarks on an extended stroll through the maze. I am pleased (Johnson 2004).

Johnson's experience, like similar anecdotes from neurofeedback patients, demonstrates just how difficult it can be, especially for novices, to control the sorts of brain activity that an EEG picks up. In addition, even though biofeedback therapy is unlikely to migrate from the fringes to the mainstream of medical acceptability, we shall see that many researchers attempting to build brain-machine interfaces are now pursuing essentially the same EEG technique. As one of the leading brain-machine interface researchers told the *New Yorker* in 2003, his work could rightly be called "biofeedback"—but he doesn't want anyone to confuse it with that "white-robed meditation crap" (Parker 2003).

## 7.3 Into the Brain, Into the Mind

While EEG provides a kind of confused, collective sense of the brain's electrical activity, there is a much more direct way to tap into the brain: stick a wire into it. This approach allows not only for the measurement of electrical activity in parts of the brain, but also for the direct electrical stimulation of the brain.

The forerunners of today's brain implants can be found in the nineteenth century efforts to map different brain functions by shocking different parts of the brains of anesthetized or restrained animals. These efforts continued for decades, yielding a picture of the brain that was stupefyingly complex. However, this great body of work revealed very little about the brain's electrical activity during normal behavior, since there were practically no attempts to put electrodes in the brains of animals that were not drugged or restrained. Stanley Finger's *Origins of Neuroscience* tells of a little-known German professor named Julius R. Ewald "who put platinum 'button' electrodes on the cortex of [a] dog in 1896," then walked the dog on a leash and "stimulated its brain by connecting the wires to a battery" (Finger 1994). Finger notes that Ewald "did not write up his work in any detail, but a young American who visited Germany extended Ewald's work and then published a more complete report of these experiments" in 1900 (Finger 1994).

The first scientist to use brain implants in unrestrained animals for serious research was a Swiss ophthalmologist-turned-physiologist named Walter Rudolf Hess. Starting in the 1920s, Hess implanted very fine wires into the brains of anesthetized cats. After the cats awoke, he sent small currents down the wires.

This experiment was part of Hess's research into the autonomic nervous system, work for which he was awarded a Nobel Prize in 1949 (sharing the prize with the Portuguese neurologist Egas Moniz, the father of the lobotomy). In his Nobel lecture, Hess described how his stimulation of the animals' brains affected not merely their motions and movements, but also their moods:

> On stimulation within a circumscribed area … there regularly occurs namely a manifest change in mood. Even a formerly good-natured cat turns bad-tempered; it starts to spit and, when approached, launches a well-aimed attack. As the pupils simultaneously dilate widely and the hair bristles, a picture develops such as is shown by the cat if a dog attacks it while it cannot escape. The dilation of the pupils and the bristling hairs are easily comprehensible as a sympathetic effect; but the same cannot be made to hold good for the alteration in psychological behavior (Hess 1949).

In the decades that followed, a great many researchers began to use implanted brain electrodes to tinker with animal and human behavior. Three individuals are of particular interest: James Olds, Robert Heath, and José Delgado.

James Olds was a Harvard-trained American neurologist working in Canada when, in 1953, he discovered quite by accident that a rat seemed to enjoy receiving electric shocks in a particular spot in its brain, the septum. He began to investigate, and discovered that the rat "could be directed to almost any spot in the box at the will of the experimenter" just by sending a zap into its implant every time it took a step in the desired direction (Thompson and Zola 2003). He then found that the rat would rather get shocked in its septum than eat—even when it was very hungry. Eventually, Olds put another rat with a similar implant in a Skinner box wherein the animal could stimulate itself by pushing a lever connected to the electrode in its head; it pressed the lever again and again until exhaustion.

Thus was the brain's "pleasure center" discovered—or, as Olds came to describe it later because of its winding path through the brain, the "river of reward" (Hooper and Teresi 1992). It was soon established that other animals, including humans, have similar pleasure centers. Countless researchers have studied this area over the years, but perhaps none more notably than Robert Galbraith Heath. A controversial neuroscientist from Tulane University in New Orleans, Heath in the early 1950s became the first researcher to actually put electrodes deep into living human brains. Many of his patients were physically ill, suffering from seizures or terrible pain. Others came to him by way of Louisiana's state mental hospitals. Heath tried to treat them by stimulating their pleasure centers. He often met with remarkable success, changing moods and personalities. With the flip of a switch, murderous anger could become lightheartedness, suicidal depression could become contentment. Conversely, stimulating the "aversive center" of a subject's brain could induce rage.

By the 1960s, Heath had begun experimenting with self-stimulation in humans; his patients were allowed to trigger their own implants in much the same way as Olds's rats. One patient felt driven to stimulate his implant so often—1,500 times—that he "was experiencing an almost overwhelming euphoria and elation, and had to be disconnected, despite his vigorous protests," Heath wrote (Moan and Heath 1972). The strange story of what happened to that patient next, in an experiment so thoroughly politically incorrect that it would never be permitted today, is recounted in Judith Hooper and Dick Teresi's outstanding book *The Three-Pound Universe*:

> [The patient] happened to be a schizophrenic homosexual who wanted to change his sexual preference. As an experiment, Heath gave the man stag films to watch while he pushed his pleasure-center hotline, and the result was a new interest in female companionship. After clearing things with the state attorney general, the enterprising Tulane doctors went out and hired a "lady of the evening," as Heath delicately puts it, for their ardent patient.
>
> "We paid her fifty dollars," Heath recalls. "I told her it might be a little weird, but the room would be completely blacked out with curtains. In the next room we had the instruments for recording his brain waves, and he had enough lead wire running into the electrodes in his brain so he could move around freely. We stimulated him a few times, the young lady was cooperative, and it was a very successful experience." This conversion was only temporary, however (Hooper and Teresi 1992).

Another brain-implantation pioneer, José Manuel Rodríguez Delgado, described how he induced the same effect in reverse: when a particular point on a heterosexual man's brain was stimulated, the subject expressed doubt about his sexual identity, even suggesting he wanted to marry his male interviewer and saying, "I'd like to be a girl" (Delgado 1969).

That experiment is described in Delgado's riveting 1969 book, *Physical Control of the Mind* (Delgado 1969). A flamboyant Spanish-born Yale neuroscientist, Delgado, like Heath, began exploring in the 1950s the electrical stimulation of the reward and aversion centers in humans and animals—what he called "heaven and hell within the brain" (Delgado 1969). Like Heath, Delgado tells stories of patients whose moods shifted after their brains were stimulated—some becoming friendlier or flirtatious, others becoming fearful or angry. He describes artificially inducing anxiety in one woman so that she kept looking behind her and said "she felt a threat and thought that something horrible was going to happen" (Delgado 1969). In other patients, Delgado triggered hallucinations and déjà vu.

Delgado invented a device he called the "stimoceiver," an implant that could be activated remotely by radio signal. The stimoceiver featured prominently in the experiment for which Delgado is best known, in which he played matador, goading a bull into charging him, only to turn off the bull's rage with a click of the remote control at the last instant. The bull had of course had a stimoceiver implanted in advance.

This is just one of a great many bizarre animal experiments detailed in Delgado's brilliant, absurd, coldhearted, sickening book. It is not for the squeamish. A weird menagerie of animals with brain implants is shown in the book's photographs. One little monkey is electrically stimulated so that one of its pupils dilates madly. Friendly cats are electrically provoked to fight one another. Chimpanzees "Paddy" and "Carlos" have massive implants weighing down their heads. One rhesus monkey is triggered 20,000 times so that the scientists can observe a short ritual dance it does each time; another loses its maternal instinct and ignores its offspring when triggered; yet another is controlled by cagemates that have learned that pressing a lever can bring on docility.

The creepiest revelation of Delgado's book is how easily the brain can be fooled into believing that it is the source of the movements and feelings actually induced by electrical implants. For instance, a cat was stimulated in such a way that it raised one of its rear legs high into the air. "The electrical stimulation did not produce any emotional disturbance," Delgado writes, but when the researchers tried to hold down the cat's leg, it evinced displeasure, "suggesting that the stimulation produced not a blind motor movement but also a desire to move" (Delgado 1969). A similar effect was noticed with humans, as in the case of a patient who was stimulated in such a way that he slowly turned his head from side to side:

> The interesting fact was that the patient considered the evoked activity spontaneous and always offered a reasonable explanation for it. When asked, "What are you doing?" the answers were, "I am looking for my slippers," "I heard a noise," "I am restless," and "I was looking under the bed." In this case it was difficult to ascertain whether the stimulation had evoked a movement which the patient tried to justify, or if an hallucination had been elicited which subsequently induced the patient to move and to explore the surroundings (Delgado 1969).

When other patients had their moods suddenly shifted by stimulation, they felt as though the changes were "natural manifestations of their own personality and not … artificial results of the tests" (Delgado 1969). One might reasonably wonder what such electrical trickery and mental manipulability suggest about such concepts as free will and consciousness. To Delgado, these are but illusions: he speaks of the "so-called will" and the "mythical 'I'" (Delgado 1969). It is not surprising, given his totally physicalist views, that Delgado should end his book with a call for a great program of researching and altering the human brain with the aim of eliminating irrational violence and creating a "psychocivilized society" (Delgado 1969).

There was admittedly some interest in such ideas for a short while in the late 1960s and early 1970s, primarily among those hoping to study and rehabilitate prison inmates by means of electrical implants. The chief byproduct of these efforts, it would seem, was the creation of a lasting paranoia about U.S. government plans to control the population with brain implants. (Do an Internet search for "CIA mind control" to see what I mean.) But in reality, of course, the vast social program Delgado envisioned never came to pass, and in some ways, research into the manipulation of behavior through electrical stimulation of the brain has not gone very far beyond where Delgado, Heath, and their contemporaries left it. Consider, for example, the brain-controlling implant that received the most attention in the past few years: a 2002 announcement by researchers at the SUNY Downstate Medical Center that they could control the direction that rats walk (through mazes, across fields, and so forth) by remotely stimulating the pleasure centers in their brains. The scientists claimed that this disturbing research might eventually have practical applications—like the use of trained rats in search-and-rescue operations. But the media excitement about these "robo-rats" obscured the fact that this remote-controlled rodent perambulation was barely an advancement over the work James Olds first did with rats a half-century ago.

## 7.4 The Brain Pacemaker

In his 1971 novel *The Terminal Man*, Michael Crichton imagined the first-ever operation to insert a permanent electrical implant into the brain of a man suffering from psychomotor epilepsy (Crichton 1972). In the story, the patient's seizures and violent behavior are repressed by jolts from the implant. Relying on the best available prognostications about how such futuristic technology could work, Crichton meticulously described every detail: the surgery to insert the 40-electrode implant; the implant's long-lasting power pack; the tiny computer inserted into the patient's neck to trigger the implant when a seizure was imminent; and the testing, calibration, and use of the implant. In true Crichton style, things go awry soon after the surgery and the patient runs away from the hospital and starts killing people.

Just a few years later, similar surgeries were being carried out in real life in the United States. Perhaps the first was an operation to insert an implant designed by Robert Heath and his colleagues, a permanent version of the implants Heath had used in the previous decade. The patient was a mentally-retarded young man prone to fits of terrible violence. Some of the things Crichton predicted hadn't yet been developed—so, for example, Heath's real-life implant didn't have a tiny computer telling it when to zap the brain, it just zapped on a regular schedule, much as an artificial pacemaker sends regular electrical impulses into the heart. And instead of a small power pack under the skin, Heath's implant was connected by wire to a battery outside the skin.

The operation had the desired effect and the patient became sufficiently calm to go back home—until, as in Crichton's story, something went wrong and the patient abruptly became violent and tried to kill his parents. But unlike Crichton's implantee, who met with a bloody end, the real-life patient was captured and safely returned to the hospital, where Heath promptly discovered that the problem was caused by a break in the wires connecting the implant to its battery. The battery was reconnected and the patient went back home.

In all, Heath and his colleagues inserted more than 70 similar implants in the 1970s, with some of the patients seeing dramatic improvements and about half seeing "substantial rehabilitation," according to Hooper and Teresi.

In the years that followed, the study of brain implants stalled. Then, in the 1980s, after French doctors discovered that an electrode on the thalamus could halt the tremors in a patient with Parkinson's disease, researchers began to focus on the feasibility of treating movement disorders with electrical implants. Previously, the only treatments available to Parkinson's patients were drugs (like levodopa, which has a number of unpleasant side effects) and ablative surgery (usually involving either the intentional scarring or destruction of parts of the brain). Other hoped-for cures, like attempts to graft dopamine-producing cells from kidneys or even fetuses into the brains of Parkinson's patients, weren't panning out. But subsequent research confirmed the French doctors' discovery, and a company called Medtronic—one of the first producers of cardiac pacemakers in the late 1950s—began work on an electrical implant for treating Parkinson's patients. After several years of clinical investigation, the Medtronic implant was approved in 1997 by the U.S. Food and Drug Administration (FDA) for use in treating Parkinson's disease and essential tremor; in 2003, it was approved for use in treating another debilitating movement disorder called dystonia.

The implant is often called a "brain pacemaker" or a "neurostimulator"; Medtronic uses brand names like Soletra and Kinetra. The treatment goes by the straightforward name "Deep Brain Stimulation." In the procedure, a tiny electrode with four contacts is permanently placed deep in the brain. It is connected by subcutaneous wire to a device implanted under the skin of the chest; this device delivers electrical pulses up the wire to the electrode in the brain. (Many patients are given two electrodes and two pulse generating devices, one for each side of the body.) The device in the chest can be programmed by remote control, so the patient's doctor can pick which of the four contacts get triggered and can control the width, frequency, and voltage of the electrical pulse. Patients themselves aren't given the same level of control, but they can use a handheld magnet or a remote control to start or stop the pulses.

The first thing that must be said about Deep Brain Stimulation is that it really works. There is always a risk of complications in brain surgery, and even when the operation is successful, there is no guarantee that the brain pacemaker will bring the patient any relief. But more than 30,000 patients around the world with movement disorders have had brain pacemakers implanted, and a majority of them have apparently had favorable results. Some of the transformations seem miraculous. The *Atlanta Journal-Constitution* tells of Peter Cohen, a former lawyer whose dystonia robbed him of his livelihood and left him stooped, shaking, and often stretched on the floor of his home; less than 2 years after his operation, he was off all medication, walking about normally without attracting the stares of passersby, and hoping to resume his law career (Guthrie 2004). A single mother told the *Daily Telegraph* of London how her tremors from Parkinson's abated after she received her implant; before the surgery she had been taking "forty or fifty pills a day," but afterwards she was off all medication and feeling "like a normal mum again" (Doyle 2011). Tim Simpson, a top professional golfer, had to quit playing after a series of health problems, including essential tremor; after he had his brain pacemaker implanted, his hand steadied and he has since returned to pro golf, according to a profile in the *Chicago Sun-Times* (Ziehm 2005). The *San Francisco Chronicle* describes a family in which three generations have had the implants: a mother and her elderly father with essential tremor have gotten over their trembling, while her teenage son with dystonia has regained the ability to walk (DeFao 2005). Thousands of similarly treated patients have come to regain normal lives.

The second thing that must be said about Deep Brain Stimulation is that nobody knows how it works. There are many competing theories. Perhaps it inhibits troublesome neural activity. Or maybe it excites or regulates neurons that weren't firing correctly. Some researchers think it works at the level of just a few neurons, while others think that it affects entire systems of neurons. That it works is undeniable; how it works is a puzzle far from being solved.

Given the mysteriousness of Deep Brain Stimulation, it should come as no surprise that the implants seem to be capable of much more than just stopping tremors. According to various sources, scientists are investigating the use of the implants for treating epilepsy, cluster headaches, Tourette's syndrome, and minimally-conscious state following severe brain injury. What's more, since at least the 1990s it has been clear that the implants can affect the mind, and in the past few years they have been used experimentally to treat a few dozen cases of severe depression and obsessive-compulsive disorder—cases where several other therapies had failed. Establishing experimentally whether such treatments will work is tricky business, since there can be no animal tests for these mental illnesses and since it's all but impossible to conduct blind studies for brain implantation surgeries. But the evidence from the small pool of such patients treated so far seems to show that several have been helped, although none has been cured.

The evidence also suggests that these implants affect mood and mind more subtly than those used by Delgado and Heath more than a generation ago. Consider this 2004 testimony from G. Rees Cosgrove, a Harvard neurosurgeon, to the President's Council on Bioethics:

> So we have four contacts in [the brain], and Paul Cosyns, who is one of the investigators in Belgium, relates this very wonderful anecdote that one of the patients [he] successfully treated has, you know, their four contacts, and she says, "Well, Dr. Cosyns, when I'm at home doing my regular things, I'd prefer to have contact two [activated], but if I'm going out for a party where I have to be on and, you know, I'm going to do a lot of socializing, I'd prefer contact four because it makes me revved up and more articulate and more creative."…

> We have our own patient who is a graphic designer, a very intelligent woman on whom we performed the surgery for severe Tourette's disorder and blindness resulting from head tics that cause retinal detachments, and we did this in order to try and save her vision. The interesting observation was that clearly with actually one contact we could make her more creative. Her employer saw just an improvement in color and layout in her graphic design at one specific contact, when we were stimulating a specific contact (Cosgrove 2004).

These stories suggest that brain implants could be used intentionally to improve the mental performance of healthy minds with less imprecision than mind-altering drugs and less permanency than genetic enhancement. But that possibility is remote. For the foreseeable future, there is no reason to believe that any patient or doctor will attempt to use Deep Brain Stimulation with the specific aim of augmenting human creativity. The risks are too high and the procedure is too expensive. But even if the surgery were much safer and cheaper, we know so little about how these implants affect the mind that any such attempt would be as likely to dull creativity as to sharpen it.

More significant is the possibility that implants will, in time, move into the mainstream of treatment for mental illness. Not counting the tens of thousands of motion-disorder patients with brain pacemakers, Deep Brain Stimulation has so far only been tried on a few severe cases of mental illness. However, there is another technique that involves the stimulation of the vagus nerve in the neck; it is mainly used in the treatment of epilepsy, but for the past few years has been used in Canada and Europe to treat the severely depressed. In July 2005, the FDA approved it for use as a last-resort treatment for depression. According to a 2005 article in *Mother Jones*, Cyberonics, the company that makes the vagus nerve stimulator, "has hired hundreds of salespeople to chase after the 4 million treatment-resistant depressives that the company says represent a $200 million market—$1 billion by 2010" (Slater and Jeffrey 2005). Consider this recipe for a new industry: ambitious companies eager to break into a new market, vulnerable consumers looking for pushbutton relief, and growing ranks of neurosurgeons with implant experience. How long before patients pressure their doctors to prescribe an implant? How long before the defining-down of "last resort"? How long until brain stimulation becomes the neuromedical equivalent of cosmetic surgery—drawing upon real medical expertise for non-medical purposes?

Of course, this may never come to pass. Implants may stay too dangerous for all but the worst cases—or implant therapy for mental illness might be outpaced and obviated by improved psychopharmacological therapies. But there are those who would like to see brain implants become a matter of choice, even for the healthy. David Pearce, a British advocate of transhumanism, has argued that implants in the brain's pleasure centers should be one technological component of a larger project to abolish suffering. His musings on this subject are outlined in an intriguing, if laughably idealistic, manifesto called "The Hedonistic Imperative" (Pearce 2011). (On one of his websites, Pearce offers this quote purportedly from the Dalai Lama: "If it was possible to become free of negative emotions by a riskless implementation of an electrode—without impairing intelligence and the critical mind—I would be the first patient" (Pearce 2011).) Pearce's idealism may seem, on the surface, to be the antithesis of Delgado's dreams of an imposed "psychocivilized society." But they are of a piece. Enamored of the possibilities new technologies open up, unsatisfied with given human nature, and unburdened by an appreciation for the lessons of history, they both forsake reality for utopia.

## 7.5 One Letter at a Time

The most compelling research being done on brain-machine interfaces is as far from utopia as can be imagined. It is in the hellish reality of a trapped mind.

Modern medicine has made it possible to push back the borders of "the undiscover'd country" so that tiny premature babies, the frail elderly, and the gravely sick and wounded can live longer (Shakespeare). One consequence has been the need for new categories that would have gone unnamed a century ago—"brain death" (coined 1968), "persistent vegetative state" (coined 1972), "minimally conscious state" (coined 2002), and so on (U.S. Congress, Office of Technology Assessment 1987; Jennet and Plum 1972; Giacino et al. 2002). Perhaps the most terrifying of these categories is "locked-in syndrome" (coined 1966), in which a mentally alert mind is entrapped in an unresponsive body (Plum and Posner 1968). Although the precise medical definition is somewhat stricter, in general usage the term is applied to a mute patient with total or near-total paralysis who remains *compos mentis*. The spirit is willing, but the flesh is weak. The term is sometimes used to describe a patient who retains or regains some slight ability to twitch and control a finger or limb, but locked-in patients can generally only communicate by blinking or by moving their eyes—movements that must be interpreted either by a person or an eye-tracking device. Sometimes they lose even that ability. Swallowing, breathing, and other basic bodily functions often require assistance. In a word, it is the greatest state of dependency an awake and sound-minded human being can experience.

Locked-in syndrome can develop inexorably over time as the result of a degenerative disease like ALS (Lou Gehrig's disease), or it can be the sudden result of a stroke, aneurysm, or trauma. Misdiagnosis is a frequent problem; there have been documented cases of locked-in patients whose consciousness went unnoticed for years; in more than a few cases, locked-in patients have reported the horror of being unable to reply when people within earshot debated disconnecting life support.

Reliable statistics are nonexistent, but there are surely thousands, and perhaps tens of thousands, of locked-in patients (depending on how broadly the term is defined). Their plight has received attention in recent years partly because of a number of books and articles written by locked-in patients, painstakingly spelling out one letter at a time with their eyes. A Cornell student paralyzed by a stroke at age 19 described her fears and frustrations in her 1996 book, *Locked In* (Mozersky 2000). A former publishing executive in France defiantly titled his 1997 memoir of locked-in syndrome *Putain de silence* (*F***ing Silence*; the English version was given the sanitized title *Only the Eyes Say Yes*) (Vigand and Vigand 2000). A young rugby-playing New Zealander left locked-in by strokes described in a 2005 essay in the *British Medical Journal* how he "thought of suicide often" but "even if I wanted to do it now I couldn't, it's physically impossible" (Chisolm and Gillett 2005). By far the most famous account of a locked-in patient is *The Diving Bell and the Butterfly*, a bestseller written by French magazine editor Jean-Dominique Bauby (and turned into a film in 2007) (Bauby 1997). Bauby spent less time locked-in than the other patient-authors—his stroke was in December 1995, he dictated his book in 1996, and he died 2 days after it was published in 1997—but his account is the most poignant and poetic. The book's title refers to his body's crushing immobility while his mind remains free to float about, flitting off to distant dreams and imaginings. He describes the love and the memories that sustain him. In addition, he tells of the times when his condition seems most "monstrous, iniquitous, revolting, horrible," as when he wishes he could hug his visiting young son (Bauby 1997).

Brain-machine interfaces are likely to make it easier for patients with locked-in syndrome to communicate their thoughts, express their wishes, and exert their volition. Experimental prototypes have already helped a few locked-in patients. With sufficient refinement, brain-machine interfaces may also make life easier for patients with less total paralysis—although for years to come, any patient retaining command of a single finger will likely have more control over the world than any brain-machine interface can provide.

The concept behind this kind of brain-machine interface is simple. We know that electrical signals from brains can be detected by electrode implants or by EEG. However, what if the signals were sent to a machine that does something *useful*? Although most of the serious research in this area goes back only to the 1980s, there are some earlier examples. Perhaps the first is a 1963 experiment conducted by the eccentric neuroscientist and roboticist William Grey Walter. Patients with electrodes in their motor cortices were given a remote control that let them advance a slide projector, one slide at a time. Grey Walter did not tell the patients, though, that the remote control was fake. The projector was actually being advanced by the patients' own brain signals, picked up by the electrodes, amplified, and sent to the projector. Daniel Dennett describes an unexpected result of the experiment in his 1991 book *Consciousness Explained*:

> One might suppose that the patients would notice nothing out of the ordinary, but in fact they were startled by the effect, because it seemed to them as if the slide projector was anticipating their decisions. They reported that just as they were "about to" push the button, but before they had actually decided to do so, the projector would advance the slide—and they would find themselves pressing the button with the worry that it was going to advance the slide twice! (Dennett 1992)

That odd effect, caused by the delay between the *decision* to do something and the *awareness* of that decision, raises profound questions about the nature of consciousness. However, for the moment, let us just note that patients were able to control a useful machine with their brains alone, even if they did not realize that is what they were doing.

Researchers are divided on the question of which is the better method for getting signals from the brain, implanted electrodes or EEG. Both techniques have adherents. Both also have shortcomings. Implants can detect the focused and precise electrical activity of a very small number of neurons, while EEG can only pick up signals *en masse* and distorted by the skull. EEG is noninvasive, while implanted electrodes require risky brain surgery. The two schools of thought coexist and compete peaceably for headlines and limited grant money, although there is some ill will between them and badmouthing occasionally surfaces in the press.

The EEG-based approach dates back at least to the late 1980s, when Emanuel Donchin and Lawrence Farwell, the erstwhile collaborators now on opposite sides of the "brain-fingerprinting" controversy, devised a system that let test subjects spell with their minds. A computer would flash rows and columns of letters on a screen; when the row or column with the desired letter flashed repeatedly, a P300 wave was detected; this process was reiterated until the user had whittled the options down to one letter—and then the whole process would begin anew for the next letter. Donchin and Farwell found that their test subjects could communicate 2.3 characters per minute.

While that system clearly worked, it was indirect—that is, it relied on the uncontrollable P300 wave rather than on the user's willful control of brain or machine. Most subsequent EEG-based researchers have sought control that is more direct. For example, the work of Gert Pfurtscheller, head of the Laboratory of Brain-Computer Interfaces at Austria's Graz University of Technology, emphasizes the motor cortex, so the computer reacts when a subject imagines moving his extremities. A multinational European project, headed by Italy-based researcher José del Rocío Millán, has been working on a system called the Adaptive Brain Interface: the user's brain is studied while he imagines performing a series of pre-selected activities (like picking up a ball); the brain pattern associated with each imagined activity then becomes a code for controlling a computer with one's thoughts. Jonathan Rickel Wolpaw of the Wadsworth Center in the New York State Department of Health, the leading American authority on EEG-based brain-machine interfaces, told *Technology Research News* in 2005 that using his cursor-controlling system "becomes more like a normal motor skill"; the relationship between thought and action becomes even more direct (Patch 2005).

The best-known European researcher who works on EEG-based brain-machine interfaces is University of Tübingen professor Niels Birbaumer. In 1995, he won the Leibniz Prize, a prestigious German award, for a successful neurofeedback therapy, he devised to help epileptics control their seizures. With the prize money, he was able to fund his own research into the use of EEGs for brain-machine interfaces. He was soon testing what he called the "Thought Translation Device" on actual paralyzed patients and reporting impressive successes. One patient, a locked-in former lawyer named Hans-Peter Salzmann, was able, after months of training, to use the device to compose letters, including a thank-you note to Birbaumer published in *Nature* in 1999 (Birbaumer 1999). In the following years, Salzmann's system was connected to the Internet, so he could surf the Web and send e-mails. Here is how Salzmann, in a 2003 interview with the *New Scientist* magazine, describes the mental gymnastics needed to control the cursor:

> The process is divided into two phases. In the first phase, when the cursor cannot be moved, I try to build up tension with the help of certain images, like a bow being drawn or traffic lights changing from red to yellow. In the second phase, when the cursor can be moved,

> I try to use the tension built up in the first phase and kind of make it explode by imagining the arrow shooting from the bow or the traffic lights changing to green. When both phases are intensely represented in my head, the letter is chosen. When I want to not choose a letter, I try to empty my thoughts (Spinney 2003).

Although Birbaumer has reportedly had good results with some of the more than a dozen other patients he has worked with, none has been as successful as Salzmann, and even he has off-days.

Birbaumer's most astonishing case has been that of Elias Musiris, the owner of factories and a casino in Lima, Peru. ALS left Musiris totally locked in by the end of 2001, unable even to blink or control his eyes. A profile of Birbaumer in *The New Yorker* in 2003 describes the scientist's visit with Musiris in the summer of 2002 and how, after several days of practice and training, Musiris was able to answer yes-or-no questions and to spell his own name with the Thought Translation Device (Parker 2003). He had been unable to communicate for half a year. No fully locked-in patient—incapable even of blinking or eye motion—had ever communicated anything before.

## 7.6 Mind Over Matter

Birbaumer thinks the implant approach to brain-machine interfaces is less practicable than the EEG approach, even though the latter is slower. He says that his patients prefer sluggish communications over having a hole in the head.

But a few patients have said yes to a hole in the head, in hopes of controlling machines with their brains. The first were patients of Philip R. Kennedy, an Emory University researcher who, in the 1980s, invented and patented an ingenious new neural electrode. Even setting aside the many health risks of having an electrode surgically implanted in your brain, there were a host of technical problems associated with previous brain implants. Sometimes scar tissue formed around them, reducing the quality of the electrical signals they picked up. Sometimes the electrodes would shift within the brain, so they no longer picked up signals from the same neurons. Kennedy's new design solved some of these problems. The tip of his electrode was protected in a tiny glass cone; once it is implanted, neurons in the brain actually grow into the cone and reach the electrode. The electrode is thus sheltered from scarring and jostling.

After experiments with rats and monkeys, Kennedy obtained FDA permission in 1996 to test his implant in human patients. The first patient, a woman paralyzed by ALS and known only by the initials M.H., could change the signals the electrode detected by switching her mental gears; there was a distinct difference between when she concentrated furiously and when she let her mind idle. Unfortunately, she died two and a half months after the surgery.

Kennedy's second patient was Johnny Ray, a Vietnam vet and former drywall contractor locked in by a stroke. He received his implant in March 1998, and over the next few months learned to move a cursor around a screen by imagining he was moving his hand. By the time the press was informed in October 1998, Ray was able to move the cursor across a screen with icons representing messages—allowing him to indicate hunger or thirst, and to pick from among messages like "See you later." After months of further practice he was able to spell, using the cursor to hover toward his desired letter and then twitching his shoulder—one of the few residual muscles he could control—to select it, like clicking a computer mouse.

When asked what he felt as he moved the cursor, Ray spelled out "NOTHING" (Naam 2005). This could not have been strictly true: it is clear that moving the cursor was exhausting work. However, the doctors interpreted this to mean that Ray no longer had to imagine moving his hand. That intermediate step became unnecessary; he now just *thought* of moving the cursor and it responded.

Ray died in 2002, but Kennedy and his colleagues have carried on their work with several other patients. In his more recent studies, Kennedy has increased the number of electrodes he implants, giving him access to a richer set of brain signals. But the number of electrodes Kennedy implants is dwarfed by the number of electrodes on the implants used by the only other brain-machine interface researchers to put long-term electrodes into humans. That team, led by Brown University professor John P. Donoghue, has conducted two clinical implant studies—one on paralyzed patients, the other on patients with motor neuron diseases like ALS. Their system, called BrainGate, uses 96 tiny electrodes arrayed on an implant the size of an M&M. Seen magnified, the implant looks like a bed of nails.

The first patient to have a BrainGate implant inserted in his head was Matthew Nagle. He was stabbed in the neck with a hunting knife during an altercation at an Independence Day fireworks show in 2001. His spinal cord was severed, leaving him quadriplegic. Although communication wasn't a problem for Nagle—he could talk, and gave interviews and testified at his attacker's trial—he agreed to participate in the BrainGate study, and was surgically implanted in June 2004. The 96 electrodes in his head were estimated to be in contact with between 50 and 150 neurons, and signals from about a dozen were used to give him the same sort of cursor control Johnny Ray had. Nagle's computer was also hooked up to other devices, so he could use it to change the volume on a television and turn lights on and off (Nagle died in 2007).

Several researchers have also done impressive work with electrodes in animals. The leaders in this field are unquestionably Duke University neurobiologist Miguel A. L. Nicolelis and State University of New York neurobiologist John K. Chapin. In 1999, they demonstrated that rats could control a robotic lever just by thinking about it. The rats had been trained to press a bar when they got thirsty; the bar activated a little robotic lever that brought them water. Electrodes in the rats' heads measured the activity of a few neurons, and the researchers found patterns that occurred whenever the rats were about to press the bar. The researchers then disconnected the bar, turning it into a dummy, and set up the robotic lever to respond whenever the right brain signals were present—much as Grey Walter had used a dummy remote control with his slide projector. Some of the rats soon discovered that they didn't have to press the bar, and they began to command the robotic lever mentally.

Nicolelis, Chapin, and their colleagues quickly extended the experiment, and within a couple of years reported successes in getting monkeys to control a multi-jointed robotic arm. To be precise, the monkeys didn't *know* they were controlling a robotic arm: they were trained, with juice as a reward, to use a joystick to respond to a sort of video game, while unbeknownst to them the joystick was controlling the robotic arm. Their brains' electrical signals were measured and processed and interpreted. The researchers then used the brain signals to control the robotic arm directly, turning the joystick into a dummy. Eventually the joystick was eliminated altogether. As a bit of a stunt, the researchers even sent the signals over the Internet, so that a monkey mentally controlled a robotic arm hundreds of miles away—unwittingly, of course.

Other scientists have improved and varied these experiments further still. Andrew B. Schwartz, a University of Pittsburgh neurobiologist who has for more than two decades studied the electrical activity of the brains of monkeys in motion, has trained a monkey to feed itself by controlling a robotic arm with its mind. In video of this feat available on Schwartz's website, the monkey's own limbs are restrained out of sight, but a robotic arm, with tubes and wires and gears partially covered by fake plastic skin, sits beside it. A gloved researcher holds a chunk of food about a foot away from the monkey's face, and the arm springs to life. The shoulder rotates, the elbow bends, and the claw-hand takes the chunk of food, then brings it back to be chomped by the monkey's mouth. The researcher holds the chunk closer, and the monkey changes his aim and gets it again. The whole time, the back of the monkey's head, where the electronic apparatus protrudes, is discreetly hidden from view.

No one can deny that these are all breathtaking technical achievements. Neither should anyone deny that there are a number of major interlocking obstacles that must be overcome before implant-based brain-machine interfaces will be feasible therapeutic tools for the thousands of people who could, in theory, benefit from their use.

The first problem relates to implant technology itself. Implant design is rapidly evolving. Newer implants will have more electrodes. New materials and manufacturing processes will allow them to shrink in size. And implants will likely become wireless. These advances will carry with them new problems to be solved; wireless implants, for example, might cause thermal effects that weren't a problem before.

Second, even though biocompatibility is always considered when designing and building brain implants, most implants don't work very well after a few months in a real brain. There are exceptions—electrodes in a few test animals have successfully picked up readings for more than 5 years—but in general, implant longevity is a problem. One way around it might be to use electrodes capable of moving small distances to get better signals (a notion proposed by Caltech researcher Richard A. Andersen in 2004).

Third, there is still much disagreement about which spots in the brain give the most useful signals for brain-machine interfaces. And much work needs to be done to improve the "decoding" of those signals—the signal processing that seeks meaning in the measurements.

Finally, as the technology moves slowly toward commercial viability for treating patients, standard practices, procedures, and protocols will have to be established, and there will be challenges from government regulators on issues like safety and consent.

In time, the technology will improve, and implant-based brain-machine interfaces will be worthwhile for some patients, perhaps many. However, as things stand today, they make sense for almost no one. They involve significant risk. They are expensive, thanks to the surgery, equipment, and manpower required. They can be exhausting to use, they generally require a lot of training, and they are not very accurate. Only a locked-in patient would benefit sufficiently, and even in some locked-in cases, it would not make sense. For all the *technical* research that has been done, there has been very little *psychological* research, and we still know very little about the wishes and aspirations of severely paralyzed patients.

## 7.7 Artificial Limbs

Experiments allowing animals to mentally move robotic arms raise the question: To what extent will brain-machine interfaces allow paralyzed humans to regain mobility?

One sure bet is that some paralyzed patients will be able to control their own hands and arms, at least in a rudimentary fashion. A little-known but remarkable technology that has been used clinically for more than two decades can restore very basic control to paralyzed muscles. The technology is called Functional Electrical Stimulation (FES). It uses electrical impulses, either applied to nerves or directly to muscle, to jumpstart paralyzed muscles into action. FES has become an important physical therapy tool for some paralytics, allowing them to exercise muscles they can't control. But it can do much more: It has been used to give paralyzed patients new control over their bladder and bowels; it has been used to help several 100 paraplegics stand and haltingly walk with a walker; it has even been used in a number of cases to give quadriplegics a semblance of control over their arms and hands. The first patient to use FES to control his own hands was Jim Jatich, a design engineer left quadriplegic by a 1978 diving accident. In 1986, he had stimulating electrodes implanted into his hands; he can control those implants with a sort of joystick technology manipulated by his chin. Thanks to this system, Jatich and hundreds like him can use computers, write with pens, groom themselves, and eat and drink on their own.

It takes no great leap of the imagination to see how this approach might work in conjunction with the cursor-controlling systems, and indeed, researchers at Case Western Reserve University reported in 1999 that they had already combined the two technologies. A test subject who used FES to open and close his disabled hand was first trained to move a cursor using an EEG-based brain-machine interface. Then the EEG signal was connected to the FES, so that when he controlled his brain waves he could open and close his hand. A more recent study by researcher Gert Pfurtscheller used a similar approach, finding that a patient who triggered his FES by changing his EEG activity "was able to grasp a cylinder with the paralyzed hand" (Pfurtscheller et al. 2003). Researchers using brain implants have taken notice, too. "Imagine if we could hook up the sensor directly to this FES system," implant pioneer John Donoghue told *The Scientist* in 2005. "By thought alone these people could be controlling their arm muscles" (Constans 2005).

But FES doesn't work for everyone. Patients with many kinds of nerve and muscle problems can't use FES—and, needless to say, amputees cannot use it either. Such patients might instead turn to robotics. Donoghue has already shown that his BrainGate system can be used for basic robotic control: His patient Matthew Nagle was able to open and close a simple robotic hand using his implant. That sort of robotic hand is increasingly available to amputees, replacing the older mechanical prostheses normally controlled by cables. The newer robotic prostheses are usually controlled by switches, or by the flexing and flicking of muscles in the amputee's stump. And some more advanced models respond to electromyographic activity—that is, the electrical activity in muscles.

Consider the case of Jesse Sullivan. A bespectacled average Joe in his 50s, Sullivan was fixing electrical lines for a Tennessee power company in 2001 when he was badly electrocuted. Both his arms had to be amputated and he was fitted with mechanical prostheses. Then, researchers led by Todd A. Kuiken of the Rehabilitation Institute of Chicago replaced Sullivan's left prosthetic arm with a robotic arm he can control through nerves grafted from his shoulder to his chest. This lets him move his robotic arm just by thinking where he wants it to go, and according to the institute's website, "today he is able to do many of the routine tasks he took for granted before his accident, including putting on socks, shaving, eating dinner, taking out the garbage, carrying groceries, and vacuuming" (Introducing Jesse Sullivan 2011).

It will be many years before any locked-in patient can control a robotic limb that fluidly. The brain-machine interfaces that let patients slowly and sloppily move a cursor today might be able to control a simple and clunky claw, but nothing that matches the complexity of Jesse Sullivan's new arms. And even Sullivan's high-tech robotic limbs don't come near to rivaling the versatility of the real thing. A real human arm has seven degrees of freedom and a hand has 22 degrees of freedom. While robotic limbs will surely be built with at least that level of complexity, capable of imitating or surpassing all the billions of positions that a human arm and hand can take, it is hard to see how such complex machines can ever be controlled by either the muscle signals that Jesse Sullivan uses or by a descendant of today's brain-machine interfaces. There is just too much information required for dexterous control. Born as infants "wired," so to speak, with countless neuronal connections in our limbs, it takes us years to master our own bodies. No artificial appendage will get that intimate and intricate a connection.

Of course, paralyzed patients and amputees do not necessarily need full equivalency; even partial functionality can dramatically improve their quality of life. Moreover, there is no reason why a patient would have to control every aspect of a prosthetic limb—computers built into the prosthesis itself could do some of the mental heavy lifting. Therefore, while a patient might use a brain-machine interface to tell an artificial hand to grasp a cup, the hand itself might use computerized sensors to tweak the movements and adjust the firmness of the grasp. As one of the researchers already working on this concept told *The Scientist*, this "shared control" idea "seems to make the tasks a lot more reliable than having solely brain-based control" (Constans 2005).

This concept can be extended even further. Wheelchairs controlled by EEG or brain implants are plausible—although patients using breath-control devices can operate their wheelchairs more adroitly than would be possible with any of today's brain-machine interfaces. In addition, if brain-controlled wheelchairs might someday be available, why not other machines? Several research teams around the world have been working for years on exoskeletons. While these robotic suits are generally intended for use by soldiers, the elderly, or the disabled, there are many other possible applications, as a recent article in *IEEE Spectrum* points out: "Rescue and emergency personnel could use them to reach over debris-strewn or rugged terrain that no wheeled vehicle could negotiate; firefighters could carry heavy gear into burning buildings and injured people out of them; and furniture movers, construction workers, and warehouse attendants could lift and carry heavier objects safely" (Guizzo and Goldstein 2005). Making exoskeletons work with brain-machine interfaces for severely paralyzed patients is a distinct, if distant, possibility.

## 7.8 The Higher Senses

Unsurprisingly, much of the funding for research on brain-machine interfaces has come from the United States military, to the tune of tens of millions of dollars. Most of this funding has come through DARPA, the Pentagon's bleeding-edge R&D shop, although the Air Force and the Office of Naval Research have also chipped in substantially. (DARPA's British and Canadian counterparts have also, to a lesser extent, funded brain-machine interface work over the years.) DARPA's interest in robotics and brain-machine interfaces is quite broad—according to its website, the agency would like to find ways to "seamlessly integrate and control mechanical devices and sensors within a biological environment" (Department of Defense Fiscal Year 2007).

DARPA is also interested in a less sophisticated form of mental control over military aircraft—one in which aircraft are made more responsive to the needs and wishes of pilots and aviators by closely monitoring them with sensors and adapting accordingly. This approach—given names like the "cognitive cockpit" and "augmented cognition" (augcog)—would rely on EEG and other indicators (Adams 2005). An *Aviation Today* article explains it this way: "Instead of merely reacting to pilot, sensor, and other avionics inputs, the avionics of tomorrow could detect the pilot's internal state and automatically decrease distractions, declutter screens, cue memory or communicate through a different sensory channel—his ears vs. his eyes, for example. The system would use the behavioral, psychophysiological, and neurophysiological data it collects from the pilot to adapt or augment the interface to improve the human's performance" (Adams 2005).

It will be years before that sort of adaptive cockpit is regularly implemented. Moreover, even that is very different from the idea of direct mental control of airplanes. This is sometimes called the "Firefox" scenario, after the 1982 action film *Firefox*, in which Clint Eastwood was ordered to steal a shiny new Soviet fighter jet specially rigged to read and obey the pilot's mind, to save him the milliseconds it would take to actually press buttons. There is a hilarious catch, though: the brain-reading technology, which is built into Eastwood's flight helmet, can only read thoughts mentally expressed *in Russian*. At the film's climax, Eastwood must destroy a pursuing fighter, but the mission is almost ruined when he forgets to mentally fire his missiles in Russian. At the last moment, he remembers, the missiles fire, and the day is saved. In real life, it is hard to see why brain piloting of a fighter jet would ever be necessary or desirable, especially given the advances in remotely controlled unmanned aircraft.

If brain-machine interfaces are to advance sufficiently for people to control robots with their minds or for the severely disabled to interact normally with the world around them, researchers will have to improve not just the ability to detect and decode the brain's commands, they will also have to improve the feedback that users get. A locked-in patient moving a cursor on a screen can see the results of his mental exertions, but it would be much harder for him to tell, for example, how tightly a robotic hand is grasping a Fabergé egg.

Normally, our senses give our brains plenty of feedback, especially our senses of hearing, vision, touch, and proprioception (balance and orientation). For patients who are deaf, blind, or disabled, researchers and therapists have long sought methods by which one sense could be substituted for another. Haptic technology, by which sensory information is translated into pressure on the skin, has been around for decades; it is central to telerobotics and it has even been used to give some blind patients a very crude kind of "vision" by translating camera images into tactile sensations. It is also of consummate interest to researchers and theorists working on virtual reality. Some basic version of haptic technology, one that puts pressure somewhere on a locked-in patient's skin, would be a simple way to give at least a little non-visual feedback for controlling robotic devices.

In addition, what if it were possible to create the illusion of tactile sensation by directly stimulating the nervous system? Such illusions have long been created haphazardly by researchers stimulating nerves and the brain with electrodes, but what if they could be produced in an organized way, to correspond with the motions of a prosthetic device? A team led by University of Utah bioengineering professor Kenneth W. Horch has reportedly been able to do just that. They ran wires from a robotic arm to nerves in the forearms of amputees. The wires sent nerve signals to the robotic arm, giving the amputees control over the robot. However, the wires also carried electrical impulses back into the amputees' nerves, giving them feedback from the arm. According to a 2004 article in *The Economist*, this enabled the patients "to 'feel' natural sensations as though through the device's fingers," and "made it possible for them to gauge how much pressure to apply when commanding the motors to grip. In addition, position sensors in the robot's joints were translated into 'proprioception' signals that enabled the subjects to feel the arm's position, even when their eyes were closed" (Once Again With Feeling 2004). This technology is still quite far from practical use, but it is a very impressive proof-of-concept.

There have also been advances in the electric creation of other perceptions. Cochlear implants, devices which can restore hearing in patients with certain kinds of deafness or hearing loss, have constantly improved since they went on the market in the early 1980s. These implants, which circumvent the natural hearing mechanism by electrically stimulating nerves in the ear, have now been used by more than 80,000 patients. In those rarer cases of deafness caused by damage to the auditory nerve, cochlear implants are not an option; in some of these cases—a few 100 around the world so far—researchers have begun putting implants directly on the auditory brainstem. The sound quality of auditory brainstem implants is greatly inferior to even that of cochlear implants, but still valuable to those with no other options.

To a lesser extent, there has been progress in artificial vision. Several research teams have been working on implants that could replace *some* of the functionality of *some* of the retina of patients whose own retinas are degenerating. Each team has its own approach. Some researchers are working with simple implants that go just behind the retina, using sunlight to stimulate degenerating retinal cells; they need no wires or batteries. Others have been experimenting with bulkier systems that send data from external cameras to chips implanted in front of the retina. In clinical trials in the past few years, both approaches have succeeded in improving vision in a handful of patients—minimal improvements, but improvements nonetheless. Moreover, for patients whose vision problems are unrelated to the retina, researchers have been working on tapping into the optic nerve or directly stimulating the visual cortex. One of the pioneers of the latter technique, maverick scientist William Dobelle, died in 2004; even though he had some successes with patients, it isn't clear whether any other researcher will emulate either Dobelle's technique (sending signals from cameras to the surface of the visual cortex) or his style (taking patients to Portugal so as to skirt the FDA approval process).

It must be emphasized that artificial vision research is today utterly primitive. Human vision is made possible by an almost unimaginably complex biological system involving millions of photoreceptors, billions of cells, and methods for processing information that researchers can still only guess at. Likewise, the technologies for controlling robotic limbs and restoring hearing will take many more years to mature. Our ignorance is awesome. But we have made progress; real patients have had lost powers restored to them. In our lives, the blind have received their sight, the lame have walked, and the deaf have heard. These are miraculous times.

## 7.9 The New Brain Science

The future of brain-machine interfaces will depend, in part, on several overlapping areas of research now in their infancy. Some of them will fade from view, forgotten footnotes even historians will ignore. Others will likely rise in importance in the years to come, shaping how we think about minds and computers.

First, the study of neurobiology and neurochemistry is progressing rapidly, and scientists are learning ever more about how the brain and nervous system function. The full import of some of the revolutions in brain science is only now beginning to be understood. Chief among these revolutions is the overthrow of the notion of the static brain. Thirty years ago, scientists believed the adult brain was "hardwired"—immutably fixed like electronic circuits. They now know that the brain is flexible, adaptive, and resilient. Understanding the extent of this *plasticity*—the ability of neurons to form new connections and to strengthen or weaken existing connections—is central to understanding how our brains grow, heal, and age.

A second area of research might make brain-machine interfaces unnecessary for some patients: biological solutions for some kinds of paralysis. Although brain damage due to strokes can't be undone, and no magical stem-cell-derived cures for spinal cord injuries should be expected anytime soon, it is worth remembering that biomedical research is progressing simultaneously with the research on brain-machine interfaces. Indeed, the last few years have seen remarkable advances in growing, moving, and manipulating nerve cells in ways that could benefit paralyzed patients.

A third subject that interests scientists is a new way of manipulating the brain called Transcranial Magnetic Stimulation (TMS). Originally developed in the 1980s as part of the growing arsenal of brain-mapping techniques, the fundamental idea in TMS is this: Because the activity of the brain and nervous system is electrochemical in nature, very intense magnetic fields can alter the way the brain functions. This may sound like the sort of charlatanry that has been around for ages—from Mesmer's *baquet* to the "magnetotherapy" cures advertised in late-night infomercials—but TMS is the real deal. Scientists have found that TMS can alter mood. It can reduce hallucinations and treat some migraines. It has been used by one researcher to generate an illusion of what he calls "sensed presence," which he hypothesizes might explain the paranoid and the paranormal. Another researcher has shown that TMS can improve creativity, although only in some people and only very slightly and briefly; he theorizes that by temporarily disabling some of our normal brain functions, TMS might be used to turn us into temporary savants. DARPA has apparently considered TMS as a tool to help soldiers perform well without sleep, and TMS has been clinically tested as a replacement for electroshock in the treatment of depression.

Investigators are also exploring other noninvasive techniques for mind-influence. Studies have recently shown that a specialized technique related to MRI called echo-planar magnetic resonance spectroscopic imaging (EP-MRS) can temporarily improve the mood of patients with bipolar disorder. Another technique called direct current polarization—the use of a battery to send a very tiny electrical current through the front of the head—supposedly can slightly improve verbal ability, according to researchers.

As with electroshock therapy and Deep Brain Stimulation, no one is certain how TMS, EP-MRS, and direct current polarization produce the effects they do. It still isn't clear whether they can affect neurons deep in the brain or just those closer to the surface. Although none of these techniques is painful, little is known about their health risks or long-term effects. And these outside-the-skull techniques are blunt;

they don't give fine control over the mind; they're more like a cudgel than a scalpel. But, taken together, these developments may presage a new interest in the use of machines to influence the mind.

A fourth area of research involves the study of disembodied brains and neurons in combination with silicon chips. The gory aspects of some of this research seem intended chiefly to attract the attention of editors and headline writers. One can be forgiven for wondering what real scientific value there is in removing the brains from a lamprey (an eel-like fish) and connecting them to a little robot. Or in wiring up a mass of disembodied rat brain cells to a robotic arm holding colored markers, so that the ex-brain blob's electrical activity is turned into "art." Or in the simple experiment that prompted absurdly exaggerated headlines like "Brain Grown from Rat Cells Learns to Fly Jet" (Sherwell 2004).

When you look behind the hype, some of this research has serious scientific value: it could improve our understanding of neurochemical processes, it could teach us about how neurons interact, and it could help in the design of better electrodes. It isn't obvious where this gruesome combining of silicon and neurons will lead, but it is clearly a growing area of research that cuts across several disciplines.

Fifth, and finally, a few scientists believe that computer chips could perform some of the higher functions of the brain. As we have seen, the vast majority of the research on brain-machine interfaces and neural prosthetics relates to the body's motor functions and sensory systems. But what if an implant could assist in the brain's cognitive functioning?

To date, the only serious effort to create a "cognitive prosthesis" is the work of a team of researchers headed by University of Southern California professor Theodore W. Berger. This team is attempting to create a computer chip that can do some or all of the work of the hippocampus, part of the brain critical for the formation of long-term memories. Damage to the hippocampus has been connected to amnesia; degeneration of the hippocampus is associated with Alzheimer's disease. As Berger and his colleagues described their hopes in *IEEE Engineering in Medicine and Biology*, if artificial "neurons" can approximate the functions of biological neurons, and ultimately replace damaged neurons, then we will see the rise of "a new generation of neural prostheses" that "would have a profound impact on the quality of life throughout society; it would offer a biomedical remedy for the cognitive and memory loss accompanying Alzheimer's disease, the speech and language deficits resulting from stroke, and the impaired ability to execute skilled movements following trauma to brain regions responsible for motor control" (Berger et al. 2005).

These are grand ambitions. Berger and his team have planned a gradual research program, starting by "reverse-engineering" the hippocampus—thoroughly analyzing the electrical functions of thin slices of rat brain—and then moving on to designing and testing microchips that can replicate those functions. Eventually those chips will be connected to living animal brains for testing.

Will Berger's approach work? And if so, will it someday lead to cognitive prostheses capable not only of restoring damaged brains but of doing much more, such as connecting brains telepathically or giving us editorial control over our memories? Speculation abounds, sometimes rooted in fantasies of the imagination rather than in the best scientific evidence.

## 7.10 Beyond the Cyborg

For the past many decades, serious brain-machine science has evolved alongside popular dreams and nightmares about the meaning of merging men and machines. These visions of the future make incremental advances in the laboratory seem like the slow March toward an inevitable age of cyborgs.

In the summer of 1947, the brilliant American mathematician Norbert Wiener coined the term *cybernetics*—derived from the Greek for "steersman"—to describe the study of "control and communication theory, whether in the machine or the animal" (Wiener 1950). He considered cybernetics to be a vitally important new discipline, and he explained it in two books (*Cybernetics*, 1948, and *The Human Use of Human Beings*, 1950) that are surprisingly humanistic, in light of the subject matter and the author's impeccably technocratic credentials (Wiener 1948, 1950). For a short while, cybernetics aroused significant academic interest—at the intersection of physiology, computers, engineering, philosophy, economics, psychology, and sociology. However, eventually, its ideas were so fully absorbed into the disciplines that much of cybernetics came to seem obvious.

In 1960, at the height of interest in cybernetics, the word *cyborg*—short for "cybernetic organism"—was coined by researcher Manfred E. Clynes in a paper he co-wrote for the journal *Astronautics* (Clynes et al. 1960). The paper was a theoretical consideration of various ways in which fragile human bodies could be technologically adapted and improved to better withstand the rigors of space exploration. (Clynes's co-author said the word cyborg "sounds like a town in Denmark" (Clark 2003).) Around the same time, Jack E. Steele, a polymath doctor-engineer-neuroanatomist serving in the U.S. Air Force, coined the word *bionics* for the use of principles derived from living systems to solve engineering and design problems.

These words and concepts soon entered the popular imagination, starting with a 1972 science fiction novel, *Cyborg*, that became the basis for the 1970s TV show *The Six Million Dollar Man* about the "world's first bionic man," and then the spin-off *The Bionic Woman* (Caidin 1972). This was followed by decades of movies, TV shows, books, comics, and video games with cops, criminals, soldiers, and aliens who were cyborgs—from Darth Vader in the 1970s to Robocop in the 1980s to the Borg in the 1990s. The flood of cyborgs in pop-culture caught the attention of academics, and soon anthropologists, philosophers, and literary theorists were offering up unreadable piles of "cyborg scholarship."

In popular usage, the term "bionic" now refers to any kind of electronic implant or prosthesis, and so several different people—including amputees with robotic arms, like Jesse Sullivan—have been dubbed "the world's first" bionic man or woman. Similarly, the term "cyborg" has been overextended to the point of meaninglessness. Who was the first human cyborg? Maybe it was Johnny Ray, the first brain-machine interface implantee to control a computer cursor with his thoughts. Or maybe the Australian performance artist Stelios Arcadiou—called STELARC—known for his decades of grisly forays into high-tech body modification. Or perhaps Steve Mann, the Canadian wearable computer pioneer, who since the 1980s has spent most of his waking hours viewing the world through little screens in front of one or both of his eyes. Or Kevin Warwick, the professor in England whose audacious showmanship in having chips implanted in his body has brought him tremendous publicity despite the total lack of scientific merit in his stunts.

But to the most ambitious and most radical advocates of merging brains and machines, such advances are mere child's play. These so-called transhumanists long for an age when human beings will leave the miseries and limits of the body behind, and achieve new ecstasies of freedom and control: We will send feelings and conscious thoughts directly from brain to brain; we will delete unwanted memories at will; we will master complex subjects by "downloading" them directly into our minds; we will "jack in" to virtual realities; and eventually, we will be able to "upload" our personalities into computers or robots, where the self will live on indefinitely.

Such fantasies are staples of much science fiction, including cyberpunk novels like William Gibson's *Neuromancer* (Gibson 1984) and Neal Stephenson's *Snow Crash* (Stephenson 1992), and movies like *Total Recall*, *Johnny Mnemonic*, *Strange Days*, *The Final Cut*, and the Matrix trilogy. These books and movies are mostly dystopian visions of the future, or tales in which things have gone terribly awry—crime, cruelty, and mass delusion. However, advocates of transhumanism, like Ramez Naam, author of the 2005 book More Than Human, are far more optimistic:.

> With neural prosthetics, information from the emotional centers of someone else—say, a loved one—could be piped straight to your empathy center. So rather than having to guess what your spouse or child is feeling, you would simply be sensing it via the wireless link between your brains.... The end result might be just like having an unusually keen sense of how others are feeling, with the option to dial that sense up or down in intensity according to whatever criteria you choose (Naam 2005).

Naam imagines how that sharing of feelings might be taken further:

> You send your spouse what you see and hear and feel.... That night, you and your spouse make love while opening all of your senses and emotions to each other. The intimacy is beyond anything you have known (Naam 2005).

And further still:

> In principle we could do this for all the senses—record not just what you see, but also what you hear, taste, smell, and feel, all at the level of your brain. Playing back such an experience would be a little like reliving it. You might even be able to play that kind of sensory recording back for someone else, turning the experience you had into a set of nerve impulses that could be sent into the other person's brain, allowing him or her to experience at least the sensory parts of an event from your perspective....
>
> When sensations, emotions, and ideas become digital, it's as easy to share them with a dozen friends, or a 1,000 strangers, as it is to send them to one person.... We'll be able to broadcast the inner states of our minds (Naam 2005).

What an unattractive vision of the future—this world in which you can snoop on your children's feelings, feel what it's like to have sex with yourself, and broadcast the full sensory experience of your sexual encounters to the world. These are shallow, solipsistic aspirations, utterly divorced from the hopes and fears of mature human beings. The transhumanist fantasy is surely not the best guide for thinking about the genuine ethical dilemmas we now face at the dawn of the Age of Neuroelectronics.

## 7.11 A True Humanism

Without question, there are genuine human benefits to be gained if brain-machine technology advances in a sober, limited way. People with motor diseases or severe mental disorders can be helped with brain implants, amputees might find a new freedom and mobility in the use of mind-controlled prosthetics, the blind might get new electronic eyes and the deaf new ears, and even severely paralyzed patients might someday be "unlocked." Yet it is also possible to envisage a world where these new technologies are used for less noble purposes—from the next generation flight into alternative reality to the active manipulation and control of innocent subjects to the self-destructive pursuit of neurological perfection.

In the short term, brain implants probably shouldn't worry us: they are still very crude and the risks of brain surgery make them worthwhile only for those without other options. And we probably should not exert much energy fretting about the transhumanist future, which requires a level of scientific sophistication so far removed from the present that making predictions about its plausibility is a fool's errand. The greatest questions lie in the middle range—that time, some years hence, when today's techniques are vastly improved, when brain surgery becomes safe enough and implants become effective enough for the electronic alteration of the brain to move from desperate therapy to mainstream enhancement. Not long ago, the prospect of manipulating our minds with machines would have been universally disquieting. Now, after decades of "softening up" by advances in science and science fiction, far fewer people find the notion of neuro-enhancement troublesome. Its potential clients are not just the radicals who long for a post-human future, but ordinary people who grew up in an age of transplants and implants, of fictional bionic men and vivid cyborg fantasies.

The obvious temptation will be to see advances in neuroelectronics as final evidence that man is just a complex machine after all, that the brain is just a computer, that our thoughts and identity are just software. But in reality, our new powers should lead us to a different conclusion: even though we can make the brain compatible with machines to serve specific functions, the thinking being is a being of a very different sort. We are neither machines nor ghosts, but psychophysical unities—finite yet creative, embodied yet spiritual, cognitive yet not cognitive alone. No machine, however sophisticated, seems likely to duplicate or surpass that improbable mix of excellence, depravity, dignity, and uncertainty that is our human lot. On this score, the machine makers of the future still have much to learn from the myth makers of the past. And even as we seek to improve human life by improving the brain, we should be careful not to make ourselves into something worse in the effort to become something better.

## References

Adams, Charlotte. 2005. Q&A: Lt. Cmdr. Dylan Schmorrow: Empathetic avionics: Beyond interactivity. *Aviation Today*. http://www.aviationtoday.com/av/issue/feature/891.html. Accessed 25 Feb 2011.

Bauby, Jean-Domonique. 1997. *The diving bell and the butterfly*. New York: Vintage Books.

Berger, Theodore W., Ashish Ahuja, Spiros H. Courellis, et al. 2005. Restoring lost cognitive function. *IEEE Engineering in Medicine and Biology* 24(5): 30.

Birbaumer, N. 1999. A spelling device for the paralyzed. *Nature* 398: 297–298.

Brain Fingerprinting Testing Ruled Admissible in Court. http://www.brainwavescience.com/Ruled%20Admissable.php. Accessed 25 Feb 2011.

Caidin, Martin. 1972. *Cyborg*. New York: Arbour House.

Caton, Richard. 1875. The electrical currents of the brain. *British Medical Journal* 2: 278. qtd in Stanley, Finger. 1994. *Origins of neuroscience: A history of explorations into brain function*. New York: Oxford University Press.

Chisolm, Nick, and Grant Gillett. 2005. The patient's journey: Living with locked-in syndrome. *British Medical Journal* 331: 94.

Clark, Andy. 2003. *Natural-born cyborgs: Minds, technologies, and the future of human intelligence*. New York: Oxford University Press.

Clynes, Manfred E., and Nathan S. Kline. 1960. Cyborgs and space. *Astronautics*: 29–33.

Constans, Aileen. 2005. Mind over machines. *Scientist* 19(3): 27.

Cosgrove, G. Rees. 2004. Session 6: Neuroscience, brain, and behavior V: Deep brain stimulation. Testimony before the President's Council on Bioethics. http://bioethics.georgetown.edu/pcbe/transcripts/june04/session6.html. Accessed 25 Feb 2011.

Counterterrorism Applications. http://www.brainwavescience.com/counterterrorism.php. Accessed 25 Feb 2011.

Crichton, Michael. 1972. *The terminal man*. New York: Knopf.

DeFao, Janine. 2005. Woman's brain surgery benefits entire family; Her son, father also undergo procedure to help with ailments. *San Francisco Chronicle*: B1.

Delgado, José Manuel Rodríguez. 1969. *Physical control of the mind: Toward a psychocivilized society*. New York: Harper & Row.

Dennett, Daniel. 1992. *Consciousness explained*. Boston: Little Brown & Co.

Department of Defense Fiscal Year 2007 Budget Estimates. Feb 2006. http://www.darpa.mil/Docs/FY07_Final.pdf. Accessed 25 Feb 2011.

Doyle, Christine. An answer to Parkinson's? *The Daily Telegraph*. http://www.telegraph.co.uk/health/3307658/An-answer-to-Parkinsons.html. Accessed 25 Feb 2011.

Finger, Stanley. 1994. *Origins of neuroscience: A history of explorations into brain function*. New York: Oxford University Press.

Giacino, J.T., S. Ashwal, N. Childs, et al. 2002. The minimally conscious state: Definition and diagnostic criteria. *Neurology* 58(3): 349–353.

Gibson, William. 1984. *Neuromancer*. New York: Ace Books.

Guizzo, Erico, and Harry Goldstein. 2005. The rise of the body bots. *IEEE Spectrum* 42(10): 50–56.

Guthrie, Patricia. 2004. Rewiring a brain. *The Atlanta-Journal Constitution*. 1A.

Hess, Walter Rudolf. 1949. The central control of the activity of internal organs. http://nobelprize.org/nobel_prizes/medicine/laureates/1949/hess-lecture.html. Accessed 25 Feb 2011.

Hooper, Judith, and Dick Teresi. 1992. *The three-pound universe*. New York: Putnam.

Introducing Jesse Sullivan, the World's First 'Bionic Man.' Rehabilitation Institute of Chicago. http://www.ric.org/research/accomplishments/Bionic.aspx. Accessed 25 Feb 2011.

Jennet, B., and F. Plum. 1972. Persistent vegetative state after brain damage: A syndrome in search of a name. *The Lancet* 7753: 734–737.

Johnson, Steven. 2004. *Mind wide open: Your brain and the neuroscience of everyday life*. New York: Scribner.

Kaempffert, Waldemar. 1941. Studies of epilepsy: Science and seizures. *The New York Times*: BR29.

Kneeland, Timothy W., and Carol A.B. Warren. 2002. *Pushbutton psychiatry: A history of electroshock in America*. Westport: Praeger.

Millett, David. 2001. Hans berger: From psychic energy to the EEG. *Perspective in Biology and Medicine* 44(4): 522.

Moan, Charles E., and Robert G. Heath. 1972. Septal stimulation for the initiation of heterosexual behavior in a homosexual male. *Journal of Behavior Therapy and Experimental Psychiatry* 3: 23–30.

Mozersky, Judy. 2000. *Locked in*. Ontario: Dundurn Press.

Naam, Ramez. 2005. *More than human: Embracing the promise of biological enhancement*. New York: Broadway Books.

Once Again With Feeling. 2004. *The Economist*. http://www.economist.com/node/2724499. Accessed 25 Feb 2011.

Parker, Ian. 2003. Reading minds. *The New Yorker* 78(43): 52–63.

Patch, Kimberly. 2005. Brainwave interface goes 2-D. *Technology Research News*. http://www.trnmag.com/Stories/2005/020905/Brainwave_interface_goes_2D_020905.html. Accessed 25 Feb 2011.

Pearce, David. http://www.wireheading.com/dalai-lama.html. Accessed 25 Feb 2011.

Pearce, David. The hedonistic imperative. http://www.hedweb.com/hedab.htm. Accessed 25 Feb 2011.

Pfurtscheller, Gert, Gernot R. Müller, Jörg Pfurtscheller, Hans Jürgen Gerner, and Rüdiger Rupp. 2003. 'Thought'-control of functional electrical stimulation to restore hand grasp in a patient with tetraplegia. *Neuroscience Letters* 351(1): 33–36.

Plum, Fred, and Jerome B. Posner. 1968. *The diagnosis of stupor and coma*. Philadelphia: F.A. Davis Co.

Shakespeare, William. *Hamlet*, 3.1.77–3.

Sherwell, Philip. 2004. Brain grown from rat cells learns to fly jet. *Sunday Telegraph*: 31.

Slater, Lauren, and Clara Jeffrey. 2005. Who holds the clicker? *Mother Jones* 30(6): 62–71.

Some Plain English on Epilepsy. 1944. *The New York Times*: E9.

Spear, Joseph H. 2004. Cumulative change in scientific production: Research technologies and the structuring of new knowledge. *Perspectives on Science* 12(1): 55–85.

Spinney, Laura. 2003. Hear my voice. *New Scientist* 2383: 36–39.

Stephenson, Neal. 1992. *Snow crash*. New York: Bantam Books.

Thompson, Richard F., and Stuart M. Zola. 2003. Biological psychology. In *Handbook of psychology*, History of psychology, vol. I, ed. Donald K. Freedheim and Irving P. Weiner, 47–66. Hoboken: Wiley.

U.S. Congress, Office of Technology Assessment. 1987. *Life sustaining technologies and the elderly, OTA-BA-306*. Washington, DC: U.S. Government Printing Office. 59.

Vigand, Philippe, and Stephane Vigand. 2000. *Only the eyes say yes: A love story*. New York: Arcade Publishing.

Wiener, Norbert. 1948. *Cybernetics: Or control and communication in the animal and the machine*. Cambridge: MIT Press.

Wiener, Norbert. 1950. *The human use of human beings: Cybernetics and society*. Boston: The Riverside Press.

Ziehm, Len. June 5, 2005. Shakes, battle, and roll; Tim Simpson is riding Golf's comeback trail after battling 14 years of health problems. *Chicago Sun-Times*: S92.

Zimmer, Carl. 2004. *Soul made flesh: The discovery of the brain—And how it changed the world*. New York: Free Press.

# Chapter 8
# The Cochlear Implant Controversy: Lessons Learned for Using Anticipatory Governance to Address Societal Concerns of Nano-scale Neural Interface Technologies

**Derrick Anderson**

## 8.1 Introduction

A shortcoming of public policy, according to Jeffery Greene (2005), is that it tends to be reactive and not proactive. That is to say, public policies tend towards addressing looming and existent public problems rather than preventing or foreseeing and circumventing future public problems. While the "reactive" nature of public policy might be contested in some scholarly or professional circles it is safe to say that the prevailing public policy making structures in modern democracies are often more inclined to respond to public problems rather than attempt to anticipate, circumvent, or prevent them.

This shortcoming is exemplified by the policies that shape and guide the American scientific enterprise, which lacks systematic ways of addressing the safety and appropriateness of revolutionary advances. The increased pace of innovation and dependency on scientific and technological outputs make this condition problematic. As such, efforts have emerged to consider the prospects of shaping more proactive rather than reactive public policies to govern the activities within this scientific enterprise. Central to these efforts is the idea of "anticipatory governance" (Guston et al. 2007; Guston 2007, 2008). Guston (2007) describes anticipatory governance as a capacity extended through a community that can manage emerging knowledge-based technologies while such management is still possible. Anticipatory governance operates under the premise that "the scientific enterprise we have—its foci, productivity, contributions, strengths and shortcomings—is at least as attributable to the governing forces of personalities, policies, and institutions as it is to the autonomous play of researchers" (Guston 2008, 940). Under these circumstances, the value of upstream policy engagement of advancements in science and technology becomes increasingly

D. Anderson (✉)
Department of Public Administration and Policy School of Public and International Affairs, The University of Georgia, 204 Baldwin Hall, 355 South Jackson Street, GA 30602, Athens
e-mail: demander@uga.edu

S.A. Hays et al. (eds.), *Nanotechnology, the Brain, and the Future*, Yearbook of Nanotechnology in Society 3, DOI 10.1007/978-94-007-1787-9_8,

clear, especially in light of common characterizations of policy making as an iterative process involving refinement over time, or in this case, as key scientific advancements become more developed. Indeed, early involvement can help towards ensuring that policies are appropriately formulated before crisis materializes.

This chapter considers anticipatory governance of one area of scientific advancement: nano-scale neural interface technologies (nanotechnologies that interface with a body's nervous system, especially the brain). A focus is placed on the application of anticipatory governance to address the potential ethical, legal and social controversies associated with nano-scale neural interface technologies. One element of anticipatory governance is the consideration of relevant and related technologies. Reflection on these technologies can help researchers and policy makers better understand the anatomy of the challenges posed by selected scientific advancements. In this case, it is relevant to not only reflect on nano-enabled devices but also existing devices that interface with the nervous system because lessons can be learned that highlight, in particular, the unique societal challenges associated technologies that enhancement the nervous systems.

In this light, cochlear implants are chosen as a relevant and related technology for this chapter's consideration of anticipatory governance of nano-scale neural interface technologies. The purpose here is to highlight some of the ethical, legal and social controversies that emerged with the development of cochlear implants and then to reflect on what these controversies could mean for governance of nano-neural interface technologies.

## 8.2 Nano Neural Interface Technologies

Before shifting the focus toward cochlear implants it is necessary to discuss some of the technical features of nano-neural interface technologies. In comparison to cochlear implants, much less is known about nano-neural interface technologies since they are largely in the early stages of development. Nano-neural interface technologies apply nano-scale engineered materials to the central nervous system, (the brain), or peripheral nervous systems, (spinal cord and nerves throughout the body). One way to think about these technologies is to consider both existing neural interface technologies and promised characteristics of nanotechnologies; nano-scale neural interface technologies are likely to exist at the convergence of these two areas.

First, consider existing neural interface technologies. A great deal of work is currently underway in applying electronics to the brain. These include invasive and noninvasive technologies. Invasive technologies are those that requiring implantation of devices into the body and noninvasive technologies that interface through electronic signaling. An example of an invasive technology is the BrainGate™ Neural Interface System, a device that requires implantation of a computer chip into the brain (BrainGate 2010a). This technology, designed to help patients who are paralyzed or have lost limbs, is described as having the ability to:

> [S]ense, transmit, analyze, and apply the language of neurons. [It] consists of a sensor that is implanted on the motor cortex of the brain and a device that analyzes brain signals. The underlying principle behind BrainGate™ is that with intact brain functionality, brain

> signals are generated even though they are not sent to the arms, hands, and legs. The signals then can be interpreted and translated into cursor movements, offering the user an [additional technology] to control a computer with thought, just as individuals who have the ability to move their hands use a mouse. (BrainGate 2010b).

At the cutting edge of current brain interface technologies, BrainGate™ nicely represents the fundamentals of invasive neural interface technologies: a surgically implanted device that observes and interacts with brain activity. In this case, brain signals are used to help paralyzed patients move computer cursors.

While invasive technologies are far from common, some non-invasive technologies are verging on ubiquitous. Commonly used imaging technologies such as magnetic resonance imaging (MRI) represent non-invasive interface technologies. Their traditional applications have be primarily oriented towards medical diagnostic practices, however, increasing attention has been placed on expanding their scope of application to include observing brain activity in, for example, forensics (Schauer 2010). This expanded use of readily available imaging technologies represents the fundamentals of non-invasive brain interface technologies: technologies that observe and interact with brain activity without requiring surgical implantations.

The next step in moving closers towards a working characterization of nano-neural interface technologies is to apply the characteristic promises of nanotechnology to the above description of invasive and non-invasive brain interface technologies. In general, it is anticipated that nanotechnology will make existing technologies more accurate, affordable and safe (for examples, see: Roco and Bainbridge 2001, 2005; National Nanotechnology Initiative 2007). For example, the challenges of invasive technologies such as biological rejection of the implanted devise could be dramatically mitigated through nanotechnology. Similarly, the size and cost of non-invasive brain interface devises make them prohibitively expensive and cumbersome for personal and widespread consumer use. However, the promise of nanotechnology suggests that these technologies (or ones with similar capacities) could mitigate these disqualifying characteristics. In light of this, some important ethical issues emerge about having these technologies widely accessible and safe. For the most part, these nanotechnology-enhanced products are not at the point of wide public use. Thus, the appropriateness of anticipatory governance is twofold: there is both time and need for implementation of appropriate policy measures.

## 8.3 Cochlear Implants

The remainder of this chapter focuses on cochlear implants and what they teach us about nano-neural interface technologies. Today's cochlear implants represent some of the most advanced technological achievements in modern medicine. Consisting of multiple interactive parts, cochlear implants are designed to replicate hearing in thousands who experience permanent profound or complete sensorineural hearing loss—hearing loss rooted in damage or failure of the nerves in the inner ear, vestibular nerve or sound processing centers in the brain. Figure 8.1 provides a schematic for the basic operating parts of a typical cochlear implant.

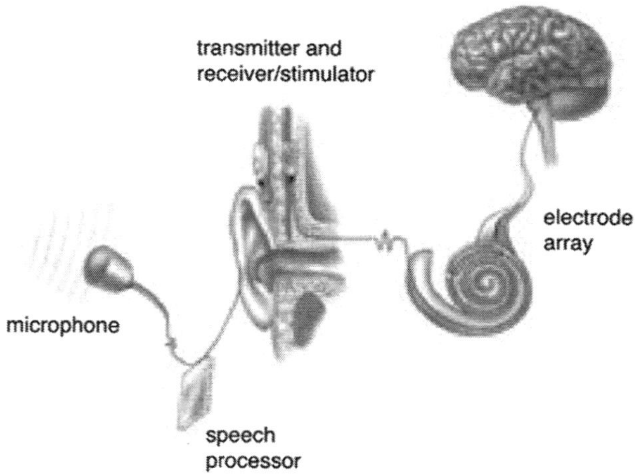


**Fig. 8.1** Working parts of a cochlear implant (Image credit: National Institutes of Health. Information about Hearing, Communication, and Understanding. http://www.ncbi.nlm.nih.gov/bookshelf/br.fcgi?book=curriculum&part=A513)

External to the body is a microphone that picks up sound waves or signals. These sound waves are converted to electronic signals in the speech processor, a pocket sized computer that filters out inaudible and audible sounds while prioritizing the audible ones for transmission to the remaining components of the device. The processor sends the audible sound in the form of electronic signals to the transmitter, a small removable device that is magnetically fixed behind the ear and is just a bit larger than a coin. The transmitter sends the electronic signals to the internal components of the implant, beginning with the receiver. The receiver is comparable in size to the transmitter and is anchored in below the skin just behind the ear in a recipient's skull. Receivers work in harmony with stimulators to transfer the electronic signals to an array of electrodes that wind their way through the head into the cochlea. To date, over 20,000 adults and 15,000 children have received implants in the United States (NIDCD 2009). Recipients are both pre- and post-lingually deaf meaning some were deaf before they learned to talk and others development deafness later in life. Getting an accurate sense of recipients' satisfaction with the device is difficult (Ou et al. 2008). However, Klop et al. (2007) were able to observe a high degree of satisfaction with the device by prelingually deaf adults. While a range of factors—including patient physiology, postoperative therapy, and the type of device used—contribute to success of the device it is generally thought that ideal patients will progress to develop normal hearing (Advanced Bionics 2010; Chorost 2005; FDA 2010). Adults can sometimes experience immediate hearing improvement while children generally progress more slowly (FDA 2010).

Despite general success and adoption by tens of thousands in the United States, a number of interrelated challenges face cochlear implants. These challenges have contributed to social concerns and controversies. Some of these concerns have highly technical foundations. For example, there are several known health risks associated with cochlear implants (FDA 2010). Biocompatibility of the electrodes (Adunka et al. 2004) and increased risk of meningitis following implantation (FDA 2003) represent some of these technical challenges. On the other hand, additional challenges have cultural foundations. For example, the Deaf community (represented by the National Association of the Deaf) responded adversely to FDA approval of cochlear implants in children (NAD 2000). A major impetus for this reaction is this idea that cochlear implants represent a threat to the Deaf community—their way of life, identity, language and legitimacy (Lane and Grodin 1997). This issue has been address widely in ethics literature and the media. For example, *The Sound and the Fury*, a 2001 Academy Award nominee for Best Documentary Film, address this issue exclusively (PBS 2010).

While, the interrelated technical and cultural challenges facing cochlear implants are very interesting and worthy of additional consideration, lengthy elaboration on them is beyond the scope of this chapter. Rather, these challenges are presented as underlying motivations behind a discussion of several "lessons" to be learned from cochlear implants. An active recognition that cochlear implants are associated with challenges and controversy demonstrates the relevancy of cochlear implants to this exercise of anticipatory governance of nano-neural interface technologies. The following sections elaborate on this.

## 8.4 Case Study Framework

This chapter considers a few specific lessons learned from the development of cochlear implants that could have implications for the anticipatory governance of nano-scale neural interface technologies. While this is by no means an exhaustive set of the lessons learned from this technology, it does represent a few of the more relevant ones. There are many things to consider when looking at the development of cochlear implants. However, the underlying policy orientation of this chapter necessitates a particular focus on the social challenges facing cochlear implants. Each of the four lessons provides a description of events that occurred in the development of cochlear implants. A short discussion of what each lesson could mean for nano-neural interface technologies concludes each section. After presenting all four lessons, this paper finishes by bringing them all together for a final reflection on what can be expected as policies attempt to address the challenges of nano-neural interface technologies.

### *8.4.1 Lesson One: Development Takes a Long Time*

Forays into stimulating the auditory system with electricity have a rich heritage. Brackman (1976) points out that the idea of stimulating the auditory system with electricity dates back to the beginnings of electricity itself. Alessandro Volta, in 1790, reportedly inserted electrodes into his ears, connected them to a battery then described the resultant sensation as a "blow to the head followed by a sound like the boiling of a viscous liquid" (Brackman 1976, 374). Fortunately, technologically and scientifically sound leaps towards cochlear implants emerged in the late 1950s. For example, by 1957, Charles Eyries of France had implanted a primitive form of the device into a male patient who claimed to have experienced some hearing that increased his capacity to lip read (Brackman 1976). Initial attempts like that of Eyries' was largely concept building, eventually leading to systematic research and development programs by the 1970s. Here, substantial work towards designing and manufacturing a commercially available device was the primary goal. Two research teams emerged as pioneers in cochlear implant research and development: William House of the House Ear Institute in California and Graem Clark of the University of Melbourne in Australia. Small-scale commercialization of cochlear implants began in 1978 when House entered into an agreement with 3 M. Clark's research group began a similar relationship with a different firm the next year (Garud and Rappa 1994).

Though proof of concept occurred in the 1950s, device development occurred though the 1970s and commercialization began in 1978, it was not until 1980 that the Food and Drug Administration (FDA) began regulating cochlear implants in the United States.[1] Then, it took another 4 years of a device to be approved. Even the, the 1984 approval was only for limited use in adult patients in the United States. Six years later, after 5 years of clinical trials, the FDA approved implantation for children two and older. In 1998, the age was lowered to 18 months and, again, in 2002 it was lowered to 12 months. Today, nine devices are approved by the FDA, including the original 1984 design manufactured by the House/3 M group.

This fairly lengthy time delay between development activities and devices availability to consumers can be attributed to a variety of forces, a few of which I will introduce. To begin, the pace of innovation is, in part, subjected to the internal dynamics of the scientific or technical community researching and developing devices. Raghu Garud and Michael Rappa offer an excellent illustration of how this prolonged the commercialization process of cochlear implants (1994). In this case, as they point out, development was driven by divergent expectations of what the technology can and should to. Specifically, the team lead by House saw cochlear implants as a tool to help recipients better experience their environment. They envisioned a device that would help recipients hear loud noises in their surroundings but not necessarily the relatively soft sounds of speech. Additionally, they preferred a minimally invasive "single-channel" implants with limited stimulating capacity but less risk of postoperative nerve damage. On the other hand, Clark's team saw the technology as a tool for helping recipients hear and understand speech. This more aggressive expectation necessitated a slightly more invasive "multi-channel" implant. The divergence in vision proved challenging for development were regulatory and licensing processed preferred a singular, well articulated set of design specifications and anticipated benefits for end users. Although the precise effect that this divergence had on the commercialization timeline is unclear, it is safe to assume that it inhibited development.

[1] Regulatory information on cochlear implants can be found at the United States Food and Drug Administration's website. See the following: http://www.fda.gov/MedicalDevices/ProductsandMedicalProcedures/ImplantsandProsthetics/CochlearImplants/ucm062823.htm

Support of this assumption is seen in the eventual intervention made by the FDA to bring the two perspectives to a consensus. Eventually, in 1988, the FDA brought researchers together for a consensus conference on the subject and a preference was made for multi-channel devices (Garud and Rappa 1994).

For the purposes of this exercise, it is safe to assume that internal conflict amongst key research and development players can lengthen the development timeline for nano-neural interface technologies as it did for cochlear implants. This point has particular saliency for the case of nano-neural interface technologies when considered through Garud and Rappa's (1994) framework of socio-cognitive model of technology evolution. According to this model, developers' expectations of the technology play a role in how it is developed or not developed. While this obviously played a role in cochlear implants, one might imagine the effects of this process being amplified in the case of nanotechnologies, where developers' expectations for technologies must navigate the uncertainties of nanoscale materials in addition to conceptualizations of how the technology can and should benefit the user.

Regulatory structures can also inhibit development timelines. Before cochlear implants could be delivered to consumers an array of tests had to be conducted. These tests, primarily oriented towards consumer health and safety (Van de Ven and Garud 1993), are difficult to expedite and mandatory for any sort of implant. For example, it took 5 years of clinical trials to broaden the approved use of the House/3 M device, which had already been cleared for adults, to include children two and older. While the pace of technological innovation might wax and wane overtime, there is no *a priori* reason to suggest that the existing regulatory processes will get faster over time. Thus, for the purposes of this exercise, it is safe to assume that the regulatory process will be a limiting factor in the delivery of these technologies.

If emerging nano-scale neural interface devices are to be subjected to the same regulations we could expect lengthy development phases as well. However, it is important to note that the nature of these devices clearly indicates that some will not, for example, require implantation and as such could experience quicker development to consumer pathways. For example, nanotechnology could enhance the BrainGate™ technology such that implantation is no longer necessary. If this were to be the case, it would not take much to imagine that the capacities of BrainGate™ could be available in a store-bought hardware/software package. Under these circumstances, the tone of the anticipatory governance discussion shifts away from the safety risks associated with brain implants and towards the equity implications of distribution based on price restraints. This distribution consideration makes a nice transition into the second lesson: market penetration patterns.

### *8.4.2 Lesson Two: Disparities in Market Penetration*

As previously mentioned, cochlear implants are used in many countries around the world. Similarly, they are manufactured and have been developed in several different countries, primarily the US and Australia. The global nature of use, development, and manufacturing highlights the inevitability of looming differences in the ways that cochlear implants are adopted. Guston et al. (2007) argue that the distribution of nano-enabled human enhancement technologies is likely to flow along existing wealth patterns. Cozzens and Wetmore (2010) further support this observation, suggesting that nanotechnologies are likely to enhance existing global disparities between rich and poor. This has been the pattern with cochlear implants and concern has been expressed regarding the capacity of residents of developing countries to afford cochlear implants (cf. Zeng 2004).

It seems reasonable to assume that two important forces contribute to the adoption of these technologies by communities: health care systems and social forces. First, consider health care system structures. In the case of cochlear implants, where devices can cost US$20,000 and require another US$20,000 for post operative therapy, it becomes obvious that the majority of recipients capable of affording the devise are those with health insurance or substantial government support. Garber et al. explored the relationship between health insurance coverage and access to cochlear implants. The results of their analysis were, in their words, "troubling" (2002, 1151). Their study supports observations that access to sufficient coverage plays a major role in adoption and, conversely, gaps in coverage—for example, only partial coverage of the device or therapy—contribute to decisions against implantation.

The second variable contributing to adoption are social forces. Here, the primary focus, in the case of cochlear implants, has been the Deaf community. In the United States alone there are millions of deaf people, the Deaf community is a subset of deaf people with a strong sense of identity that is derived from being deaf. The Deaf community is mostly prelingually deaf children and adults who communicate through sign language and consider deafness an "experience" rather than a disability (Ladd 2003). Because their cultural identity is centered on being deaf, the Deaf community advocates against cochlear implants and, more specifically, considering cochlear implants a "cure" for deafness (NAD 2000). This coupled with the fact that many of today's deaf children are born into the Deaf community contributes significantly to the adoption practices of the community.

From these two variables, we could conclude that emerging nano-scale neural interface devices are likely to be adopted through existing wealth channel. That is, the wealthiest countries and the wealthiest communities are most likely first to adopt these technologies. There is, however, reason to suggest that in the case of some enhancements, select communities will electively reject adoption and even advocate broadly against it, especially when specific technologies challenge a communities' sense of identity or suggest they are inferior and needing technological intervention in the form of therapy or treatment.

### 8.4.3 Lesson Three: The Importance of Players

As illustrated in the previous lesson, there are a number of key players or stakeholders in the cochlear implant development process. With new and emerging technologies, and sometimes with well-established technologies, cultural and social opposition is deeply associated with issues of health and safety or risk. One could hardly count the number of consumer advocacy groups that are focused on raising awareness of the negative health consequences of readily available and well-known products. Such is also the case with cochlear implants. Similarly, such is also the case with the products based on new and emerging technologies. To be sure, issues of health and risk are on the minds of consumers in the case of cochlear implants. For example, while audiologists recommend implantation early in childhood, parents of prospective recipients, knowing that implant surgery destroys residual hearing, must subject the risk and reward of cochlear implants to the whims of imperfect hearing screenings. Additionally, the risk of meningitis is ever looming (FDA 2009a, b). As is the risks associated with skull fracture during early childhood or exposure to toxic battery materials (Hansson 2005; Weinberg 2005).

In addition to these concerns regarding health and safety risks, a unique and significant opposition emerged that had little to do with health and safety. The locus of this opposition was in America's Deaf community, who claimed that support of cochlear implants was analogous to cultural genocide (Lane and Grodin 1997). In 1991, the National Association of the Deaf (NAD) issued their first statement on cochlear implants claiming they were scientifically, procedurally and ethically unsound (NAD 2000). Nearly 10 years later a new statement was issued, adding nuance to their claim by declaring implants appropriate in some instances but never a "cure" for deafness (NAD 2000)

In the United States, the participation of the Deaf community has played a central role in guiding adoption of cochlear implant technology. To be sure, this role is, in most cases, secondary, at best, to others including health care providers, educators and manufactures. Nonetheless, its important to note the existence of the Deaf community as a key player in framing the issue of the controversy over cochlear implants. Their participation fueled the emergence of the "socio-cultural" and "pathological" perspectives of deafness (Weinberg 2005). The socio-cultural perspective considers deafness a characteristic or quality whereas the pathological perspective considers deafness a medical condition, eligible for therapy, treatment, and correction.

In the case of nano-scale neural interfaces technologies, we can easily assume that, like cochlear implants, apprehensions over adoption could emerge as a result of a technology's role in redefining people with certain characteristics, especially if new definitions include prescriptions for treating, correcting, or curing these characteristics.

### *8.4.4 Lesson Four: Reach of Implications Beyond Health*

A key player in the debate over the appropriate use of cochlear implants has been the Deaf community, who argue that cochlear implants represent a threat to their identity (Lane and Grodin 1997). The legitimacy of this claim has saliency when viewed from a variety of perspectives. To be sure, the case of cochlear implants shows that adoption of a new technology can have the potential to impact, for better or worse, entire communities. If this prognosis might seem far-fetched, a more tangible reality is that, at the very least, a community can become dramatically and seriously divided.

The Deaf community has developed a nuanced perspective on the issue of cochlear implants (NAD 2000). This perspective's official face has been slightly modified over time, initially being completely against the use of implants to, now, considering implants appropriate in some instances but never a "cure" for deafness (NAD 2000). Members of today's Deaf community hold positions on this issue that range somewhere between, near and around these two official statements. The emergence of cochlear implants has created a range of activist groups associated with these slightly varying perspectives of cochlear implants. What makes this important is that none of these groups would have existed had the technology not been developed.

This final lesson is of particular importance for the case of nano-neural interface technologies. Here, the prospects of nanotechnology include alleviating many of the exogenous health and safety concerns of existing technologies. In light of this, questions emerge about the traction that opposition groups might have in the absence of substantive health and safety complaints. While some might assume that elimination of the health and safety concerns provides a channel for adoption without barriers, the case of cochlear implants illustrates otherwise. To be sure, many oppose the technology irrespective of health and safety concerns.

## 8.5 Conclusion

The need for appropriate public policy to address societal concerns of nano-neural interface technologies becomes increasingly important as the technologies become more of a reality. An unfortunate predisposition towards reactive policymaking limits the ability of appropriate policy formulations to fully address problems and, thus, benefit and protect society. Under these conditions, a primacy should be placed on proactive approaches to policy making (Greene 2005). Because widespread use of nano-neural interface technologies is a reality of the distant future the possibility of proactive policy intervention is still available. Anticipatory governance is one such approach. This chapter is an exercise in one principle of anticipatory governance of nano-scale neural interface technologies: consideration of relevant and related technologies. Reflection on cochlear implants through this lens of anticipatory governance offers important lessons.

First, development of these technologies is likely going to take considerable time because of regulatory barriers. However, in the event that regulatory barriers can be circumvented, it is reasonable to assume that they will be. If this were to occur, the policy discussions are likely to shift away from health and safety concerns and more towards concerns of distributional equity.

Second, there are a variety of factors that collectively determine how these technologies will be adopted. In the case of nano-neural interface technologies, it can be assumed that distribution will flow along existing wealth patterns. Such was and remains the case with cochlear implants. This pattern of distribution raises the same equity concerns seen in cochlear implants, namely, why is adoption centered on ability to pay and not need.

Third, as with cochlear implants, nano-neural interface technologies are likely to develop in a context involving many players. In the case of cochlear implants, some of these players were involved in development discussion while others—especially the Deaf community—felt disenfranchised. If a lesson is to be learned in this case for the purposes of anticipatory governance, it is that appropriate policies can be formulated through broadening the range of considered stakeholders in decision-making processes.

Finally, discussions of appropriateness of medical technologies are often eclipsed by health and safety considerations. This is especially interesting in the case of nano neural interface technologies, which are seen to alleviate many health and safety concerns of existing technologies. While some might assume that nano-neural interface technologies will therefore be void of any peripheral discussions of appropriateness, the experience of cochlear implants suggests otherwise. To be sure, there are many concerns that reach well beyond the realm of health and safety. The case of cochlear implants illustrates that nicely. This exercise of anticipatory governance nicely highlights this condition.

In general, the principles of anticipatory governance nicely complement a desire to leverage public policies to better protect, promote and articulate interests of the public. Perhaps many of the potential problems that emerge with nano-neural interface technologies can be circumvented through adopting principles of anticipatory governance.

## References


Adunka, O., J. Kiefer, M.H. Unkelbach, T. Lehnert, and W. Gstoettner. 2004. Development and evaluation of an improved cochlear implant electrode design for electric acoustic stimulation. *The Laryngoscope* 114(7): 1237–1241.

Advanced Bionics. 2010. Your journey to hearing. http://www.advancedbionics.com/CMS/Your-Journey-to-Hearing/FAQ.aspx. Accessed Oct 2010.

Brackman, D.E. 1976. The cochlear implant: Basic principles. *The Laryngoscope* 86(3): 373–388.

BrainGate. 2010a. Corporate website of the BrainGate Company. http://www.braingate.com/. Accessed Oct 2010.

BrainGate. 2010b. Corporate website of the BrainGate Company. http://www.braingate.com/thought.html. Accessed Oct 2010.

Chorost, M. 2005. *Rebuilt: How becoming part computer made me more human*. Boston: Houghton Mifflin.

Cozzens, Susan E., and Jameson M. Wetmore. 2010. Equity. In *Encyclopedia of nanotechnology in society*, ed. D. Guston. Thousand Oaks: Sage.

FDA—United States Food and Drug Administration. 2003. Public health notifications. http://www.fda.gov/MedicalDevices/Safety/AlertsandNotices/PublicHealthNotifications/ucm064526.htm. Accessed Oct 2010.

FDA—United States Food and Drug Administration. 2009a. Information on cochlear implants. http://www.fda.gov/MedicalDevices/ProductsandMedicalProcedures/ImplantsandProsthetics/CochlearImplants/ucm062823.htm. Accessed Nov 2009.
FDA—United States Food and Drug Administration. 2009b. Information on cochlear implants. http://www.fda.gov/MedicalDevices/ProductsandMedicalProcedures/ImplantsandProsthetics/CochlearImplants/ucm062892.htm. Accessed Oct 2010.
FDA—United States Food and Drug Administration. 2010. Information on cochlear implants. http://www.fda.gov/MedicalDevices/ProductsandMedicalProcedures/ImplantsandProsthetics/CochlearImplants/ucm062843.htm. Accessed Oct 2010.
Garber, S., M.S. Ridgely, M. Bradley, and K.W. Chin. 2002. Payment under public and private insurance and access to cochlear implants. *Archives of Otolaryngology – Head and Neck Surgery* 128(10): 1145–1152.
Garud, R., and M.A. Rappa. 1994. A socio-cognitive model of technology evolution: The case of cochlear implants. *Organization Science* 5(3): 344–362.
Greene, Jeffery D. 2005. *Public administration in the new century: A concise introduction.* Belmont: Thomson Wadsworth.
Guston, David H. 2007. Toward anticipatory governance. http://nanohub.org/resources/3270.
Guston, David H. 2008. Innovation policy: Not just a jumbo shrimp. *Nature* 454: 940–941.
Guston, David H., John Parsi, and Justin Tosi. 2007. Anticipating the ethical and political challenges of human nanotechnologies. In *Nanoethics: The ethical and social implications of nanotechnology*. Hoboken: Wiley.
Hansson, S.O. 2005. Implant ethics. *Journal of Medical Ethics* 31: 519–525.
Klop, W.M.C., J.J. Briaire, A.M. Stiggelbout, and J.H.M. Frijns. 2007. Cochlear implant outcomes and quality of life in adults with prelingual deafness. *The Laryngoscope* 117(11): 1982–1987.
Ladd, P. 2003. Understanding deaf culture: In *Search of deafhood*. Clevedon: Multilingual Matters.
Lane, H., and M. Grodin. 1997. Ethical issues in cochlear implant surgery: An exploration in disease, disability, and the best interests of the child. *Kennedy Institute of Ethics Journal* 7(3): 231–251.
National Association of the Deaf. 2000. NAD position statement on cochlear implants. http://www.nad.org/issues/technology/assistive-listening/cochlear-implants. Accessed Oct 2010.
National Nanotechnology Initiative. 2007. *Strategic plan 2007*. Washington, DC: Committee on Technology/National Science and Technology Council.
NIDCD—National Institute of Deafness and other Communication Disorders. 2009. Primer on cochlear implants. http://www.nidcd.hig.gov/health/hearing/coch.asp. Accessed Sept 2009.
Ou, H., C.C. Dunn, R.A. Bentler, and X. Zhang. 2008. Measuring cochlear implant satisfaction in postlingually deafened adults with the SADL inventory. *Journal of the American Academy of Audiology* 19(9): 721–734.
PBS—Public Broadcasting System. 2010. The sound and fury: About the film. http://www.pbs.org/wnet/soundandfury/film/index.html. Accessed Oct 2010.
Roco, M.C., and W.S. Bainbridge. 2001. *Societal implication of nanoscience and nanotechnology*. Boston: Kluwer Academic.
Roco, M.C., and W.S. Bainbridge. 2005. Societal implications of nanoscience and nanotechnology: Maximizing human benefit. *Journal of Nanoparticle Research* 7: 1–13.
Schauer, F. 2010. Neuroscience, lie-detection, and the law: Contrary to the prevailing view, the suitability of brain-based lie-detection for courtroom or forensic use should be determined according to legal and not scientific standards. *Trends in Cognitive Sciences* 14(3): 101–103.
Van de Ven, A.H., and R. Garud. 1993. Innovation and industry development: The case of cochlear implants. *Research on Technological Innovation, Management and Policy* 5: 1–46.
Weinberg, A. 2005. Pediatric cochlear implants: The great debate. *Penn Bioethics Journal* 1(1): 1–4.
Zeng, F.G. 2004. Trends in cochlear implants. *Trends in Amplification* 8(1): 1–34.

# Chapter 9
# Healing the Blind: Perspectives of Blind Persons on Methods to Restore Sight

**Arielle Silverman**

## 9.1 Introduction

According to the American Foundation for the Blind, there are approximately ten million visually impaired persons in the United States. Approximately 1.3 million (five per 1,000 Americans) are legally blind, meaning that they have a visual acuity of 20/200 or worse in the better eye with correction, or a visual field of 20° or less. Of these, 80% reported having some useful vision, while the remaining 20% have light perception or less. Half of all visually impaired Americans are over the age of 65 (American Foundation for the Blind 2007). Due to the aging of the population and improvements in technology to preserve the lives of premature infants, the prevalence of blindness and visual impairment has increased in recent decades and is likely to continue increasing.

Historically, blindness has been associated with dependence and is often depicted in ways that elicit fear and pity (Ferguson 2001). Despite technological advances and changes in public attitudes that allow blind persons to participate in all aspects of society, blindness is still the third most feared "disease" after AIDS and cancer (Dobelle 2000). Furthermore, the blind still face disadvantages in educational attainment and employment outcomes (Statistical Snapshots 2008). For these reasons, it is not surprising that developing techniques to prevent and treat vision loss has become a high research priority.

Vision loss and blindness can result from anomalies in any part of the visual system, but blindness resulting from retinal damage is especially common, both in infancy and because of aging. Approaches to sight restoration that address retinal causes of blindness have focused on techniques that either electronically mimic retinal functioning or attempt to regenerate damaged retinal tissue. Both types of interventions have shown promise in animal studies, and they have been linked to modest improvements in visual

A. Silverman (✉)
Department of Psychology and Neuroscience, 345 UCB, University of Colorado, CO 80309-0345, Boulder
e-mail: arielle.silverman@colorado.edu

S.A. Hays et al. (eds.), *Nanotechnology, the Brain, and the Future*, Yearbook of Nanotechnology in Society 3, DOI 10.1007/978-94-007-1787-9_9,

functioning when tested in humans. However, a great deal of research with human subjects is still needed to determine how much functional vision can be restored, especially for those who have no vision at all or who have always been blind.

Two sight restoration procedures have been receiving particular attention. The first, known informally as a "bionic eye," employs a silicon chip or a series of ceramic sensors implanted into the retina to compensate for inadequate photoreceptor function. A bionic eye pioneered by the Second Sight company consists of a "receiver" implanted on the retinal surface that receives visual information from a camera mounted in a pair of eyeglasses. The most advanced version of the device, called Argus II, is being studied in ongoing clinical trials throughout the United States, Mexico, and Europe. Patients who had become totally blind from retinal degeneration have gained enough functional vision from the microchip to visually track complex stimuli such as a basketball hoop and to recognize simple objects, and a later version of the microchip is intended to enable visual face recognition. However, the full extent of functional vision that the chip can enable, especially in those who have been totally blind since birth, is still not known (Gawel 2007; Medgadget 2008).

Since many forms of retinal degeneration can be attributed to genetic mutations, scientists have also invested a great deal of effort into discovering gene therapies that can revitalize damaged retinal tissue. In particular, a gene therapy targeting mutations in the retinal pigment epithelium-65 (RPE-65) gene has shown some promise. Both canine and early human studies have shown that the introduction of healthy RPE-65 genes into retinal cells can slow photoreceptor death and restore function in remaining rod and cone cells, and the viral vector used to transfer replacement genes does not provoke a dangerous immune response (Acland et al. 2005; Ghosh 2007; Clinical Trials 2009). Ongoing Phase I and Phase II clinical trials in the United States, England, and Israel continue to test the RPE-65 therapy in both adults and children aged 8–18 (Clinical Trials 2009).

Two additional procedures, the Cortical Visual Stimulator (CVS) and stem cell transplantation into the retina, have shown promise as well. The CVS is similar to the retinal microchip, except that the transducing device is implanted directly into the visual cortex of the brain. There, it receives signals from a camera mounted on a pair of glasses and conveys the signals to the visual cortex to bypass a malfunctioning retina. The CVS has been studied in limited numbers of human patients, who reported modest gains in functional vision such as a greater ability to locate objects in space visually and an enhanced ability to recognize large letters (Merabet et al. 2005; Dobelle 2000). Finally, just as gene therapy can potentially restore the function of a retina affected by faulty genes, stem cells can be transplanted into the retina to regenerate healthy photoreceptor cells. This procedure has been both safe and effective in rats, but has not yet been tested in humans (Guanting et al. 2005).

The initial success of experimental methods to restore sight has fueled interest in the scientific community, but ongoing investigation is necessary to ascertain the functional utility and long-term safety and efficacy of these techniques. Particularly little is known about whether meaningful sight can be enabled in those blind since birth, for instance, and experimental procedures today are limited in the degree of visual function that they can facilitate. Recipients of visual therapies may still need to use nonmedical methods such as Braille and the white cane to perform daily activities, and the vision regained may or may not be stable over time (Dobelle 2000; Buffoni et al. 2005).

In order to answer these questions and prepare sight-enabling therapies for mainstream application, a great deal of money and scientific effort have been invested in these ongoing clinical trials, including Phase I trials involving minors (Talk with RAC 2005; Clinical Trials 2009). Research reports about these experiments, both in the popular media and in scientific journals, often couple descriptions of findings with blatantly negative statements about the condition of blindness. A news report about a bionic eye prototype, for example, opens by featuring a formerly successful man whose life was "devastated" by retinitis pigmentosa. The article explains that blindness caused him to move about as if she were intoxicated, "required" him to retire from her job, and "forced him to rely on a cane and the help of his daughters" (Foundation for Retinal Research 2009). Approval of a Phase I gene therapy trial in children is justified by the remarks of a blind infant's father, who testifies that "All our dreams for his future were taken the day we found he has LCA" [a genetic condition causing insufficient retinal development] and that "This is why I urge you to let these wonderful doctors perform their trial for gene therapy of LCA in children…They [the children] deserve their sight". (Talk with RAC 2005). An empirical paper by Veraart et al. (2003) acknowledges the ethical concerns of involving human subjects in an "immature project" such as Phase I research. Yet they assert that, "On the contrary, excessive carefulness could hold back progress and dismiss the expectations of the blind community" (Veraart et al. 2003).

In these cases, then, restoring sight is seen as a medical imperative of sorts, justifying the risks and costs of continued investigation and early involvement of child subjects. This view has been espoused by researchers, by sensationalized media reports, and by some parents of blind infants, but not by those who have been blind for a substantial period of their lives. On the contrary, the views of blind people themselves have been largely absent from the discussion. Several searches of a collection of research databases, including PubMed, Medline, Academic Search Premier, and PsycINFO, failed to locate any empirical studies of blind people's attitudes about gaining sight or about specific devices currently in development to restore vision. There have, however, been a number of studies examining blind people's attitudes about vision loss and its psychological impact upon their lives, which suggest that attitudes about becoming sighted are complex and variable enough to warrant more thorough investigation. It is crucial that researchers and policymakers consider the actual beliefs and reactions of blind individuals, who will participate in clinical trials and who will ultimately constitute the market for sight-enabling procedures.

There is little doubt that the newly blinded and their families frequently have a negative reaction to vision loss and, at least in the early stages of blindness, regard their condition as a medical disorder that should be cured. Sight loss has been linked to depression and tendencies toward suicide (De Leo et al. 1999), and young people who have recently lost vision tend to develop a more negative self-concept (Roy and MacKay 2002). The interpretation of vision loss as an incomplete or inferior state of existence that deserves medical intervention, frequently held by those who are newly blinded and by representations of blindness in popular culture, represents a "medical model" of blindness (Schroeder 1996; Higgins 1999).

Yet there is some evidence, both empirical and anecdotal, suggesting that blindness can become a neutral characteristic or, in some cases, can comprise a positive aspect of identity. Schroeder (1996) conducted qualitative interviews of eight legally blind Americans to gauge their attitudes toward Braille and its applicability in their lives. He found that persons who use Braille extensively also identify strongly with blindness and consider Braille to be a positive symbol of their blindness. Conversely, blind persons who do not use Braille tend to self-identify as "persons with sight problems" rather than "blind persons"; they prefer to use their remaining vision whenever possible; and they are reluctant to use non-visual techniques to complete tasks even if they cannot complete them with vision. Schroeder asserts that blind persons who have fully accepted their blindness readily self-identify as blind, embrace symbols of blindness, (such as Braille), and derive self-esteem from the competent use of blindness techniques (Schroeder 1996). Newly blinded individuals do not always experience depression and hopelessness if they have opportunities to learn relevant skills and regain former roles and responsibilities (Dodds et al. 1991). From this view, it follows that the disadvantages faced by many blind people originate from societal attitudes and prejudices rather than from inherent limitations caused by vision loss (a "social model" of blindness; Higgins 1999).

This alternative view of blindness has important implications for sight restoration. It implies that some blind persons may regard sight restoration as a loss of a valued identity rather than a gain of independence and freedom. A parallel debate rages in the deaf community concerning cochlear implants; in Canada, some organizations of the deaf have openly opposed the funding of cochlear implant research, arguing that deafness is a culture, rather than a medical condition (Swanson 1997). Although the term "blind culture" is rarely, if ever, used, the strong identification that some have with blindness and Braille may lead to similar sentiments, if less intense. Concerns about losing one's "blind identity" have been expressed informally by blind persons, such as Rebecca Atkinson, a blind woman who stated that "Saying yes to seeing again, even for someone who wasn't born blind, isn't easy. The repercussions would ripple beyond my eyes into my friendships, my work, my relationship…" For persons who have always been blind, such as Atkinson's boss, the blind identity is even more deeply ingrained; "As the blind-from-birth son of blind parents, I am, in part of my soul, defined by my blindness." (Atkinson 2007). In light of these remarks, it cannot be taken for granted that all the blind wish to see.

The literature also casts doubt upon the acceptability of medical procedures that can enable only partial vision. In their study of visually impaired college students, Roy and MacKay (2002) found that people with partial sight (who often self-identify as "low vision" rather than blind) are less content with their visual impairment, on average, than those with little or no residual vision. Schroeder (1996) further explains that people with more residual vision are often less willing and able to use alternative techniques or to embrace their blindness, leading to a condition of being unable to complete tasks either visually or non-visually. Therefore, while the medical model of blindness treats visual techniques as always superior to non-visual ones, the alternative model postulates that the competent use of skills is more important than the amount of sight that one has (Schroeder 1996). Finally and perhaps most importantly, blind persons who regard their blindness as neutral or positive are likely to assess the risk-benefit ratio of sight-enabling interventions less favorably than the medical model indicates. In other words, someone who is blind but successful, independent, and physically healthy may regard having a life-altering surgical procedure as more of a liability than an asset, especially if the process of "learning to see" is likely to be long and complex. A person in this situation must weigh the gains of sight against the hardships of taking time off from work, providing for family members, and undergoing the medical risks of surgery.

In short, before research on sight-enabling technologies continues on a larger scale, scientists and policymakers must understand the range of needs, wants, and attitudes of the blind community regarding artificial vision. This includes knowing who is likely to be most or least interested in becoming sighted, what visual functions are most important to enable, and other features of medical interventions that are considered acceptable and unacceptable by blind consumers. Only then can artificial vision researchers make progress that is most helpful to the blind in the long term.

The present study is designed to empirically assess the attitudes of members of the blind community toward both the general idea of gaining sight and toward the specific technologies described above. Three overarching questions are asked: (1) What factors contribute to a blind person's willingness to gain sight?; (2) In those who do wish to gain sight, what features of interventions are most important to them?; and (3) What risks and limitations will they consider acceptable in an intervention?

It is predicted that those who identify most strongly with blindness will be least open to gaining sight, but that a variety of other factors will contribute to variability in attitudes, such as the age of onset of blindness, degree of vision loss, and attitudes toward technology in general. It is also predicted that participants will be most open to interventions that can restore complete or nearly complete visual function and that pose relatively little risk of adverse health effects.

## 9.2 Methods

A Web-based survey was administered to 281 adults, all self-identifying as blind or visually impaired and recruited from a blindness-related email listserv network. Various divisions and affiliates of the National Federation of the Blind sponsored most of these listservs, although a significant portion of participants indicated membership in other blindness consumer organizations. The group consisted of 123 females, 126 males, and 32 individuals of unspecified gender; participants ranged in age from 18 to 72 years. A majority of the participants (77.9%) indicated that they were of Caucasian descent. All but one of the participants was a native English speaker. Although the survey never directly asked participants to list their nationality, most participants indicated that they belonged to American blindness consumer organizations, although a few stated that they were members of groups based in Australia or New Zealand. All participants were invited to enter a raffle drawing for one of four cash prizes (US$250, $50, $50, and $50).

The survey began with a series of questions assessing participants' attitudes toward blindness, (i.e. "I feel good about blind people"); perceptions of public attitudes toward blind people, (i.e. "Society views blind people as an asset"); and the centrality of blindness in participants' identities (i.e. "I feel a strong attachment to other blind people"). Next, participants were asked about their degree of identification and affiliation with the blind community. Participants listed the types of activities in which they participated with other blind people, how often they spent time with other blind people, and the percentage of their close friends who were blind. The third section evaluated participants' experiences with both medical and assistive technology. Participants indicated how often they had been hospitalized in the last 10 years, how often they use assistive technology in their daily lives, and what types of assistive technology (such as portable note takers, closed-circuit televisions [CCTVs] and computers) they used regularly.

Then, participants answered questions dealing with visual prosthetics. First, they listed the types of visual functions that they believed were important for a vision-enhancing procedure to provide, ranging from mere light perception to the ability to drive a car. This was designed to assess blind consumers' specific priorities concerning the types of functions that they would most like to perform visually if they were to gain sight. Next, participants were asked to state, on a seven-point scale, how likely they would be to take a hypothetical pill that would confer guaranteed 20/20 vision without risks or side effects. This question was designed to measure participants' interest in becoming sighted at all, independent of the risks and benefits of specific procedures. Then, participants read brief descriptions of four sight-enhancing procedures (CVS, the bionic eye, tissue transplantation, and gene therapy) and were asked how interested they would be in both participating in a clinical trial of the procedure and receiving the procedure after it is FDA-approved. The descriptions were edited and approved by experts at the Foundation for Retinal Research (www.tfrr.org). Finally, participants were given the chance to provide comments explaining their opinions.

The questionnaire ended with items asking participants to specify their cause of blindness, age at becoming blind, level of residual vision, and demographic characteristics. All responses were anonymous and participants entered information for the raffle on a separate page to protect the anonymity of their survey responses.

## 9.3 Findings and Implications

When asked how likely they would be to take a "magic pill" that would restore perfect sight with no risks or side effects, most participants (72%) said that they would be at least somewhat likely to take the pill, and more than half said they would be very likely to take it. However, about a fifth of respondents (21%) indicated that they were at least somewhat unlikely to take the pill. Participants who both reported spending time with other blind people frequently and who showed strong positive attitudes about blindness and blind people were least interested in the prospect of gaining sight, while those who either spent little time with other blind people, were less positive about blindness, or both tended to express strong interest in the pill. Additionally, those who had been totally blind since birth were least interested in the pill. People who had been fully sighted as children, as well as those who have always had a good deal of usable vision, were generally interested in the pill. When invited to provide comments, participants often cited their positive identification with blindness as a reason for not wanting to see, making statements such as "I think blindness makes me who I am. Why would I want to change that?" Others simply stated that their lives were comfortable now and they did not wish to subject themselves to the process of adjusting to new sight.

Not surprisingly, participants expressed more hesitation about the prospect of receiving therapies with real risks and limitations. The two biologically based therapies (tissue transplantation and gene therapy) received substantially more support than the two "electronic" therapies, especially for those who had been blind since birth, but only about 40% of participants expressed interest in the biological treatments. This was partially because some participants were blind from non-retinal causes and therefore could not personally benefit from any of these technologies. However, even those participants who were interested in a "market" version of a therapy were significantly less interested in receiving an experimental version of the therapy. Interestingly, this effect was strongest for younger participants. In addition, people who believe that society views the blind in a more positive manner, while being slightly more interested in the pill, are often substantially more interested in specific visual prosthetics. It is possible that people who express more optimism about the public's perceptions of blindness also tend to feel more optimistic about the potential benefits of visual prosthetics, especially when the exact risk-benefit ratio is uncertain.

Finally, it appears that the majority of consumers expect sight-enabling therapies to restore a good deal of functional vision in order to be worthwhile. Specifically, 69% specified that artificial vision would need to be good enough to allow them to read standard print in order to be worthwhile, and 59% listed the ability to drive as an important visual function.

This study is most likely the first to directly investigate blind persons' attitudes and expectations regarding sight restoration. Results suggest that there is not a clear consensus in the blind community on this issue, and that the anticipated benefits of artificial vision are much less clear than popular rhetoric would indicate. A substantial minority of participants, especially those who have been totally blind all their lives and those who feel strongly affiliated with the blind community, express reservations about becoming sighted even in an ideal situation. Consumers' assessment of the risks and benefits of particular therapies vary depending on their attitudes toward blindness, the functional vision that the therapy can restore, and what responsibilities and roles, (e.g. employment, childrearing, etc.), might be at stake.

Future research will need to examine these questions in a more representative sample of blind persons, including those who are less acquainted with technology in general and those who identify less strongly with blindness and the blind community. In particular, the nature of the "blind identity" that appears to be driving some people's reluctance to become sighted should be explored in more detail. These findings will continue to inform researchers and policymakers as availability of approved therapies becomes more and more of a reality.

## References


Acland, G.M., et al. 2005. Long-term restoration of rod and cone vision by single dose rAAV-mediated gene transfer to the retina in a canine model of childhood blindness. *Molecular Therapy* 12(6): 1072–1082.

American Foundation for the Blind. 2007. Blindness Statistics. http://www.afb.org/Section.asp?SectionID=15.

Atkinson, R. 2007. The chance to see again…Would she take it?. *The Guardian*, July 17.

Buffoni, L., J. Coulombe, and M. Sawan. 2005. Image processing strategies dedicated to cortical visual stimulators: A survey. *Artificial Organs* 29(8): 658–664.

Clinical trial: Gene therapy for leber congenital amaurosis caused by RPE65 mutations. http://www.clinicaltrials.gov/ct2/show/NCT00821340?term=RPE65&rank=1.

De Leo, D., et al. 1999. Blindness, fear of sight loss, and suicide. *Psychosomatics* 40(4): 339–344.

DoBelle, W.H. 2000. Artificial vision for the blind by connecting a television camera to the visual cortex. *American Society for Artificial Internal Organ* 46(1): 3–9.

Dodds, A.G., P. Bailey, A. Pearson, and L. Yates. 1991. Psychological factors in acquired visual impairment: The development of a scale of adjustment. *Journal of Visual Impairment and Blindness* 85: 306–310.

Ferguson, R.J. 2001. *We know who we are: A history of the blind in challenging educational and socially constructed policies: A study in policy archeology*. San Francisco: Caddow Gap Press.

Foundation for Retinal Research. 2009. [Online] From www.tfrr.org.

Gawel, Richard. 2007. FDA approves study of next-generation retinal implant. *Electronic Design* 55(6): 21.

Ghosh, P. 2007. Gene therapy first for poor sight. [Online] From http://news.bbc.co.uk/1/hi/health/6609205.stm.

Guanting, Q., et al. 2005. Photoreceptor differentiation and integration of retinal progenitor cells transplanted into transgenic rats. *Experimental Eye Research* 80(1): 515–525.

Higgins, N. 1999. The O&M in my life: Perceptions of people who are blind and their parents. *Journal of Visual Impairment and Blindness* 93(9): 561–578.

Medgadget.com. 2008. Argus II retinal prosthesis implanted into first two patients in Europe. [Online] From http://medgadget.com/archives/2008/04/argus_ii_retinal_prosthesis_implanted_into_first_two_patients_in_europe.html.

Merabet, L.B., et al. 2005. What blindness can tell us about seeing again: Merging neuroplasticity and neuroprostheses. *Nature Reviews* 6(1): 71.

Roy, A., and G. MacKay. 2002. Self-perception and locus of control in visually impaired college students with different types of vision loss. *Journal of Visual Impairment and Blindness* 96(4): 254–266.

Schroeder, F.K. 1996. Perceptions of Braille usage by legally blind adults. *Journal of Visual Impairment and Blindness* 90(3): 210–218.

Statistical Snapshots: American Foundation for the Blind. 2008. [Online] From http://www.afb.org/Section.asp?SectionID=15.

Swanson, L. 1997. Cochlear implants: The head-on collision between medical technology and the right to be deaf. *Canadian Medical Association Journal* 157(7): 929–932.

Talk with RAC, December 13, 2005. [Online] From http://www.webconferences.com/nihoba/440661.html.

Veraart, C., et al. 2003. Pattern recognition with the optic nerve prosthesis. *Artificial Organs* 27(11): 996–1004.

# Chapter 10
# Nanotechnology, the Brain, and Personal Identity

**Stephanie Naufel**

## 10.1 Introduction

Revolution ruptures the status quo and introduces a new era with little room to look back. Whether by way of war and political uprising or via new inventions, this vehicle of change is ever present in history. Today's technological revolution has allowed scientists and engineers to sprint toward new advancements, but not without pause for ethical concerns. In an age where new knowledge is growing exponentially, it is important to consider the societal implications of up-and-coming innovations. Especially with regard to biomedical devices, progress has been rapid and the possibilities of the future seem boundless. However, with so much potential, it is important to ask, "Just because we can, does that mean we should?"

This paper offers a discussion of how implementing nanotechnology applications or nanotechnology-enabled applications in the brain will influence human personhood. I will first provide a brief overview of select neurotechnologies, and will discuss the potential of nanotechnology in the future of these technologies. My considerations will include cochlear implants and other brain-computer interfaces. The discussion of personhood will highlight factors that may be seen to affect the line between human and machine: permanence of the technology, affect on personality, and degree of invasiveness. Central to this paper, however, will be conjecture on how ideas of personhood and moral status will be affected by nanotechnology in the brain. This will serve as a response to Grant Gillett's question,

> If my brain functions in a way that is supported by and exploits intelligent technology both external and implantable, then how should I be treated and what is my moral status – am I a machine or am I a person? (Gillett 2006, 79).

S. Naufel (✉)
Robert R. McCormick School of Engineering and Applied Science, Northwestern University, 2145 Sheridan Road, IL 60208, Evanston
e-mail: Steph.Naufel@gmail.com

S.A. Hays et al. (eds.), *Nanotechnology, the Brain, and the Future*,
Yearbook of Nanotechnology in Society 3, DOI 10.1007/978-94-007-1787-9_10,

In his paper, "Cyborgs and Moral Identity," Gillett talks about the "increasing reality of cyborgs – integrated part human and part machine complexes" (Gillett 2006, 79). Already, individuals may be considered cyborgs in many ways, evidenced in part by the integration of heart pacemakers and many prosthetic devices in the body. However, when these assimilations involve the brain, considered by many to be fundamental to sense of self, new concerns arise. This essay attempts to address such considerations.

## 10.2 A Century of Neurotechnology

It was undoubtedly a stifling day in Iraq when an American missile exploded in Mustafa Ghazwan's neighborhood. It was June of 2007, the last time baby Mustafa would hear anything for a year and a half. The explosion had left him deaf, but now, the 3-year-old has been reintroduced to sound. At the beginning of 2009, Ghazwan was brought to San Francisco through an organization called No More Victims. A 90-min surgery provided Ghazwan with a cochlear implant, enabling him to hear. The child, who has not said a word since the accident, can now listen to his father express his happiness at the situation. He is expected to start speaking soon thanks to sessions of speech and listening therapy (Fernandez 2009).

> Today, the blind can see, the deaf can hear, and the lame can move. Arm amputees can reach, and leg amputees can walk. These are not fantasies; they are the realities of science. Neuroprosthetic development in recent decades has allowed for the attainability of the seemingly impossible. From cochlear implants to prosthetic arms, there have been substantial strides in the prosthetic world. Regardless of time and place, people throughout history have been affected with disease and tragedy that have stripped them of capabilities. However, rising to the challenge, neuroscientists and engineers have teamed up to provide and restore functionality to those desiring prosthetic relief.

In the 1960s, British-born Giles Brindley established himself as a pioneer of neuroprosthetics. Inspired by earlier work to stimulate visual cortex, Brindley set out to create an array of electrodes for his own experiments. He had read about Otto Foerster's report that stimulation had resulted in subjects seeing fixed spots of light, and wanted to use this discovery to create sight for the blind. Previously, he had tested electrodes in the motor cortices of monkeys before receiving approval to implant in the visual cortex of a human (Chase 2006, 58–60)

The year was 1967 when Brindley stood by neurosurgeon Walpole Lewin in Cambridge as he implanted his 81-electrode creation into the visual cortex of a blind 54-year old woman. Brindley's electrodes – 50 of which were functional – were implanted into the right hemisphere of her cortex. The woman reported seeing spots of light on the left side of her field of vision, and each spot corresponded to a different electrode. Brindley had created a visual implant that could allow rudimentary vision, and this inspired others to make their own attempts at creating visual prostheses (Chase 2006, 60).

Across the Atlantic, American researchers were initiating their own work to develop prosthetic devices. In 1970, only a few years after Brindley's experiment, the Neural Prosthesis Program was formed at the National Institutes of Health (NIH)

(Waltzman and Roland 2006, 7). Early work focused on continuing the development of a viable visual prosthetic. However, NIH grants have funded many projects on neuroprosthetic research throughout the years, from robotic arms to neural implants and beyond.

The 1970s also saw progress in the development of the cochlear implant. While hearing aids amplify sounds, cochlear implants differ, in that they actually provide hearing through stimulation of the nerves inside the ear. This allows the brain to receive signals that can be processed as sound, and this technology is reserved for the completely or severely deaf. By the mid-1970s, the United States was home to 13 individuals with functioning cochlear implants. Through NIH funding, a study was done to evaluate the performance of these implants under the direction of Dr. Robert Bilger. One key conclusion was that the devices helped lip-reading and recognition of environmental sounds. The published results, referred to as the Bilger Report, spurred the NIH to further invest in the development of these implants (Wilson and Dorman 2008, 5). From there, the pace quickened as improvements were made and the number of human subjects rose. In 1984, the Food and Drug Administration approved the first cochlear implants for adults of at least 18 years of age. In 1990, approval allowed implantation in children over the age of two. In 1998, the age limit was lowered to 18 months, and then to 12 months in 2002 (Anderson 2008). As of May 2008, there are over 120,000 users of cochlear implants worldwide (Wilson and Dorman 2008, 6).

In addition to cochlear implants, "brain pacemakers" are also a viable neurotechnology that are commercially widespread. Used for deep brain stimulation (DBS), these devices are analogous to heart pacemakers in that they deliver pulses of electricity to restore normal functioning. In the case of DBS, the brain pacemaker is implanted in the chest and connected by an extension to an electrode implanted in the brain. The electrode is positioned to reach the area of the brain that will be stimulated as treatment. Deep brain stimulation can treat ailments such as Parkinson's disease, depression, and obsessive-compulsive disorder. The device came about in the 1980s, and received regulatory approval in the United States in 1997 (Coffey 2009, 210). Its potentials are still being explored, and the future may see greater possibilities available through this therapy.

More recently has been the development of the BrainGate Neural Interface System. In 2004, Matthew Nagle became the first subject in the pilot trial for this system. The 25-year old had become paralyzed from the shoulders down a few years prior, but this new technology would allow him to manipulate the world with the power of his mind. The system consists of an electrode array that is implanted in motor cortex. The electrodes connect to a computer, which translates the neurons firing in motor cortex into action. In this manner, Nagle was able to control a cursor on a computer screen using just his thoughts, and this enabled him to work a television, switch lights on and off, and check email (Chase 2006, 213–216). The full potential of the system extends to providing for respiratory control and manipulation of prosthetic limbs. However, the future of research on this system, which is not commercially available, is uncertain as the company that owns it, Cyberkinetics Neurotechnology Systems, has shut down (Roush 2009). Regardless, the technology will still be considered in this discussion.

## 10.3 Nanotechnology

As neural devices grow in sophistication, nanotechnology will likely become a key player in interfacing technology and brain. Consider the size of early computers that would take up the entire wall of a room, and how these same computers have been modified into laptop and even cellular phone form. Often in the case of technology, smaller is better, and for many reasons the nanoscale is the next frontier for neurotechnology.

Francois Berger et al. (2008) speculate that nanotechnology will improve neuro-implants in their recent paper, "Ethical, Legal and Social Aspects of Brain-Implants Using Nano-Scale Materials and Techniques." His claim is that researchers will look to nanotechnology for the greater biocompatibility, smaller size, enhanced integration into tissue, and the advanced electronics that will accompany such development (Berger 2008, 241; see also Robert 2008, 228–229). Indeed, tapping into the brain has required fragile and careful work on the part of scientists and engineers, and nanotechnology may offer valuable improvements.

A common concern for introducing foreign medicines and technologies into the brain has been getting past the blood-brain barrier (BBB). This barrier of endothelial cells lines the blood vessels of the brain, protecting it from harmful substances (Berger 2008). The limitations accompanying the selectivity of this barrier have long been recognized, and a solution is offered through the implementation of nanotechnology. One need not even do an intensive search through scientific databases to discover the promises nanotechnology brings to crossing the BBB; a simple Google search will yield a large amount of articles on the topic. As Michael Berger explains, nanotechnology is ideal because it can slip through the BBB by masquerading as a compound that would normally be let through: nonionic and low molecular weight molecules. With this success, nanotechnology can slip into the brain and carry out its intended task (although this possibility is precisely the reason why many are fearful of a relationship between nanotechnology and the brain).

With specific regard to neuroprosthetics, nanotechnology shows notable potential. A successful neuroprosthetic requires a safe and feasible interface between device and brain, and nanotechnology will be valuable in providing this quality (Stieglitz 2007). Smaller technology can be expected to provide a surface pattern similar enough to the surface of the brain that will make for a compatible connection. The more natural the interface, the less of a reaction the body will have to the technology, and biocompatibility will be optimized.

Most obvious of all, however, is the fact that nanotechnology will provide for the development of smaller neural devices overall. As of now, neurotechnologies are relatively large, and could stand to decrease in size. Cochlear prosthetics, for example, have both internal and external components, and the external components are sizeable and noticeable. This way they are more exposed to the elements, and could be considered an aesthetic burden. Brain pacemakers for deep brain stimulation could also benefit from a reduction in size. A smaller pacemaker would allow surgeons to make more minimal cuts for implantation, and would likely cause less of a disruption in the body. As for the BrainGate™ system, which has required a substantially sized plug and cord sprouting out of the head, it would most likely be difficult to find a potential user who would object to the prospect of a smaller, less unwieldy device.

## 10.4 Technology and the Brain

Sensory and motor nerves are electric messengers of the body, reaching every corner and relying on the brain and spinal cord as headquarters. The brain presides over the constant exchange of messages, and is the control seat of functioning, taking charge of capabilities such as reasoning, decision-making, and communication. For this reason, introducing technology to the brain generates more unease than would arise if the nervous system were not involved as explored by Gillett in, "Cyborgs and moral identity," (2006).

The subject of human versus machine brings up the question of what it means to be a person. Whereas machines and humans may be seen more as natural categories, personhood is a moral and legal notion. While humans are paradigmatic persons, personhood has been extended to other sorts of entities, as well – most notably, corporations, to hold them responsible in legal proceedings. However, what of humans with nanotechnology-enabled brain function? Are they persons? Indeed, exactly what it means to be a person is unclear. The law, philosophers, ethicists, and others have attempted to provide an answer, but consensus has been difficult to find. Regardless, it remains important to deliberate on how the relationship between technology and the brain will affect personhood, and what implications this will have on society.

### *10.4.1 Personhood*

Throughout history, societies have allowed atrocities to be committed against various groups of individuals. Many times this has been done under the justification that these beings were not "persons" worthy of respect. Slavery and genocide are poignant examples. Today, the concept of personhood has taken on new meaning as medicine and technology advance. The moral status of humans influenced by technology has become an issue as the line between human and machine (and between human and non-human animals – Baylis and Robert 2003) becomes blurred.

Various notions of personhood exist, but this essay does not intend to redefine or create new criteria for personhood. The focus here is instead to analyze how different factors and technologies will affect the line between human and machine.

The intention is not to decisively define this line, but rather explore the shades of gray where the user of a device may reconsider his or her sense of self as it relates to personhood.

Yet the question remains, why is personhood an important concept to consider? Returning to Gillett, it is important to realize that with this question, he raises countless others. The reality of humans admixing with machines has many implications for society, and supplemental to Gillett's stated concern are the following:

How will I see myself?
How will I be seen in the eyes of my family?
How will I be seen in the eyes of God?
How will others see me in the eyes of the law?
How will others see me in the eyes of morality?

These questions are important to the consideration of personhood because society's treatment of individuals depends on its view of their moral status as persons. As Farah and Heberlein explain in their paper on personhood, it is important to understand who qualifies as a person because persons have moral responsibilities that non-persons do not have. In their words, "if a person works hard and accomplishes something good, we give that person moral credit…if a falling tree branch kills someone, we do not regard the branch or its behavior morally wrong" (Farah and Heberlein 2007, 38).

Although it is clear that a human is a person and a tree is not, this distinction becomes blurred when one considers a human whose functioning is closely influenced by particular technologies. Especially with respect to the brain, where technology intimately affects the person at the location of their decision-making, personhood becomes an issue. For example, if the user of a brain pacemaker is influenced by stimulation to become more impulsive, and he or she commits a crime, how should the user be treated? Is he or she a full person who is expected to uphold standards of society by not breaking the law? Alternatively, should society give this individual leniency because the technology aspect of his or her being cannot be held morally accountable?

## 10.5 Factors Influencing the Line Between Human and Machine

### *10.5.1 Permanence of the Technology*

The more permanent a technology, the more it may seem a fundamental part of one's body. Permanent retainers, for example, feel foreign until a few days pass, and the tongue no longer hesitates over the groove between metal and tooth. Contact lenses soon feel natural, as do knee replacements that persist under the skin and offer a healthy hinge between thigh and calf. In addition, just as heart pacemakers are able to successfully integrate into the body, revolutionary neurotechnologies can be incorporated into the nervous system.

If a technology is permanently interacting with an individual's brain, it is set up to become an established component of the body. In the specific case of the brain, the technology will be connected to the seat of reason and decision-making, and this involvement will be one intended to last. Therein lay the source of uneasiness. For the most part, people like and hope to remain in control of their actions. This is not to say that many will not acquiesce under the strict orders of another, or allow themselves to be influenced by drugs and alcohol. However, it is human nature to hold on to volition. That said, if a device is developed to permanently affect the brain, the user would not have the ability to remove it, and this may be seen as an attack on the amount of his or her control over the device. Instead of being able to decide the extent of the device's involvement in the body, the user will have to accept that the device will be there to stay. An inability to easily remove a neural device may leave the user feeling eternally allied with technology, a sentiment that could leave the user feeling less human, and more cyborg. Nanotechnology, once implanted in the brain, will most likely be designed for permanence. The technology will be so small that to isolate it for removal would not only be tedious but might not be worth the effort. This situation leaves an individual with a technology that is forever implanted, thus solidifying the bond between human and machine.

### 10.5.2 Effect on Personality

In the children's book series *The Tripods*, author John Christopher creates a science-fiction world of mind control via technology developed by sophisticated aliens. In this post-apocalyptic series, society is under the authority of these aliens, who subject humans to cranial implants from the age of 14. These cap-like devices put the humans under the volition of their foreign masters, in part by hindering their capacity for creativity and questioning.

However futuristic and imaginative this plot may be, the premise is not as far-fetched as it initially seems. The brain's functioning can already be altered through several media. Drugs, disease, and now neurotechnology can drastically change behaviors. The main neurotechnology that has proven to affect personality in multiple ways is deep brain stimulation. Serving as treatment for depression among other disorders, DBS is able to alter moods and affect personality. Cases have also arisen where enhanced capabilities have been observed. In one example, creativity in the user of a brain pacemaker was seen to increase (Drago et al. 2009).

With such potential to influence the brain, deep brain stimulation is a therapy that should be kept under scrutiny. Therefore, the case must be with other technologies that produce similar effects. Future developments that will involve nanotechnology may be able to integrate so effortlessly in the brain that upcoming devices will seamlessly integrate into the body and wield a consistent sway on the mind. This can be seen to inherently affect personhood because it will fundamentally change the individual. Of course, drugs and disease, as aforementioned, can produce similar outcomes. However, in the case of neural devices, the technology aspect creates a gray area around the line between human and machine that merits contemplation.

### *10.5.3 Degree of Invasiveness*

Brain technologies vary in degree of invasiveness, from magnets and electrodes on a skullcap to electrodes surgically implanted in the brain. How invasive a device is may have an effect on how people perceive its influence because of the intimate association between device and body. General reasoning teaches that more often than not, it is easier to wield direct power from near than from far, and this way of thinking may lead individuals to believe that an invasive device poses a greater threat to personhood than a non-invasive one.

There is also the issue of internal versus external technologies. An important consideration is that humans are often inclined to judge based on sight. Therefore, an external technology may inspire more pause than an internal technology that can easily be "out of sight, out of mind." Consider two men walking down the street. One has a prosthetic hand that is not neurotechnological, but rather more rudimentary. The second individual has an internal and hidden retinal prosthetic that is providing eyesight for up to 3 miles away. Upon initial glance, it may seem that the first individual is the one with the most potential to be a human-machine hybrid. However, it is the second individual who is not only seeing when he was previously blind, but also has enhanced vision. His life is more influenced by technology, and yet even when this is explained, one may still be more inclined to see the other individual as less human.

## 10.6 Neurotechnologies to Consider

It would be presumptuous to say for certain what the notions of personal identity are from one neurotechnology user to the next. Nevertheless, analyzing different neurotechnologies for how their capabilities and relationship with the brain will affect notions of personhood is a useful beginning. As mentioned in the initial discussion of personhood, it is necessary that society consider what it means to be a person so that it may treat individuals accordingly. But before continuing on to discuss specific technologies, it is first important to consider the words of ethicist Paul Root Wolpe who, in an effort to inform his audience of how individuals may feel seriously integrated with their technological aids, says:

> In a sense, what we are talking about [are] … technologies that are going to be incorporated into our very flesh that will become part of who we are. We are already doing it. We have prosthetic limbs that all of us know about, and yet somehow we think of those limbs as separate from ourselves, unless of course you happen to use one. And, if you talk to someone who has an artificial arm or leg or other limb that they depend on, and you ask them about their relationship to it, it's not just a piece of hardware that they strap on (Wolpe 2002).

### 10.6.1 Cochlear Implants[1]

It may be best to begin with an analysis of a relatively established neurotechnology. Cochlear implants have been available to the public since the early 1980s and have been widely successful. However, this is not to say that there has not been outcry from the deaf community over these implants. Some critics have argued that cochlear implants are a means to eliminate the society of the deaf. Protest has gone so far as to suggest that providing this technology to the deaf is comparable to genocide (Balkany et al. 1996, 748; see also Anderson, this volume).

With such objections, it may be determined that these individuals feel every bit as capable as hearing individuals. To extend this idea, they regard themselves to be "persons" of equal worth as their hearing counterparts, as they should. Gregor Wolbring offers a clear validation of this idea in his paper, "Confined to Your Legs" (Wolbring 2004, 139–156). He contends that society tries to "fix" the disabled when many times the truth is that the disabled do not see themselves as such, and have proven themselves to be completely capable without "fixing".

What these arguments offer to the discussion of personhood is a look at how the deaf define what it means to be a person. As has been argued, many deaf individuals feel fulfilled without being able to hear, and will therefore regard the cochlear implant as an unnecessary accessory. Many will believe that they are already valuable as persons without this technology, and this goes to show that perhaps society should take a step back before impressing technologies on those who may not feel that they need to be "helped" by a device.

However, for those who do embrace cochlear implants, the human versus machine contention can still be considered. The verdict may be as follows. Although the implant is permanent and invasive, the brain and technology relationship arising from this device seems minimally threatening to notions of the user's personhood. Cochlear implants can be seen mainly as a medium for turning sounds into stimulation for the brain to process. In this manner, the device is an aid to the brain, but does not directly influence decisions and reasoning. The user may therefore continue life as he or she did before implantation, only now with the benefit of enhanced communication.

Then again, now comes up the issue of enhancement. If a cochlear implant can provide "normal" hearing for an individual, it is not a far stretch to say it may also be capable of providing super hearing powers. Here is where issues may arise. Consider Mark, who has a cochlear implant that can allow him or her to hear voices a mile away. What implications does this ability have? Will he consider himself a normal person, just gifted? How will the law consider him? Because he now has advanced capabilities, will he have greater responsibilities as an individual in society?

[1] The arguments made for cochlear implants may be analogous to arguments that may be made for visual prostheses. See also Anderson, this volume, and Silverman, this volume.

For example, he could be called upon to use his powers for the benefit of his community, perhaps by using his hearing to identify dangers off in the distance. Moreover, will his family and friends view him as the Mark they always knew, or will they be wary of the affect of technology on his mind? Such are the questions that arise if the full possibilities of cochlear implants are explored.

### *10.6.2 Brain Pacemakers*

As has been mentioned, brain pacemakers are analogous to heart pacemakers because of their ability to send pulses of stimulation to the body. Another similarity between them is the uneasiness felt by many at the introduction of both of these technologies into society. The public viewed heart pacemakers with a wary eye when they first became publicly available (Liker et al. 2008, 1129). Therefore, if a device supporting the pumping of the heart originally generated concern, how much more will a pacemaker that stimulates the brain?

According to Ford and Kubu, deep brain stimulation can potentially "alter the brain itself that results in a more fundamental change in the self and being." They cite "risks to memory, executive function, language, or personality variables" as changes that may occur as the result of DBS (Ford and Kubu 2006, 107). The question to ask here, is, is the user of a brain pacemaker still a person? If a user, Patrick, has a shifted personality and functional capabilities because of his device, will he still be held up to standards expected of persons? His judgments will be influenced by stimulation, and this means that his choices will not be entirely his own. Will societal expectations be different for Patrick than for someone else? If he commits a crime, will he take on full blame, or can he blame his technological counterpart for affecting his actions? Also important is the question of how Patrick will view himself. Is he still his own person, or should he adopt the idea that he is partially controlled by stimulation, and that he and technology are wedded and geared to face life together?

### *10.6.3 BrainGate™ Neural Interface System*

If technology is interfacing with your brain to helping you move your limbs, go to the restroom, and breathe, how does this affect your sense of self? Such are the future possibilities of the BrainGate™ Neural Interface System, and its implications for ideas of personhood are considerable.

Consider an example case. After a terrible accident, Sylvie has become a quadriplegic, and is a user of this BrainGate™ system. A large plug and cord protruding from her skull serve as a visible intermediary between her brain and the computer that is translating her neural activity into action. She uses the system to move a computer cursor, thereby controlling the lights in her room, the temperature, and other regulatory factors. Thanks to the many neurons firing in her brain, she can also use to system to move her wheelchair, getting from one place to the next on the fuel of thought. Is she still a full-fledged person, even though her body is non-functional? It is true that her brain is still actively humming, but the only way she is capable of living life somewhat normally is through a close connection with a computer. Does her community see her as woman, machine, or hybrid, and how will this affect their treatment of her? Will she still be held to the same moral standards as she was before her accident? In one case, society could easily forgive her for mistakes, explaining that a substantial amount of her functioning depends on technology, and this influence may have glitches. On the other hand, she could be assigned increased responsibility; under the reasoning, that she now has to make sure that her technological components are in harmony with her human components.

### *10.6.4 Looking Forward*

Michael Chorost's take on personhood may be useful. In his book, *Rebuilt: How Becoming Part Computer Made Me More Human*, Chorost discusses his experiences with deafness and his cochlear implant. He acknowledges his human side as well as his machine side, and discusses the advantages and shortcomings of both. One take-home message from his account is this:

> My bionic hearing made me neither omniscient nor dehumanized: it made me more human, because I was constantly aware that my perception of the universe was provisional...I was not disconnected from the world, remote and uncaring in the bioelectronic shell of my skin. The very provisionality of my perception reminded me that my political perspective was provisional also, and that it was my task as a human being to strive to connect ever more complexly and deeply with the people and places of my life (Chorost 2005, 157).

Chorost may have struck upon a true benefit that neurotechnology has to offer: it keeps us alert and active in the world. Cochlear implants allow deaf individuals to enjoy music, hear the voice of a loved one, and communicate with others. Brain stimulation, although it may greatly affect moods and personality, also relieves the symptoms of terrible diseases, and may provide for a healthier lifestyle. Moreover, the BrainGate system can restore meaning to the life of paraplegics by giving them the power to manipulate the world again, freeing them from the frozen isolation of their bodies. In looking to the future of neurotechnological applications, it is important to consider the alterations of personhood that may occur, but it is also important to realize what these devices can offer to life. They can enhance the humanness of the individual by enabling a greater level of participation in the world. They provide hope and remedy, and if society takes heed to be aware of the ethical implications of these technologies, they may very well thrive without hindrance.

## References

Anderson, Derrick. 2008. The history of cochlear implants. Poster presented at the CNS-ASU All Hands Meeting, Tempe, AZ.

Balkany, Thomas, Annelle Hodges, and Kenneth Goodman. 1996. Ethics of cochlear implantation in young children. *Otolaryngology – Head and Neck Surgery* 114(6): 748–755.

Berger, Michael. 2008. Crossing the blood-brain barrier with nanotechnology. *Nanowerk*. http://www.nanowerk.com/spotlight/spotid=6269.php.

Berger, Francois, Sjef Gevers, Ludwig Siep, and Klaus-Michael Weltring. 2008. Ethical, legal, and social aspects of brain-implants using nano-scale materials and techniques. *NanoEthics* 2: 241–249.

Chase, Victor D. 2006. *Shattered nerves: How science is solving modern medicine's most perplexing problem*. Baltimore: Johns Hopkins University Press.

Chorost, Michael. 2005. *Rebuilt: How becoming part computer made me more human*. Boston: Houghton Mifflin.

Coffey, Robert J. 2009. Deep brain stimulation devices: A brief technical history and review. *Artificial Organs* 33(3): 208–20.

Drago, V., P.S. Foster, M.S. Okun, I. Haq, A. Sudhyadhom, F.M. Skidmore, and K.M. Heilman. 2009. Artistic creativity and DBS: A case report. *Journal of the Neurological Sciences* 276(1–2): 138–142.

Farah, M., and A.S. Heberlein. 2007. Personhood and neuroscience: Naturalizing or nihilating? *The American Journal of Bioethics* 7(1): 37–48.

Fernandez, Elisabeth. 2009. Iraqi boy can hear again after implant surgery. *San Francisco Chronicle*, 18 Feb 2009, News section.

Ford, Paul, and Cynthia Kubu. 2006. Stimulating debate: Ethics in a multidisciplinary functional neurosurgery committee. *Journal of Medical Ethics* 32(2): 106–109.

Gillett, G. 2006. Cyborgs and moral identity. *Journal of Medical Ethics* 32(2): 79–83.

Keiper, A. 2006. The age of neuroelectronics. *The New Atlantis* 11: 4–41.

Liker, Mark, Deborah Won, Vikas Rao, and Sherwin Hua. 2008. Deep brain stimulation: An evolving technology. *Proceedings of the IEEE* 96(7): 1129–1141.

Robert, Jason Scott. 2009. Nanoscience, nanoscientists, and controversy. In *Nanotechnology and society: Current and emerging issues*, ed. Fritz Allhoff, and Patrick Lin, 225–239. Dordrecht: Springer.

Roush, Wade. 2009. Cyberkinetics loses energy. Xconomy. http://www.xconomy.com/boston/2009/03/19/cyberkinetics-loses-energy/.

Stieglitz, T. 2007. Restoration of neurological functions by neuroprosthetic technologies: Future prospects and trends towards micro-, nano-, and biohybrid systems. *Acta Neurochirurgica. Supplementum* 97(Pt 1): 435–42.

Waltzman, Susan, and J. Thomas Roland Jr. 2006. *Cochlear implants*. Stuttgart: Thieme Medical Publishers, Inc.

Wilson, B., and M.F. Dorman. 2008. Cochlear implants: A remarkable past and a brilliant future. *Hearing Research* 242(1–2): 3–21.

Wolbring, G. 2004. Confined to your legs. In *Living with the genie: Essays on technology and the quest For human mastery*, ed. Lightman Alan, Sarewitz Daniel, and Desser Christina, 139–156. Washington, DC: Island Press.

Wolpe, P.R. 2002. Neurotechnology, cyborgs, and the sense of self. In *Neuroethics: Mapping the field*, ed. S.J. Marcus, 159–167. New York: The Dana Foundation.

# Chapter 11
# Ethical, Legal and Social Aspects of Brain-Implants Using Nano-Scale Materials and Techniques*

**Francois Berger, Sjef Gevers, Ludwig Siep, and Klaus-Michael Weltring**

## 11.1 Introduction

An especially sensitive research area from an ethical, social and legal point of view is the development of brain implants, which constitute a major opportunity to treat neurological illness. The therapy of brain diseases is difficult because of the low accessibility of the brain as well as the high functionality of most of its parts. Systemic therapy using drugs has many negative side effects due to the absence of local and functional targeting. Brain dysfunction is often located in a specific neuronal network and therefore very difficult to target from the periphery. Compared to systemic therapy local targeting provided by brain implants promises a better efficacy. Moreover, brain devices can deliver a functional and reversible effect, which dramatically increases their safety.

*Springer Science+Business Media Dordrecht/Nanoethics, 2, 2008, p. 241–249, Ethical, Legal and Social Aspects of Brain-Implants Using Nano-Scale Materials and Techniques, Francois Berger, Sjef Gevers, Ludwig Siep & Klaus-Michael Weltring, Received: 29 January 2008/Accepted: 18 September 2008/Published online: 11 October 2008, with kind permission from Springer Science+Business Media Dordrecht 2012.

F. Berger
Department of Brain Nanomedicine, INSERM, Grenoble, France
e-mail: francois.berger@ujf-grenoble.fr

S. Gevers
Department of Social Medicine, Academic Medical Center, Amsterdam, The Netherlands
e-mail: j.k.gevers@amc.uva.nl

L. Siep
Philosophisches Seminar, Westfälische Wilhelms-Universität, Münster, Germany
e-mail: siep@uni-muenster.de

K.-M. Weltring (✉)
Gesellschaft für Bioanalytik Münster e.V., Mendelstr. 11, 48149 Münster, Germany
e-mail: weltring@bioanalytik-muenster.de

S.A. Hays et al. (eds.), *Nanotechnology, the Brain, and the Future*,
Yearbook of Nanotechnology in Society 3, DOI 10.1007/978-94-007-1787-9_11,

Since the work of Alim Louis Benabid in 1987, it has been demonstrated that high frequency electro stimulation is able to do the same as an irreversible lesion by reversibly inhibiting brain functions in the specific location targeted by the electrodes. This concept proved highly efficient in the treatment of the Parkinson's disease tremor and was extended to other targets providing a major cure for previously incurable neurological illnesses such as dystonia or epilepsy. Long-term effects and safety were clearly demonstrated, with a follow-up of 20 years for older patients. More recently, therapeutic efficacy was also observed in major depression and obsessive compulsive disorders resistant to chemical therapy. Forty thousand patients all around the world were treated with the new technology (Schwalblow 2008; Voges et al. 2007).

In what sense are implants and stimulations changed by nanotechnology? In medical applications the implementation of nano-modifications on micro-devices will promise major progress, because it positively influences resistance, biocompatibility, physiological integration in the tissue and implementation of multifunctionality from diagnosis to therapy. Research will tend to decrease the size and to add nano-materials to the micro-devices.

Miniaturization is a prerequisite for the second generation of devices mandatory for improved targeting of a specific area in the brain or for a complex stimulation associating several locations at the same time to treat more complex diseases using multi-target tri-dimensional brain stimulation (Elder et al. 2008). The neuroprosthetic field is another application of brain implants which supports nano-modifications (Stieglitz 2007). First applications, which have been carried out recently in animal and humans, demonstrated that patient thoughts can be recorded in paraplegic patients to guide an external device. To do that, silicon microelectrode arrays have been developed for in vivo recording of neural activity, but this has been hampered by scar tissue formation at the site of implantation. Therefore, increasing biocompatibility and improving the interfaces between technical devices and the biological environment is a crucial research direction (Szarowski et al. 2003). The unique properties of single-wall carbon nanotubes (SWNTs) (Malarkey and Parpura 2007) or other nano-scale coatings for example may have a tremendous impact in the future developments of micro-systems for neural prosthetics (Silva 2006). Several publications have demonstrated that such well-controlled nano-scale coatings have the potential to significantly improve the compatibility and performance of these implants (Patil and Turner 2008). Another example are nano-beads which have been demonstrated to be able to reach the brain after intravenous injection. We can imagine that these beads could be used to target a specific brain area using, for example, magnetic targeting and localisation. Magnetic stimulation can be delivered outside the brain; the result of magnetic stimulation is a local current. This concept could provide a direction for the next "nano-electrodes".

The introduction of nano-modifications on the classical and validated devices used in patients for brain neuro-stimulation should support the development of methodology for in vivo to in vitro anticipation of the toxicity of nano-modifications in humans (Linkov et al. 2008). A major issue will be to validate specific tests addressing this issue as well as the indispensable methods to control the systemic diffusion of the nano-material in the body.

To sum up, nanotechnology can lead to major progress in neuro-stimulation of the brain, because nano-tools will help to improve the targeting and avoid lesions. Effects may be recorded immediately and application is tightly controlled. Thus patients are protected from unwanted or negative side effects in a better way due to the miniaturization process and to the ability to modify neuronal pathology in a functional non-lesional way. In addition, brain implants are applied to a growing number of pathologies from neurodegenerative diseases such as Parkinson disease, movement disorders such as dystonia or tremor, psychiatric disorders such as depression and obsessive disorders or pain. Further- more, the new field of prosthetic device therapy was recently introduced in human (neuro-surgery), with a first patient suffering from paraplegia treated in 2007. These are all major pathologies in European countries with a growing number of cases especially due to the aging population. Nanotechnology promises to bring considerable benefits to the growing number of patients suffering from brain diseases and degenerative processes, if the location in the brain is known.

Since help for suffering patients is an inherent duty in all moral codes[1] and theories (Beauchamp and Childress 2001), there are obviously many good ethical reasons to proceed with these technologies. However, brain surgery is already an ethically controversial form of therapy. It is, of course, highly invasive and touches a very complex and sensitive organic system, which forms the basis for the most important mental states and activities of the human being. Public concern regarding the development and application of such therapies and techniques is quite natural. These concerns are likely to increase in the case of neuro- stimulation, the more so, if it takes place by means of nano-containing implants. Every researcher should be aware of this and consider the impact of his research on some serious ethical and legal and social aspects (ELSA).

The modern use of the term "Ethics", especially in a European context, reflects the plurality of legal, moral and religious norms and convictions within a community of pluralistic democratic societies. It includes not only norms of what is legally or morally forbidden—especially moral permissions are, of course, in many cases controversial—but also the duty not to risk the advantages which may be achieved by technological progress. However, the development of safe techniques to achieve these advantages often has to pass experimental phases where individuals or groups bear risks or even suffer damage which are not outbalanced by the benefits for them. In these situations not only the legal order but also the principles of "common morality", especially autonomy, welfare, and justice come into play (Beauchamp and Childress 2001). Moreover, in the long term these techniques may change our way of life, our body and our environment in a way, which few people really want. Public concern is quite understandable although "slippery slope" arguments are often speculative. Therefore foresight and the assessment of consequences to common morality and public acceptance is a task for ethical considerations in projects dealing with the nanotechnological improvements of neuro-implants.

[1] Compare for instance the European Convention on Human Rights and Biomedicine, Preamble and Art. 3 (Convention for the Protection of Human Rights and Dignity of the Human Being with regard to the Application of Biology and Medicine: Convention on Human Rights and Biomedicine 1997).

There are many issues that can be raised in this context, both with a view to the immediate future and in a more extended time perspective. In this paper, we focus on the following questions, the first and second relating to short and medium-term, the third rather to long-term developments.

A. Are there special issues regarding the testing of micro-nano-devices in clinical trials?
B. What are the individual risks for safety, autonomy and self-understanding involved in the implantation of micro devices with nano- components into human brains?
C. Do micro-nano-implants constitute a new step on the way to enhancement beyond other forms of brain therapy (pharmaceutical or technical)?

The following sections will describe the issues related to the above questions to sensitize nanotechnologists to the ethical, social and legal frameworks they have to know and deal with once they want to apply their technologies to brain implants.

## 11.2 Issues Concerning the Testing of Nano-Enabled Brain Devices

Before a new medical product or procedure is offered to patients on a regular basis clinical research is required to evaluate its safety and effectiveness. For nano-enhanced brain implants the following three issues are important to consider:

(a) Which regulatory framework applies to nano- brain devices and to what extent?
(b) What does this imply and what does it mean for the so-called phase-1 studies of humans in particular?
(c) Patients who need brain implants may not be competent and able to give informed consent; does this exclude them from clinical research with nano-enabled devices?

To answer these questions, it is necessary to take into account the relevant national and international regulations and standards on drugs, medical devices and clinical research. Here, we will focus on international and in particular European standards.

(a) In general, ‘nanomedicine’ does not stand for a new category of health care products; it is rather a new enabling technology used in the design and production of medical devices and pharmaceuticals (European Commission, European Technology Platform on NanoMedicine 2006a). The problem is that the regulatory framework is quite different, depending on whether the product is a pharmaceutical (i.e. drug; medicinal product) or a device.[2]

[2] We are discussing here the regulations on the marketing of both kind of products; when a drug or a device has been introduced on the market and it turns out to be unsafe, the manufacturer is in both cases liable for physical injury caused by the product; this product liability is stricter than liability based on negligence.

Basically, on the basis of the relevant EU directives, pharmaceuticals need preliminary market authorisation (Directive 2001/83); this authorisation will be given only, if the manufacturer can demonstrate that the medicinal product has been investigated in a clinical trial. Devices, however, can be introduced without previous authorisation (Directive 1993/42). The manufacturers themselves are responsible for the performance and related safety of their devices on the basis of an appropriately conducted and systematic risk management procedure, taking into account all the relevant information they can gather. For devices falling in a higher risk class, like implantable devices, their conclusions must be examined and confirmed by an independent third party ('Notified Body'). Usually, implantable devices will need some kind of clinical evaluation, which may include a clinical trial, before the declaration can be issued that it meets the official requirements. It is up to the Notified Body that is responsible for that declaration to decide whether the clinical evaluation carried out by the manufacturer can be considered sufficient.

So, to some extent, the regulations of pharmaceuticals are of a different nature compared to the regulations of medical devices. While in the first the protection of the patient/consumer/research subject is predominant, in the latter the functioning of the market and the process of innovation have been taken into account to a greater extent, leading to a less restrictive regime. The result of the co-existence of these two regimes is a certain tension between the level of protection required on the one hand, and the formal requirements that are based on a presupposed dichotomy between drugs and devices on the other (whereas in reality there is a grey area between the two). The European Technology Platform document on Nanomedicine rightly points out that this situation calls for improved collaboration between regulators responsible for medical devices and medicinal products, as integration of competences might be required for complex nanotechnology based products (European Commission, European Technology Platform on NanoMedicine 2006b) (An example of this is a drug eluting stent.)

Since, from a legal point of view, it matters whether a product should be considered a medicinal product or a device, the relevant EU directives on pharmaceuticals and on medical devices contain definitions for that purpose. The definition of a pharmaceutical is very wide, and furthermore in case of doubt (when a product has both the characteristics of a device and a pharmaceutical) the rules on pharmaceuticals take precedence. So, unless a micro-implant with nano- material is obviously 'only' a device, it will be considered a pharmaceutical; the latter applies anyway, if it is presented as a pharmaceutical.

Although the relevant regulations deal in a different way with 'drugs' and 'devices' as to whether clinical research is essential, in practice that difference may not exist when nano brain implants are used. It is hard to see that such implants would be introduced into clinical practice without prior research on humans. Even if the Notified Body did not require such research, hospitals and other health care institutions would not be likely to proceed to such new and risky procedures without previous systematic testing in a clinical situation.

(b) Basically, if a clinical trial takes place, the international standards for human subject research will apply, e.g. Declaration of Helsinki, Additional Protocol to the Biomedicine Convention on Biomedical Research. On the whole, these standards are quite substantive (for example, they require approval by an ethics committee of the research protocol, on the basis of a complete description of the study, including detailed information on the risks and burdens that it may entail for research participants and on the ways participants will be selected and requested to participate). However, in the case of a drug trial the applicable regulations are much more extensive and elaborate, since clinical trials must meet all the requirements laid down under the EU clinical trial directive (2001/20) and the EU good clinical practice directive (EU 2005/28); for instance, the research protocol must also be submitted to a competent national authority which may object to the study, even if the ethics committee will approve it.

What does this mean for so-called phase-1 studies? The wording 'phase 1' has its origin in clinical trials with medicinal products/drugs, but is also often used in other research settings where a new procedure or device is applied for the first time to human beings. In general, phase-1 studies are intended primarily to assess the safety of the new product or procedure. In addition, they are expected to yield the first data about its potential effectiveness in humans. In principle, phase-1 studies should be done with healthy volunteers, unless this raises substantive problems.

It seems obvious that brain implants cannot be tested on healthy volunteers (because the burdens and risks would be disproportional), and that therefore the research subjects are likely to be patients with some kind of brain disease. As a consequence, the safety of the implant has to be tested as much as possible before by appropriate in vitro or animal testing.

If the product to be implanted is to be considered a pharmaceutical/medicinal product, the EU regulations and the documents referred to therein will require animal testing before the product is used in a clinical trial (the Investigators Brochure to be submitted to the Ethical Review Committee must contain information on pharmacokinetics and product metabolism in animals). If the product is rather a medical device according to the meaning of the relevant EU directive (1993/42), then it would seem to depend on the 'Notified Body' (see above) whether appropriate clinical evaluation has to include previous animal testing. Since nano-materials have new chemical properties one can expect that nano-devices will also require toxicology and safety test including animal models.

(c) Patients in brain implant trials might or might not be incompetent (for instance most Parkinson patients can still make decisions).

In the latter case, the general rule is that if the research intervention is 'therapeutic' (i.e. one may expect that it will directly benefit the patient), it can be carried out, provided that the legal representative of the incompetent person gives their consent. See for instance the most authoritative document on human subject research in Europe, viz. the Additional Protocol to the Convention on Biomedicine and Human Rights on Biomedical Research (Council of Europe 2004). This protocol further stipulates that 'research of comparable effectiveness cannot be carried out on individuals capable of giving consent.'

It is very likely that a research project on a brain implant will be 'therapeutic', at least if the product is implanted in a person who needs it on a medical basis, even if the study is a so called phase-1 study. If, for some reason, the intervention cannot be considered 'therapeutic', then strict requirements apply. First, incompetent patients can only be included if the study cannot be done with competent individuals. Second, as to the level of burden/risk to be allowed, if 'the research has not the potential to produce results of direct benefit to the health of the person concerned' it should entail only 'minimal risk and minimal burden' (Council of Europe 2004). The EU clinical trial directive (2001/20) is even stricter, since if there are no benefits for the research subject, there should be no risk at all; this provision only applies however, when the product that is tested must be considered a pharmaceutical or medicinal product, rather than a device.

The regulations and ethical considerations described above have to be taken into account when estimating the ethical and regulatory problems of using nano-materials in brain therapy.

## 11.3 Risks and Possible Damage(s) Involved in Using Nano-Devices in the Brain

There are two main aspects regarding this question:

(a) Problems of safety, data protection and control over one's own body.
(b) Conflicts between a possible change of personality and the patient's autonomy and informed consent

(a) The brain functions like a "holist" network and the change of specific functions may have consequences, which are hard to foresee, to control and to reverse. With reference to the techniques used so far, the problems of side effects like depression (suicide), apathy, hypomania, gambling etc. have been reported with deep brain stimulation (less than 1% of patients), mainly caused by the bad position of the devices, for example implanted outside the motor area. Beside a function in motor regulation, the subthalamic nucleus, which is the target for Parkinson disease, is also involved in mood regulation explaining some psychiatric effects (Seijo et al. 2007; Tabbal et al. 2007).

In addition, technical devices may malfunction or become overused. It has to be discussed whether nano-tools used to improve implants or instruments for neuro-stimulation can diminish these risks. On the one hand, it is easier to target and survey them. On the other hand, nano-devices may get out of control more easily than other technical devices due to their size and complexity. These problems should be taken into account in any further technical development.

Particular problems may arise for the patient's autonomy and self-determination. Risks may be higher if they transmit data and receive signals beyond the control of the "owner" (questions also discussed regarding telemedicine, (Siep 2007) theranostics etc.). There has been a good deal of discussion regarding the right to "informational self-determination" especially in the field of genetic diagnostics. If therapeutic devices and micro-transmitters using nano-particles were more difficult to control by patients and therapists, the self- determination of the patient could be endangered, not only concerning information and data processing etc. but also with regard to the control over his or her own body (European Group on Ethics in Science and New Technologies (EGE) 2005).

(b) Like other brain-implants, implants with nano- elements may pose difficult problems regarding possible personality changes in the person.[3] For instance a psychiatric patient may undergo a change from being a deeply depressed person to a extremely cheerful one. To begin with, such a change would only seem legitimate, if behaviour and personality were abnormal, and if this abnormality was beyond the threshold.

Specifying these conditions and identifying real improvements is not an easy task. It requires, for instance, the specification of a "normal" degree of cheerfulness. In addition, normality is a standard or average concept whereas personality is characterized by uniqueness. There are scales and thresholds of "abnormality" regarding rigidity, tremor, anxiety, compulsion, mania, aggression, sadness, hope, indifference, creativity, and mood. But can there be a generalization when it comes to very different characters? From the point of view of modern ethics, these questions should not be decided according to general quantitative standards but by a precise analysis of the singular case and the particular person. On the other hand, standards drawn from a broad range of cases may function as a guideline for individual cases.

Regarding the patient's autonomy the question will arise, whether the cured person is the same as before the treatment. Is she ever capable of consenting to her treatment with hindsight? This question is not limited to cases of incompetent patients. Even competent ones may not be able to decide about their future as a "different person". Which degree of paternalism is allowed in curing a person with a nano-enhanced micro-implant? Are there means of identifying in advance which degree of suffering a patient would accept to maintain his/her unique personality? How should the value of "remaining the same" be weighed against "getting rid of pains and disablement"? Is the wish to reverse a procedure a proof of autonomy if the person has changed? On the other hand, one may refer to similar cases where such a change of personality seems to be accepted, for instance cases of voluntary change of gender.

While it is not clear whether nano-tools pose special problems with regard to these questions, every researcher and therapist dealing with far-reaching changes of psychical traits in their patients should consider them seriously. Although nano-researchers may not, at first sight, relate directly to such issues, nonetheless their work may have considerable consequences in this context for instance concerning the lifetime of implants or the accessibility of brain areas. "Making aware" is an

[3] Similar questions of personality changes have been discussed regarding brain-tissue transplantation. Quante (1996) cf. also Gharabaghi and Tatagiba (2008).

important task of ethics and ethics boards or committees. Of course, such a list of ethical questions for nano-researchers may include issues, which are relevant in other fields of medical technology as well.[4]

## 11.4 Micro-implants in the Brain and the Question of Therapy vs. Enhancement

The discussion about the permissibility or "desirability" of enhancing human faculties by drugs or implants has a special emphasis on brain-enhancement.[5] Since the brain is the bodily "center of control" and the basic condition for cognitive performances, its enhancement has far-reaching consequences for human interaction. Most of these performances, e.g. the faculties of memory and attention, are already within the reach of psycho-pharmaceuticals. But the use of implants with nano- elements might improve the "fine-tuning" of these influences, not all of which might be desirable (Bruce 2006; Roco and Bainbridge 2003).

Some of these consequences are of a social character. If somebody's faculties are enhanced beyond the normal level, equal opportunity and other principles of "justice as fairness" are endangered, especially if medical techniques are expensive and not paid for by the public insurance system. Although we certainly do not live in a perfectly fair society, increasing the distance between the "improved" and the normal human beings collides certainly with accepted criteria of equality and solidarity. It may be less likely that healthy persons will use brain implants for enhancement purposes than drugs. However, if the "transhumanist" arguments[6] are increasingly accepted, we may see more interests and pressure in this direction which would have serious implications. If humans with radically improved brain capacities ever became a social fact rather than merely a future speculation, humanity might split up in different "sub- species" which could even have difficulties understanding each other. Such consequences would challenge widely accepted basic principles of equality, autonomy and even non-violence. This, of course, is a long-term perspective.

Another part of the enhancement problem has been touched on already. If a depressed person becomes a cheerful one, is the person cured or enhanced? If a person with a hearing defect is turned into an extraordinarily good hearer, is this the best outcome of the therapy? If defects can be cured, why not aim for the best possible quality of the function, which is damaged? But will the "cured" person be able to cope with her new capability? This may raise problems even for the tuning of devices: which degree of performance should be tuned in or programmed?

[4] Another, long-term issue concerns the question whether the self-understanding and self-feeling within a partly mechanised and robotised body leads to an estrangement of the traditional human self understanding. See Bruce and Ach (2009) and Siep (2009).

[5] For general problems of enhancement cf. Parens (1998), Fuchs (2002), Siep (2003).

[6] For a discussion of the transhumanist arguments see Siep (2006).

The outcome of a technical perfectionism of this sort may be problematic at least in three aspects:

(a) The person may feel estranged in her own body. She may not "recognize" herself any more and may lose her familiar sensations and behaviour.
(b) She may develop problems with her environment. Already the questions of autonomy and consent may be part of this problem. But there may be other difficulties with family and social environment.
(c) The problems of social equality mentioned above may emerge. A "run for the best brain" may be triggered off.

There are other social aspects, in particular concerning the change of acceptance and expectations. For instance, the expectations regarding the hearing qualities of elderly persons have certainly changed. This may have social consequences especially if the age of retirement is increasingly postponed. With new possibilities to improve the faculties of the aging brain, new forms of competition and resulting stress may arise. Sometimes, however, reasonable social practices are developed regarding problems which may seem impossible to solve from the present ethical perspective.

## 11.5 Final Comments

Neuro-stimulation by brain implants has been worked on for decades and it provides an alternative therapy to drug treatment of neurological diseases. The demand for such treatment will increase in the future due to the aging population, which will cause an increase in Parkinson's and Alzheimer's disease, for example. Nanotechnology is an important platform technology to assist in meeting this demand because it will add new features like improved biocompatibility, smaller size, and more sophisticated electronics to neuro-implants improving their therapeutic potential. Therefore, it would be unethical simply to stop or slow down Research and Development of nanotechnological improved neuro implants.

However, development of brain implants by itself touches many ethical, social and legal issues, which apply in a specific way to devices enabled or improved by nanotechnology. Therefore we have discussed in this paper the short term problems of testing and clinical trials (A), the short and medium- term questions of risks in the application of the devices (B) and the long-term perspectives related to problems of enhancement (C).

Researchers working with these new technologies have the obligation to thoroughly consider such issues and consequences before they start and while they carry out their projects. To discuss possible ethical implications with ELSA experts early on in a project may relieve the pressure for regulatory bodies to be proactive in response to the high speed of the development, because normally regulations are based on long term learning and experience. The consequence of an open and early communication process between researcher, regulatory bodies and ELSA committees will be a mutual learning process, which can serve to identify and handle ELSA issues as they arise and thereby accommodate the fast development. This will help to prevent ethically, socially and legally non-acceptable developments for the benefit of patients—and also the success of the European health economy.

Regulations are not the only answer, and too much regulation can even be ethically unacceptable, because it might make research too expensive. This would result in a growing dependency on big pharmaceutical companies. Therefore a social debate involving health insurance companies and patient interest groups is needed. It must tackle the problems of who finances the development and enables the companies to develop a new nano-enhanced implant, even if it is more expensive than a less effective drug.

Due to the speed of development in nanotechnology new possibilities open up fast. This raises the question of how to test the safety and efficacy of emerging new devices and therapies with the traditionally low number of patients in clinical trials for brain therapy. Improved biocompatibility will enable more applications, but it is difficult to test the risk of long-term effects. Nano- coatings or particles as such might bear a risk of toxicity. A possible solution could be the development of new in vitro test systems not only for toxicology studies of new materials but also for neuronal functions in cell cultures. More research is needed to test the value of nano-structured 2D and 3D scaffolds for neuronal cells in this regard. This would reduce the number and risk of clinical and even of animal trials which are ethically problematic as well.

Nanotechnology will almost certainly lead to more sophisticated devices which provide better targeting with fewer side effects on the one hand but also the possibility of more external control, data transmission and misuse on the other. They might also lower the barrier to enhancement, because the devices are more biocompatible and therefore seem less risky. To avoid these effects, the involvement of ELSA experts in foresight exercises and a close monitoring of the field is needed to identify such developments and to initiate a social debate concerning them. This will require the set-up of competent ELSA boards at the European level, with enough competence in the different fields required for achieving their tasks. The benefit would be a transparent, ethically and socially acceptable development of nano-enhanced brain therapies.

**Acknowledgements** This paper is in large parts the result of intensive discussions between members of the ELSA Board and Scientists of the European Network of Excellence Nano2Life.

## References

Beauchamp, T.L., and J.F. Childress. 2001. *Principles of biomedical ethics*, 5th ed. Oxford: Oxford University Press.

Bruce, D. 2006. Nano2Life ethics – A scoping paper on ethical and social issues in nanobiotechnologies. In *Nano-bio-ethics. Ethical dimensions of nanobiotechnology*, ed. J.S. Ach and L. Siep, 63. Münster: LIT.

Bruce, D., and J.S. Ach. 2009. "Improving Human Performance"? Sceptical remarks on the idea of an 'improvement' of human performance features through converging technologies. In *Ethics in nanomedicine*, ed. J.S. Ach and B. Lüttenberg. Berlin: Lit.

Convention for the Protection of Human Rights and Dignity of the Human Being with regard to the Application of Biology and Medicine: Convention on Human Rights and Biomedicine. 1997. http://conventions.coe.int/Treaty/EN/Treaties/Html/164.htm.

Council of Europe. 2004. *Protocol on biomedical research*. Strasbourg: Council of Europe.

Elder, J.B., C.Y. Liu, and M.L. Apuzzo. 2008. Neurosurgery in the realm of 10(−9), part 2: Applications of nanotechnology to neurosurgery—Present and future. *Neurosurgery* 62(2): 269–284. discussion 284–5. doi:10.1227/01.neu.0000315995. DOI:dx.doi.org 73269.c3 DOI:dx.doi.org.

European Commission, European Technology Platform on NanoMedicine. 2006a. *Nanotechnology for health-strategic research agenda*, 2 1.1. Luxembourg: Office for Official Publications of the European Communities.

European Commission, European Technology Platform on NanoMedicine. 2006b. *Nanotechnology for health-strategic research agenda*, 26 3.4. Luxembourg: Office for Official Publications of the European Communities.

European Group on Ethics in Science and New Technologies (EGE). 2005, March 16. Ethical aspects of ICT implants in the human body. In *Opinion Nr. 20*, ed. S. Rodotà and R. Capurro, Brussels. Retrieved from http://ec.europa.eu/european_group_ethics/docs/avis20_en.pdf.

Fuchs, M. 2002. *Enhancement. Die ethische Diskussion über biomedizinische Verbesserungen des Menschen*. Bonn: DRZE-Sachstandsbericht.

Gharabaghi, A., and M. Tatagiba. 2008. Neuroprothetics and cognition and personality. In *Nanobiotechnology, nanomedicine and human enhancement*, ed. Johanns Ach, and Beate Luttenberg, 77–88. Munster: Litverlag.

Linkov, I., F.K. Satterstrom, and L.M. Corey. 2008. Nano-toxicology and nanomedicine: Making hard decisions. *Nanomedicine* 4: 167–171.

Malarkey, E.B., and V. Parpura. 2007. Applications of carbon nanotubes in neurobiology. *Neurodegenerative Diseases* 4(4): 292–299. doi:10.1159/000101885.

Parens, E. 1998. *Enhancing human traits: Ethical and social implications*. Washington, DC: Georgetown University Press.

Patil, P.G., and D.A. Turner. 2008. The development of brain-machine interface neuroprosthetic devices. *Journal of the American Society for Experimental NeuroTherapeutics* 5: 137–146. doi:10.1016/j. DOI:dx.doi.org nurt.2007.11.002 DOI:dx.doi.org.

Quante, M. 1996. Hirngewebstransplantation und die Identität der Person: ein spezifisches ethisches Problem. In *Cognitio humana-dynamik des Wissens und der Werte, XVII*, ed. Ch. Hubig and H. Poser, 1459–1466. Deutscher Kongreß für Philosophie Leipzig 1996, Workshop-Beiträge Band 2. Berlin: Akademie-Verlag.

Roco, M.C., and W.S. Bainbridge. 2003. *Converging technologies for improving human performance: Nanotechnology, bio-technology, information technology and cognitive science*. Dordrecht: Kluwer.

Schwalblow, J.M. 2008. The history and future of deep brain stimulation. *Neurotherapeutics* 5(1): 3–13.

Seijo, F.J., M.A. Alvarez-Vega, J.C. Gutierrez, F. Fdez-Glez, and B. Lozano. 2007. Complications in subthalamic nucleus stimulation surgery for treatment of Parkinson's disease. Review of 272 procedures. *Acta Neurochirurgica Supplement (Wien)* 149(9): 867–875. doi:10.1007/s00701-007-1267-1. discussion 876.

Siep, L. 2003. Normative aspects of the human body. *Journal of Medicine and Philosophy* 28(2): 171–185. doi:10.1076/jmep. 28.2.171.14208.

Siep, L. 2006. Die biotechnische Neuerfindung des Men- schen. In *No body is perfect-Baumaßnahmen am menschlichen Körper–Bioethische und ästhetische Aufrisse*, ed. J.S. Ach and A. Pollmann, 21–42. Bielefeld: Transcript-Verlag.

Siep, L. 2007. *Ethik und Telemedizin, Telemedizin–Innovatio- nen für ein effizientes Gesundheitsmanagement. AnyCare Schriftenreihe*, 65–77. Stuttgart: Georg Thieme Verlag.

Siep, L. 2009. Ethical problems of nanobiotechnology. In *Ethics in nanomedicine*, ed. J.S. Ach and B. Lüttmann. Berlin: Lit.

Silva, G.A. 2006. Neuroscience nanotechnology: Progress, opportunities and challenges. *Nature Reviews Neuroscience* 7(1): 65–74. doi:10.1038/ DOI:dx.doi.org nrn1827 DOI:dx.doi.org.

Stieglitz, T. 2007. Restoration of neurological functions by neuroprosthetic technologies: Future prospects and trends towards micro-, nano-, and biohybrid systems. *Acta Neurochirurgica Supplement (Wien)* 97(Pt 1): 435–442.

Szarowski, D.H., M.D. Andersen, S. Retterer, A.J. Spence, M. Isaacson, H.G. Craighead, J.N. Turner, and W. Shain. 2003. Brain responses to micro-machined silicon devices. *Brain Research* 5–983(1–2): 23–35.

Tabbal, S.D., F.J. Revilla, J.W. Mink, P. Schneider-Gibson, A.R. Wernle, G.A. de Erausquin, J.S. Perlmutter, K.M. Rich, and J.L. Dowling. 2007. Safety and efficacy of subthalamic nucleus deep brain stimulation performed with limited intraoperative mapping for treatment of Parkinson's disease. *Neurosurgery* 61(3): 119–127. discussion 127–9.

Voges, J., A. Koulousakis, and V. Sturm. 2007. Deep brain stimulation for Parkinson's disease. *Acta Neurochirurgica Supplement (Wien)* 97(Pt 2): 171–184. doi:10.1007/978-3-211-33081-4_19.

# Part III
# Enhancing the Brain and Cognition

# Chapter 12
# The Complex Cognitive Systems Manifesto

**Richard P.W. Loosemore**

## 12.1 Introduction

In order to understand the functioning of the brain, we need to have a reasonable understanding of the connection between processes and architectures described at the cognitive level and those described at the neural level. We need to know approximately how the psychology maps onto the neural hardware. And this need for clarity in the mapping from high level descriptions to low level mechanisms is not unique to brain science: there are many systems in the world—both natural and artificial—that are so complex that different types of explanation exist at different levels, and in each case it is vital to understand how the levels relate.

This is especially true in the context of nanotechnology and the brain. We are approaching a time when nanotech tools will allow us to intervene, augment or duplicate brain systems: but the mere possession of a microscopic screwdriver is not the same thing as knowing what will happen when said screwdriver is inserted into the circuit board of a supercomputer. Wielding the screwdriver must be backed up with theoretical knowledge of how the system works.

The poster child for high-level to low-level mappings must be the science of chemistry. We have a more-or-less coherent picture of chemical processes that starts down in the quantum physics of atoms and goes all the way up to the functional properties of enzymes. But this story about different levels of description in chemistry was assembled by an extremely reductive program of research in the hard sciences, and in this paper I am going to suggest that there is a serious structural problem with using that same type of reductionist scientific methodology to understand the relationship between high and low levels in the human cognitive system.

R.P.W. Loosemore (✉)
Mathematical and Physical Sciences, Wells College,
Aurora, NY 13026, USA
e-mail: rloosemore@wells.edu

S.A. Hays et al. (eds.), *Nanotechnology, the Brain, and the Future*,
Yearbook of Nanotechnology in Society 3, DOI 10.1007/978-94-007-1787-9_12,

In spite of the seemingly rapid advances being made in neuroscience and cognitive psychology—and especially in the field of brain imaging—I would argue that progress is far less significant than it might appear because there is an unrecognized technical problem in certain aspects of our research methodology.

This methodological problem—the *complex systems problem*, or *CSP*—is hard to understand and exceptionally difficult to verify. Moreover, it seems that the only way to solve the complex systems problem is to make substantial changes to our research methodology. But because the magnitude of these changes is so large, there is a strong incentive for cognitive scientists to deny that the problem exists.

To get a bird's-eye view of what the complex systems problem involves, imagine a dynamic system that contains many interacting components, and imagine that the net result of the component interactions is that the overall behavior of the system shows some noticeable regularity. Now suppose that there *does not exist* any mathematical analysis that will allow someone to start from the component interactions and predict the appearance of that overall regularity. Under such circumstances a number of serious problems would arise if researchers were to treat the system in the same way that "regular" systems (where there is almost always an underlying mathematical model waiting to be discovered) are treated. That set of serious problems is what is meant by the *complex systems problem*.

Rather than try to define the complex systems problem in its most general and abstract form, this paper will focus on how the CSP impacts one particular aspect of cognition: how behavior at the cognitive level should be explained in terms of events at the neural level. In other words, the neural-cognitive mapping.

The neural-cognitive mapping is the backbone of all cognitive science. The mapping question is whether the basic units of thought (concepts, symbols, etc.) are implemented in the brain as single neurons, redundant clusters of neurons, distributed patterns of activation across large networks of neurons, virtual entities with no direct connection to the hardware, or some other theoretical construct. Choosing between these rival interpretations is clearly important, if for no other reason than that data from brain scanning experiments can hardly be interpreted at all without making assumptions about how the psychological level is related to neural events.

There are two main reasons why it is important to present the complex systems problem in the context of the neural-cognitive mapping. One stems from a revolution that occurred in cognitive science back in the 1980s—variously known as *connectionism*, *neural nets* or *parallel distributed processing*—which turns out to have been a textbook illustration of how the complex systems problem can undermine research. The second reason to focus on the neural-cognitive mapping is that the connectionist revolution can be adapted to yield a solution to the complex systems problem: once we understand how the connectionist ideas were compromised by the CSP, it becomes possible to correct the damage and forge a new kind of connectionism that has the potential to overcome the CSP.

In the next section I will analyze the concept of a complex system and try to understand how the properties of complex systems might have an effect on the research methodology used by (among others) the various branches of the cognitive science community. This general account of the CSP then leads to an examination of the neural-cognitive mapping issue, especially in the form that it took during the connectionist revolution. Following that is a proposal for a generalized version of connectionism that is designed to be a partial solution to the complex systems problem. Finally, I will look at some of the broader implications of the CSP: the need for new types of software tools and the problem of overcoming the academic inertia that is perpetuating the problem.

## 12.2 The Complex Systems Problem

Systems of all kinds—whether natural or artificial—appear to come in two broad categories. On the one hand there are "regular" systems, which can be defined as those in which the components interact in ways that seem tractable. What defines a regular systems is the fact that we can write down equations or algorithms that describe how the components of the system interact, and these equations or algorithms can be solved to get a prediction of the system's overall behavior. As it happens, most of the systems of interest to science belong in the regular category, because in the majority of cases we have found ways to develop a convincing, rigorous argument that takes us from a description of the system's underlying rules to a prediction of its overall behavior.

The other category of system is labeled "complex," and it can be defined by our inability to solve the system's underlying equations. Roughly speaking, a complex system has an overall behavior that can only be understood by constructing a simulation of the system. If we simulate the system's underlying mechanism and then observe that the simulation behavior corresponds to the original system behavior, we might say that we "understand" where the system's behavior comes from—but this is a very different kind of understanding than the one we get if we solve the equations and prove that the overall behavior must arise, given those underlying mechanisms. Unfortunately, if we want to get any kind of explanation for the behavior of a complex system we seem to be stuck with either a simulation or nothing.

This is an extremely simplified (even contentious) definition of complexity, so one of my first goals in this section is to analyze the concept in sufficient detail to bring out the implications that it might have. Before getting into the detailed analysis, it might be worth summarizing the shape of the full argument that will eventually emerge:

- A system that is 100% complex cannot be *reverse engineered*—which means that its underlying mechanisms cannot be discovered by looking at its behavior alone. So if the universe contains any system of interest to science that happens to be 100% complex (and if we cannot directly inspect the system's underlying mechanisms), that system will be insuperably difficult to understand.
- If a system is *partially* complex, we would expect to find some aspects of the system that suffer from the same insuperable difficulty found in those that

are 100% complex. This means that some features of the system's behavior will be caused by underlying mechanisms that look as if they could never explain the behavior.
- Because of this difficulty, the process of finding a scientific explanation for a partially complex system will exhibit a kind of *global pathology*—which is to say, if our scientific methodology is driven by a search for locally-plausible models for every aspect of the system, we will never be able to integrate these locally-plausible models into a complete explanation.
- There is evidence that cognitive systems (including the human cognitive system and all human-level artificial intelligence systems) are, in fact, partial complex systems.
- Our current scientific methodology is deeply attached to the strategy of searching for locally-plausible models: but if cognitive systems are partially complex systems this strategy will result in an endless stream of local models that never fit together.

The main conclusion to be had from this argument is that the usual divide-and-conquer approach to science will not work for cognitive systems. We would be in a position somewhat akin to that of a group of people trying to lay square tiles on a floor that appears flat if examined locally, but which is actually embedded in a curved non-Euclidean space. Each tiler can start working from one position and be convinced that everything is going well, only to discover that large irregular gaps appear when the separate patches of tile encounter one another. If the global nature of their problem is not understood, it might seem that these gaps can be resolved by adjusting any pair of conflicting tile patches. In normal space these pairwise adjustments would eventually converge on a global solution, but if the space is curvilinear, these pairwise changes will never converge on a solution.

### *12.2.1 Characterizing Complexity*

This might seem to be a good point at which to give the accepted definition for the term "complex system." Unfortunately this is a nontrivial task, because even complex systems scientists have not been able to reach a consensus definition (Mitchell 2008). In fact, some critics of the field (Horgan 1995) have used this state of confusion to argue that no proper definition will ever emerge.

In order to get past this difficulty it is necessary to find a way of defining the term that includes an explanation of why it should appear to be such a fractured idea at the moment. Reconciliation between the competing interpretations of the concept can best be achieved by clarifying why it is that there is so much competition and disagreement.

Accordingly, I will now try to develop an extended account of what complex systems are, together with an explanation for why we currently have several conflicting interpretations. Then, having laid the groundwork, we can move on to ask what effect this idea might have on the scientific methodologies we use to study such systems.

#### 12.2.1.1 Systems

A *system* consists of a number of *components*, and these components engage in certain *interactions* with one another.

Every system exhibits an overall *behavior* that is a result of the interactions between its components. Viewed from the outside, the behavior of the system is the thing that is most apparent, whereas the components and/or interactions might initially be hidden. From a cause-effect point of view the behavior is an effect, while the components and interactions are the cause.

It is often convenient to use the term *mechanism* to stand for a combination of the components and their interactions. In that case, we would say that the (observable) behavior of a system is a consequence of the (often hidden) mechanism that underlies the behavior.

Strictly speaking there are two features of a system that are caused by the mechanism: the behavior proper, and the *form* (shape, structure, or state) of the system. The first is a dynamic feature, the second is less time-dependent. For the sake of narrative convenience the term *behavior* will often be used to signify both of these. So "behavior" can be used to cover both the dynamic and static features of the system.

#### 12.2.1.2 Regularities

When we talk about a system's behavior what we usually mean is a *regularity* in the behavior. These two are not the same, because a given system can have many different regularities in its behavior, and these can exist at many levels of description. The behavior of hurricanes, for example, includes one regularity that is the spiral shape, but there are other regularities like the role played by hurricanes in the world's ecosystem, or the typical regions in which they occur, or the typical time history of a hurricane.

There is no reason why all of the different regularities that we might see in a system have to be of the same type, or follow the same rules. In fact, the concept of a regularity is observer-dependent and often quite subtle, so it is not very meaningful to talk about "all" of the regularities possessed by a given system. A regularity is a construct that we see in the behavior.

This distinction between system, behavior and regularity is important, but for convenience we often blur the distinction by using the words "system" or "behavior" to describe one particular regularity. So for example, we would say that Newton used his inverse square law of gravitation to explain the motion of planets in the solar system—but what we really mean is that he explained a certain cluster of regularities in the behavior of the solar system (namely, Kepler's Laws). Other kinds of regularity, like the pattern of temperatures on the surface of the planets, were not addressed by his theory.

A regularity is nothing more than a non-random *pattern* in the behavior of a system. The concept of a pattern is quite vague, so regularities cannot always be captured in concise laws like those discovered by Kepler. In some cases we might observe a pattern in a system's behavior but find it hard to write down an objective, closed-form description of the pattern. In spite of this, though, elusive regularities can still demand that we give them a scientific explanation.

#### 12.2.1.3 Explanation

Explaining a regularity entails much more than just writing down the correct underlying mechanism. Before Newton finished work on his law of gravitation some of his contemporaries had already suspected that there was an inverse-square force of attraction between the planets and the sun. But at that point this was just a candidate mechanism, because nobody could prove that this mechanism led unambiguously to a prediction that the orbits would follow Kepler's laws.

At the risk of laboring a point that is surely second nature to any scientist, the process of finding an explanation involves two steps, in which a candidate mechanism is first generated (the hypothesis), and then the candidate is used to construct a chain of inference that leads to a prediction of the behavior. This means that we first go "backward" from behavior to candidate mechanism (the conceptual brainstorming that Newton and others did before they guessed that there might be an inverse-square attractive force), and then we turn around and go "forward" from candidate mechanism to behavior (which in Newton's case involved the invention of the calculus, so he could solve the inverse-square force equation).

The reason for stating the obvious here is that this backward step from behavior to candidate mechanism is very significant but often underestimated. Folk wisdom portrays the art of scientific discovery as an inspiration-plus-perspiration effort, in which the initial inspiration is a blinding flash of insight that enables the scientist to come up with the (in hindsight, correct) hypothesis. Then comes the perspiration phase when the implications of the hypothesis are rigorously elaborated to show that the mechanism does lead to the observed behavior. But by shrouding that first step in the concept of "inspiration" we do a disservice to the very concrete cognitive processes at work when a hypothesis is created.

We know little about this backward pass from observation to hypothesis, but it seems safe to say that—taking Isaac Newton as an example once again—many features of the behavior of planets, moons and apples contributed to a chain of (mostly unconscious) clues that pointed toward the idea that objects falling on earth were connected to planetary orbits. Newton came up with a candidate mechanism that explained Kepler's laws not by magic, luck, divine intervention or blind guesswork, but by being sensitive to many factors that pointed toward the correct mechanism.

#### 12.2.1.4 Forward Path as One-Way Street

One surprising feature of the systems studied by scientists over the last 300 years is that in the overwhelming majority of cases there exists a rigorous chain of inference that goes from the candidate mechanism to the predicted behavior.

This may seem a trivial observation, but it only seems trivial if we assume that explanations are always there to be found. There is really nothing necessary about the existence of such proofs: it is an empirically interesting fact about the universe that so many of the systems of interest to science turn out to have concise, provable connections from candidate mechanism to behavior. There is no reason why this should always be the case: there is nothing in the structure of the universe that guarantees that every mechanism is connected by a clean proof to the behavior it gives rise to.

In particular, there is no reason to assume that *if* a regularity exists in the behavior of some system, *therefore* a deducible connection from the mechanism to the regularity can be found. Naive intuition might lead us to suppose that if a system behaves in some elegant, structured way, this is a sign that somewhere beneath the surface there exists an elegant explanation for the behavior. But however compelling this might seem—and however frequently it turned out to be true in the history of science—there is nothing logically necessary about the existence of a compact explanation, given the existence of a regularity.

Are there any examples of systems where a behavioral regularity exists, but no explanation can ever be had? This is a troublesome question, because we can never be sure that *no* explanation will ever be possible. We might suspect a system of being beyond explanation in this way, but there is always a chance that we will be surprised, tomorrow, by an unexpectedly new and elegant proof.

What we can say, however, is that to an omniscient scientist, with access to all the knowledge that could possibly exist, it might be knowable that there really are systems in this category. But with our limitations we can only say that we *suspect* some systems of having no explanation that connects the underlying mechanism to an observable behavior. As a first approximation, then, the definition of a complex system is that it *appears* to belong in this category. Another way to phrase this is that for all practical purposes we have to assume that no rigorous, analytical explanation will be possible.

If all of a system's behavior regularities appear to have no explanation, then we can say that the system is 100% complex. If some regularities are explicable in the usual way, while others seem beyond explanation, then the system is *partially complex*.

Numerous examples could be cited, but Stephen Wolfram (2002) has investigated a notably extensive set. Wolfram (2002), in fact, uses the term *computationally irreducible*, as an alternative way to refer to complex systems. This means that we cannot compute the behavior of the system by using an algorithm or equation that is more compact (more reduced) than the algorithm or equation that is encapsulated in the system's mechanism itself.

Notice that so far this definition only references the forward path of the explanation cycle: a system is complex if the route from mechanism to behavior is broken. This begs an interesting question that we now consider.

#### 12.2.1.5 A Break in the Backward Path

Suppose that a system is complex in the above sense, so there is no way for us to extract a prediction about its behavior given knowledge of its mechanism. Would it nevertheless be possible for some human (or machine) genius to look at the behavior and traverse the backward path from behavior to mechanism? Could someone intuit the correct mechanism that was behind a set of behavioral observations, even though they will never be able to develop a proof that their hypothesis was correct?

If this were possible it would significantly lessen the impact of complex systems. After intuiting the correct underlying mechanism, the scientist could then feed this into a computer simulation and use the simulation to prove that the candidate mechanism is valid. There would be no need for a mathematical proof or argument to go from mechanism to behavior.

It is difficult to collect information about whether this ever happens, because in practice the known examples of regular systems and complex systems tend to be treated differently. The regular systems that have dominated most of our science have always been subjected to that backward pass (for the obvious reason that this is an indispensable part of building an explanation). But in the case of many of the complex systems that have been studied, we have *invented* the mechanisms that define the system, so we have almost never tried to work backwards from known behavior to unknown mechanism.

We can go one step further and note that in those cases where a natural complex system has been studied, there are always some aspects of the system that are regular, so when the underlying mechanisms are not obvious, and have to be discovered, the discovery was initially done without using the complex aspects of the system. Having thus uncovered the mechanism by doing normal science on a (largely) regular system, the complex aspects of that system could then be studied in the same way that we study artificial complex systems—namely, by exploring the consequences of the mechanism using simulations.

Given this observation, and the fact that there are no known examples (at least, known to this author) of artificial complex systems whose behavior was written down first and then used to work backward to the mechanism that gave rise to the behavior, we can make the following conjecture:

- If a system does not have a logico-mathematical path leading from mechanism to behavior (if there is no "forward path" that explains the behavior), then the absence of this path means that the backward path (from behavior down to mechanism) cannot be traversed either.

What this conjecture says, in effect, is that when a scientist first encounters some observable behavior that needs to be explained, the process of generating a viable hypothesis about the cause of the behavior (the process of intuiting a candidate mechanism) will only be feasible if there exists a proof or argument that leads in the other direction, from hypothesis to behavior. If the system is such that the only way to get from candidate mechanism to behavior is via a computer simulation of the mechanism, then the subtle cognitive apparatus that scientists use to come up with a hypothesis about the system will not come into play. The process of scientific discovery cannot happen unless there is a non-simulation route that can be used to explain the behavior of the system.

Some obvious caveats need to be mentioned. If the system is simple enough, it might be possible for a simulation to be done in the head of a human scientist, and in that case the conjecture would not apply. Also, it would be feasible to work backward from behavior to mechanism in those cases where the number of possible mechanisms was sufficiently small that we could mount an exhaustive search through all the possible simulations.

This feature of complex systems is not given a great deal of attention because, as I explained above, nobody tries to invent complex systems that have a pre-ordained behavior. That kind of choose-the-behavior-first activity could be described as *reverse engineering* a complex system, and as a general rule it is simply assumed by complex systems researchers as being too obviously infeasible to be worth considering. The definition of complexity, after all, is that the behavior is emergent and therefore not what would have been expected from the mechanism—and this unexpectedness is normally assumed to imply that reverse engineering is the one thing that cannot be done.

#### 12.2.1.6 Recipe for Complexity

When using the terms "complex system" and "complexity" we need to be clear about whether we are referring to the behavior (effect), the mechanism (cause), or the connection between the two. Notice in particular that a system is never complex *because* it has certain behaviors, or certain mechanisms—rather, it is the relationship between these two that makes it complex. This is a frequent point of confusion in discussions of complexity. Calling a system "complex" is really a statement about whether we have any chance of discovering a theory that concisely describes what the behavior should arise from the mechanism.

Having said that, though, if we look at the empirically observed characteristics of known complex systems, it is possible to give a list of design ingredients, or aspects of the mechanism, that tend to make a system complex. If we see a system in which a plurality of these features occur, we could say that past experience teaches us that a complex relationship might exist between behavior and mechanism:

- The system contains large numbers of interacting computational elements.
- Simple rules govern the interactions between elements.
- There is a significant degree of nonlinearity in the element interactions.
- There is adaptation (sensitivity to history) on the part of the elements.
- There is sensitivity to an external environment.

When the above features are present and the system parameters are chosen so that activity does not go into a locked-up state or an infinite loop, then there is a high probability (though by no means a certainty) that the system will show signs of complexity.

#### 12.2.1.7 No Diagnostic Test for Complexity

One fact about complex systems is especially subtle:

- It is (virtually) impossible to find a compact diagnostic test that can be used to separate complex from non-complex systems, because the property of "being a complex system" is itself one of those behavioral regularities that, if the system is complex, cannot be derived analytically from the low-level mechanism of the system.

The "virtually" qualifier, above, refers to the fact that complex systems are not completely excluded from having behavioral features that are derivable from local mechanisms. So it is conceivable in principle that a system could have no explanation for a significant chunk of its behavior, while at the same time this lack of existence of an explanation could itself be a provable fact about the system. But although conceivable, this would be a bizarre situation—the proof would have to be rigorous in spite of the fact that it contained a concrete reference to a thing (the unexplainable regularities in the behavior) that could not be connected to the rest of the facts about the universe through any kind of formal structure or proof. It is hard to see how any proof could still count as a proof, while containing such an intangible.

This is an interesting result, because it means that complex systems are defined in such a way that the whole concept can only be coherent and internally consistent if the definition never becomes precise. One consequence is that when we debate whether a particular class of systems (e.g. intelligent systems) might be complex, the debate cannot include a demand for a definitive proof or test of complexity, because there is no such thing.

We can now begin to get some traction on the problem mentioned earlier, that different complex-systems researchers define complexity in different ways. If anyone did produce a perfect, closed-form definition of what complexity was, that definition would auto-destruct, so perhaps this impossibility of finding a perfect definition is having an effect on all attempts to build comprehensive definitions. Although this does not explain all of the variance to be seen across the different efforts to pin down the nature of complexity, it does shed some light on one source of confusion.

#### 12.2.1.8 Deniability

The fact that a complete definition of complexity is impossible leads to another consequence that has far-reaching implications. If complexity effects ever became a nuisance to some scientific community—for example, if those effects seem to imply that the community should adopt a radical change of methodology—the easiest strategy for the community to adopt is to deny the existence of the effects altogether. Denial is easy in this case because of the extraordinary difficulty of defining what is and is not a complex effect—and therefore what is or is not a *consequence* of complexity. It is always possible for the skeptic to insist that concrete proof be given that complexity effects are responsible for some situation. Then, in the absence of such a proof, the situation can instead be blamed on a mere difficulty with the understanding of a regular system.

These circumstances already seem to have arisen in economics and elsewhere (Waldrop 1992). Substantial conflicts have taken place between groups promoting the opposing viewpoints—for or against the importance of complexity—and it is arguable that these conflicts have been exacerbated by the difficulty of distinguishing complexity from regular system effects that have not been fully analyzed yet. If the above analysis is correct, and the indefinability of complexity is intrinsic to its nature, then the battle between these opposing viewpoints may be even more protracted than it usually is in a scientific paradigm conflict.

#### 12.2.1.9 Partial Complexity Versus Full Complexity

Most systems that are complex at all, are only partially complex: some aspects of their behavior can be understood as a regular consequence of some mechanism, while other aspects seem emergent.

This partial complexity can appear in many forms. One of these is that a system can have several levels of description, and some levels can be complex while others are regular. One example of a multilevel system is the well-known cellular automaton invented by J. H. Conway, known as "Game of Life" (Gardner 1970). In this system there are some very simple rules that determine whether each cell of a square grid is in the *on* or *off* state, at every cycle of a global clock. Certain patterns of initially-on cells will result in cyclic activity: the pattern repeats after a fixed number of clock cycles. There are many known patterns that have periodic behavior in this way, and from our point of view the behavioral regularities of interest would be the shape of the stable patterns and the period of each one. As far as we know, there are no forms of analysis that would allow us to input the rules of Game of Life and receive, as output, a prediction of the shape and period of all the stable creatures that can be found in this system.

At the level of the first batch of creatures to be discovered, the regularities are complex, because of their lack of derivability from the mechanism. But it is possible to use some of these basic patterns as ingredients for higher-order patterns, and those higher patterns can be constructed in such a way that they perform quite predictable, regular-system behaviors. Indeed, it has been shown that the patterns can be arranged in such a way that a complete Turing Machine can be built inside the system.

Other ways to encounter partial complexity are easier to appreciate. A system can have a number of subsystems, governed by different mechanisms, with only some of these being complex. Or, a number of different regularities can be due to the same mechanism, but with differences in their complexity. Gravitational motion in the solar system, for example, approximates very well to a regular system, but only if we ignore such effects as Pluto's occasional bursts of chaotic behavior, and the braiding patterns observed in planetary ring systems due to systematic influence of nearby moons.

Complexity exists as a continuum, so we could say that some partially complex systems are *primarily regular*, while others are *dominated by complexity*. The solar system would be primarily regular, because there are simple regularities in the orbits of the planets that enabled Newton to derive an extremely accurate account of the underlying mechanism. Whenever a system has enough regular aspects to it that we can use those regular aspects to work backward and uncover all of the underlying mechanism, we can categorize the system as primarily regular. If, on the other hand, there are significant behaviors that seem complex, and we do not have any easy way to go around to the back door (so to speak) and use regular aspects of the system to uncover the mechanisms, we would classify the system as dominated by complexity, or as containing significant amounts of complexity.

### *12.2.2 Are Cognitive Systems Complex?*

How do we decide whether intelligent systems should be treated as containing significant amounts of complexity? There are some aspects of human intelligence that seem to involve sequences of logical inference that are governed by rules, so from that point of view the system looks regular. And there are plenty of other regularities to be found, across all the paradigms of experimental cognitive psychology.

But of perhaps greater significance is the fact that the core engine of our intelligence—the mechanism that creates, develops and deploys *concepts*—is known to involve a host of subtle interactions and sensitivities. Concept construction and deployment is one of the most poorly modeled of all aspects of cognition, in the sense that we are still grasping for the correct metaphors with which to characterize them (prototypes?, exemplars, clusters of microfeatures?), we still have many choices of theory to describe their developmental aspects, and it is still not possible to build working artificial intelligence systems that construct and maintain concepts at arbitrary levels of abstraction, using only raw real-world input.

The creation and development of concepts is the place where we would most expect to see complexity, because this is where we have evidence of *intractable interactions* between the components of the system. We can combine concepts in a seemingly infinite variety of ways, and we can use them with degrees of flexibility that appear to be unbounded. Almost every time we use a concept we adapt it to the specific context in which it is used. All of this flexibility, context-dependence and combinability seems to point to a system in which component interactions are out of control.

Without pushing the case as far as it might be pushed—by listing a full catalogue of examples that seem to indicate complexity—let's step back for a moment and consider the purpose of this line of argument. Are we trying to decide whether there is conclusive evidence that a significant amount of complexity is present in human cognition? If this were the goal, it would be a risky one: we have already seen that it is impossible, in principle, to prove that a system is complex. It seems that if we had the lesser goal of showing that *there is a substantial risk that complexity is present*, we might be able to close the case immediately. I submit that this is already done, and is widely accepted by the cognitive science community. The features of concept building described above have been remarked upon throughout the history of the subject—so much so that it is almost a standing joke that when anyone tries to pin down the meaning of a concept in an algorithmically closed form, someone else will immediately produce a counterexample.

Speaking informally, cognitive scientists seem quite ready to concede that many aspects of cognition (including concept mechanisms) show evidence of complexity. They may not be willing to take the next step and admit that this has great significance, but it is enough for our purposes to note that the existence of complexity in this context is widely accepted.

#### 12.2.2.1 The Risk of Complexity

One of the main goals of this paper is to argue that complexity has much greater, damaging consequences than has been appreciated. Ideally, such an argument would be supported by a proof that cognitive systems must be complex systems, so we could then draw the obvious conclusion that these consequences will have an impact on cognitive science. But since we cannot, in principle, give a proof that cognitive systems contain significant amounts of complexity, the best we can do is show that there is a substantial risk that they do. In view of that risk, we would then need to take action.

I believe that the risk of complexity in the known properties of the concept mechanism is sufficiently high that it is imperative that we ask whether the consequences of that complexity would be severe. It is that last question that we now consider.

### *12.2.3 Complexity and Scientific Methodology*

Given all of the preceding arguments about the definition and characteristics of complex systems, what can we conclude about the way we choose to investigate systems that appear to be complex? In particular, what impact might this have on the methodology of the cognitive sciences?

If cognitive systems are partial complex systems, one practical consequence is that there are some behavioral regularities that can only be explained by mechanisms that do not look as though they would ever explain those regularities. This is just another way of saying that the link from mechanism to behavior is broken in those cases—broken in both the forward and the backward directions. If the link is broken, the mechanism must look unreasonable in some way.

But this means that if we approach the scientific analysis of a partial complex system by looking for models of local aspects of the system that are always rational and regular—models in which there is always an understandable relationship between the behavior being explained and the model being used to explain it—then we will be setting ourselves up for failure. The system cannot be entirely made from components that have a non-complex relationship to the behaviors they produce. If the system is partially complex, it must be the case that at least some of the models have a pathological (complex) relationship to the behaviors they generate. We may never know which components should have this pathology (although we can sometimes make a shrewd guess), but we can be sure that if we have prior reasons to suppose that the system is partially complex, then somewhere there will be trouble.

This innocuous-looking point has profound consequences. In a truly fundamental way our science is built on the idea that we can understand the world by using occam's razor to find the simplest, most elegant explanation for all the components of a system, then combine these separate understandings into a unified understanding of the system as a whole. But in the case of an egregiously complex system this is a doomed strategy: it will always miss the truth.

What would happen if, in spite of the danger, we simply forged ahead and applied the usual scientific strategy? The naive conclusion might be that in the case of those system components that require a complex explanation, we would see our models breaking and eventually conclude that this component was the one that needed to be treated differently. Alas, there is no reason why the locus of trouble should be so easy to pin down. More likely, we would find that we can always build models of all components of the system, but some of those models will just be locally applicable, or will reference such minute and insignificant aspects of the system that they are actually avoiding the features that are complex. In other words, local model building will not fail, it will just produce an unending stream of poor quality models.

And as this model building continues, two other processes would probably be observed. One is that each model will be extendable only at the cost of excessive complications. In order to make the model more general or apply it to more cases, it will have to be extended or elaborated with arbitrary extensions that eventually turn it into a theoretical kludge. The second process is that when two models collide, the process of integration (which means, adaptation of each to make a unified whole) will come only at great cost: again, the result will be excessive complication. More likely than the successful combination of models, though, would be their insularity: researchers in different paradigms will simply decline to integrate their model with others.

All of this can be expected to happen as a result of a situation in which the complexity of the system was being denied or not acknowledged. Local efforts at model building would work toward the goal of a complete explanation for the system that was complexity-free, but since no such non-complex, complete explanation exists, the goal would be unattainable, and the separate model-building efforts would only become more complicated and less plausible as time progressed.

This is, arguably, exactly what is happening in the cognitive sciences, both within disciplines and across them.

### 12.2.4 Solving the Problem

The main way to start solving this problem is to avoid a narrow focus on locally optimal models. To do this we need to find ways to develop the widest possible range of different models for each component of the target system, where previously we might have chosen only one. Then these models need to be tested in parallel, because it is important to assume from the outset that the proper explanation for any given behavior could be a mechanism that looks unreasonable. To some extent this involves a semi-blind search through the space of all possible explanatory models.

Inventing single models is hard enough, but inventing large sets of models to explain just one set of observations is even harder. This is a process that cries out for some kind of automation, but automation implies that we need to stop thinking of models as individual hand-crafted works of art, and instead treat them as entities that can be described in terms of *design parameters*. Then, with this new vision of a model as a commodity—as just a set of choices for the design parameters—we can start to produce large numbers of models, each of them being a single point in a multidimensional space of parameters.

This would entail a radical shift of mindset, from models to *generators* of models. Instead of thinking of models as separately interesting things, we need to think about the kind of engines that could be used to generate sets of models.

This would clearly bring a significant change to cognitive science. There would still be room for cognitive scientists to interpret the results of a particular experiment by conceiving a new model, but such an act of model creation should then lead to a mental disassembly of the model, to see whether it can be built with already existing parameters, or to see if new parameter-concepts need to be invented to capture it.

### 12.2.5 Frameworks and Paradigms

One possible approach would be to write an abstract mathematical formalization of the space of possible cognitive systems, and then devise an algorithm to explore that space. This might be called the "pure mathematics" approach to the problem. Although it might be attractive to mathematicians, such an approach seems less than optimal. After all, we do not expect every aspect of the target system to be egregiously complex, so there will be many features of the its behavior that we can deduce from regular scientific analysis. Going back to first principles and writing abstract formalisms to describe everything in the system would be equivalent to throwing away our existing knowledge.

The main alternative to the pure mathematics strategy would involve somehow collecting all of the existing knowledge about the human cognitive system into one large framework, and then using that framework as the basis for a parameterized exploration of different models and components that are consistent with the framework.

The term *framework* is used here in the particular sense that is common in the philosophy and methodology of science: a framework is a loose set of organizing ideas and assumptions that inform a cluster of theories. Individual theories derived within a framework are supposed to be subject to confirmation or refutation as the result of experimental data, but the framework itself exists at a higher level, and is not subject to direct empirical attack. The concept of a *paradigm* (Kuhn 1962) is closely related, although it could be argued that frameworks are situated at a level of generality somewhere in between paradigms and theories.

The traditional separation of power between models/theories, on the one hand, and frameworks on the other, is that the models/theories are dominant while the frameworks sit quietly in the background. Papers get published because they supply new data that is pertinent to some models or theories that are supposed to account for the data; papers rarely get published if they claim to deliver an improvement at the framework level. Success is immediately verifiable in the former case, whereas claimed improvements in frameworks are often deemed to be mere speculation.

In order to solve the complex systems problem, this situation has to be inverted. It makes no sense to *always* put a premium on models that are good at explaining particular data, or on research that discriminates between pairs of such models. Instead, what matters is our ability to generate models as commodities within a framework. So, rather than leave the framework in the background as a poorly articulated cluster of ideas, the framework needs to be brought front and center, where it can be made as explicit as possible and turned into a mechanism that generates (hopefully, in an automatic way) a very large set of candidate models that can then be implemented and tested for their fit to the data.

The problem, of course, is that saying we need fully articulated frameworks is not the same thing as actually building them. For example, what is the framework or paradigm currently accepted by cognitive psychologists as their consensus assumption about the way cognition happens? A cursory glance at any standard textbook of cognitive psychology makes it clear that the field consists of many different sets of ideas, some of which may be backed by profoundly incompatible frameworks. What, for example, are the common assumptions about cognitive processing shared by the cohort model of spoken word recognition (Marslen-Wilson 1990) and the Bruce and Young (1986) model of face recognition? If the McClelland and Rumelhart (1981) model of word recognition is considered valuable because of its sub-symbolic aspects, how does it square with the apparently symbolic process that occurs in sentence production?

### *12.2.6 Frameworks as Art*

There is an uncomfortable truth at the heart of the framework-building process. Frameworks are not constructed by cautious logical arguments and scrupulous support from empirical data. They are intuited. They represent extended judgment calls. They are born out of a collective feeling of what is elegant and right, and what most comfortably sits with the various local ideas that are working best at the moment. Sometimes a framework has to be a bold leap in the dark.

This is another way of saying that if we accept that the complex systems problem must be solved by a new emphasis on top-down design—starting with a big-picture view of the human cognitive system, turning that into a concrete framework, then exploring very large numbers of models within the framework—we must also accept that there can be no hard-and-fast rules for how to devise the framework. There is a need for a new kind of theoretical activity within cognitive science, in which frameworks are invented and described in the broadest possible terms, and where the scientific community judges them on the degree to which they appeal to a sense of what is elegant.

The most viable frameworks should then be turned into explicit software engines that enable models to be constructed and tested. But even though the process should eventually lead to proper empirical testing of the sort that science is familiar with, the initial process of framework construction needs to be given the breathing room it needs to be creative and inventive. There needs to be an attitude shift, so that a clear line exists between allowed creativity and invention on the part of the framework theorists, and down-to-earth Darwinian ruthlessness of the sort that applies when specific models are being confirmed or refuted by empirical data. The framework theorists need credibility and respect.

### *12.2.7 Connectionism and Constraints*

I will close by showing how an explicit framework can emerge from the weaknesses of an older, less explicit and non-CSP-friendly framework.

The older framework was known as *connectionism* or *parallel distributed processing* (McClelland et al. 1986a; Rumelhart et al. 1986b). The connectionist revolution that happened in the 1980s was often taken to be about neuron-like processing units, but although this was an important feature of the new ideas, the background motivation was actually more general than that. It was about finding ways to build cognitive models in which *multiple simultaneous constraint relaxation* was the dominant theme. McClelland et al. (1986b) gave a catalogue of examples in which cognition seems to involve mutual simultaneous constraints:

- Reaching and grasping
- The mutual influence of syntax and semantics
- Simultaneous mutual constraints in word recognition
- Understanding through the interplay of multiple sources of knowledge
- Stereoscopic depth perception
- Perceptual completion of familiar patterns
- Content addressability of memory

The significance of these aspects of cognition is that they seem to point toward a type of model that involves large numbers of objects connected by weak constraints, in such a way that none of the constraints is rigidly enforced, but where the system nonetheless shows intelligent behavior. The challenge faced by the early connectionists was to find ways to build such models, and what was inspiring about the connectionist revolution was that a group of early algorithms—Boltzmann machines, interactive activation, back-propagation, among others—were given as concrete examples of what could be done with networks of simple processing units that weakly constrained one another's state.

Interestingly, though, as the connectionist movement matured it started to restrict itself to the study of networks of neurally inspired units with mathematically tractable properties. Network models such as the Boltzmann machine (Ackley et al. 1985) and backpropagation learning (Rumelhart et al. 1986a) were designed in such a way that mathematical analysis was capable of describing their global behavior. So the primary characteristics of these systems were not complex.

But if the CSP is real, this reliance on mathematical tractability would restrict the scope of the field to a very small part of the space of possible systems. The original connectionist researchers could have taken their original inspiration and used it to explore large numbers of systems in which weak constraints were operating, but instead they tried to fix the known weaknesses of the early models by either looking for better (but still mathematically tractable) models, or by combining the known models into hybrid architectures.

The subsequent history of connectionism was disappointing to some. The field tended toward stagnation, with no dramatic solutions to the larger problems of cognition to follow the bold progress that occurred within a few years at the start of the revolution.

The field was driven by a strong, implicit assumption that the best (and perhaps only) place to look for models in which constraint satisfaction was the driving force, was in those models that could *provably* be shown to have the correct type of behavior. From the point of view of the complex systems problem, this aversion to complex systems was a grave mistake.

### *12.2.8 A Molecular Framework for Cognition*

So if we wanted to hit the restart button on the connectionist revolution, exploring models of cognition in which mutual simultaneous constraint relaxation played a significant role, what kind of framework might we use? There are many possibilities, of course, but in what follows I will articulate one of these possible frameworks in a little detail, both as an example of where connectionism might have gone if it had not restricted itself, and as an illustration of one direction we might go next, now that we understand the force of the complex systems problem.

This *molecular framework* (Loosemore and Harley 2010) has one simple idea at its core. When the relative strengths and weaknesses of connectionism were first debated, many of the weaknesses could be traced to the fact that the neuron-like units doing the constraint relaxation were locked into fixed positions within the system (just as real neurons are). This eventually leads to trouble, because we know that higher cognition involves clusters of concepts that can be rapidly linked together in different configurations, and fixed neurons cannot easily do this unless every one is connected to all others. These configurations of concepts are indispensible: it is not enough to know that the nodes for [girl], [bites] and [dog] are all active: what matters is whether they are arranged in the sentence "girl bites dog", or "dog bites girl". Cognition also involves situations in which there are copies of the same concept: the sentence "dog eat dog" needs to be represented with two instances of the dog concept. So, in order to model these processes, we would like to have a system in which multiple copies of concepts can be activated, with flexible, transient relationships between those concepts.

This seems to imply a return to the older, pre-connectionist ideas that involved symbol processing—where the symbols are free to be created, deleted, linked and modified in any way, and perhaps with structure inside the symbols—but with the additional feature that the symbols have the type of mutual, simultaneous constraint relaxation properties that are the hallmark of a connectionist system.

The fundamental construct of the molecular framework, then, is a symbol-like entity that is free to roam around like an atom in a molecular soup. As the atom moves around it forms temporary bonds with other atoms, because this is the only way to preserve the idea that the atoms constrain one another. This leads to a picture in which transient, ephemeral molecules are being made and unmade—each molecule resembling a structured model of some aspect of the world.

The goal now is to see whether a cognitive framework can be built around this generalized version of connectionism, in such a way that most aspects of human cognition make sense in the framework. The following is a brief sketch of some of the main features of the proposed framework.

- Symbols appear in two distinct locations: they are either in long term memory, where they do nothing, or a copy of the symbol is transferred to a special place (the *foreground*) in which the system is representing the current state of the world, and all of its currently active thoughts. To make it clear that these really are distinct, the symbol is referred to as an *element* when it is in long term memory, and as an *atom* when it is in the foreground. There is only one element for each symbol, but there can be many atoms.
- To make this distinction more concrete, we can make a tentative mapping between these abstract notions and the neural hardware of the brain. Suppose that the primary functional role of a cortical column is to host a single active atom. A secondary role of the columns is to keep a collection of elements, in such a way that each column keeps an overlapping subset of the total set of elements in the system's long term memory. A single element might be redundantly encoded in a cluster of (perhaps adjacent) columns, so it could not be destroyed by a malfunction in any one of the columns. When a particular element is called upon to deliver a copy of itself—in the form of an atom—one of these host columns produces the atom and tries to find a place for it on the nearest column that is either vacant, or which contains the weakest active atom. Thus, as time goes on the population of

atoms hosted across the entire sheet of cortical columns will change in such a way that the strongest atoms stay active and the weakest are replaced by newly activated atoms. When an atom is deactivated it is returned to the original element, where some aspects of its recent activation are used to update the element.

- The concept of working memory is therefore roughly equivalent to the current set of atoms being hosted on all cortical columns (the total number of columns being of the order of one million—although this depends on how columns are counted).
- When an atom is active its job is to look at the relatively nearby atoms and try to find ways to get support by establishing connections to them. To this end, every symbol has an internal idea of what kind of neighborhood it would like to see in the foreground. The neighborhood is not necessarily a simple list or fixed set of other atoms, but can contain slots (variables) that can be filled by atoms that themselves must satisfy certain requirements. If an atom can quickly establish links to neighbors in such a way that both it and they find the link satisfies their internal desired-neighborhood map, the atom becomes strongly supported and is less likely to be deactivated.
- One of the main sources of support for foreground atoms is the flood of information coming in from the senses. This information drives certain columns to activate certain atoms that represent the incoming information. So there would be some columns whose hosted atom is determined, not by the presence of nearby supportive atoms, but by the fact that it has a direct line to (say) a patch of color detectors in the retina.
- Activation of high-level representations would then proceed in a manner similar to that which occurs when an input pattern arrives at a neural network: neighboring, strongly supported atoms that encode low-level sensory signals get together to active candidate atoms that might represent slightly higher symbols. This bottom-up cascade of activation happens in the context of some top-down activation of atoms representing expected features.
- Notice, at this point, that the framework does *not* specify exactly how the mutual interactions between atoms will cause clusters to form, representing coherent thoughts or coherent representations of perceived objects: the exact choice of mechanisms is expected to be especially sensitive to complex system effects, so this is the area in which exploration of different possible mechanisms has to be undertaken in an empirical manner. To cover our future discoveries of the exact mechanisms, we simply summarize this aspect of the system by saying that what the foreground atoms do is engage in mutual, simultaneous, weak constraint relaxation, in such a way that the system as a whole exhibits coherent, intelligent thought processes.
- The interactions between atoms may include a number of unorthodox features (at least, unorthodox in purely connectionist terms). Some atoms may act as operators, whose only function is to transform configurations of other atoms in certain ways, then leave (there may be loosely defined set of *analogy* operators that do this on a routine basis). Also, a large number of atoms will represent *relationships* rather than features, with hierarchies of these relationship atoms

that are every bit as extensive as the hierarchies of features. And there may be some processes that are not encapsulated in atoms at all: for example, a process of elaboration that sometimes kicks in when a cluster of atoms fail to reach consistency with respect to one another: the initial failure causes the elaboration level to go up, so that all participating atoms try to surround themselves by greater-than-normal numbers of related atoms.

- The concept of an elaboration area can be used to capture the idea of the *focus of attention*. At any given time there will be a unique zone of high elaboration that is the attentional focus, with special mechanisms that cause it to move from place to place in the foreground, depending on factors like the unexpectedness or novelty of certain sensory inputs, or sequences of learned actions.
- Just as there is a flow of activation from the arriving sensory input, as more and more abstract symbols are activated to represent the input, so there is a dual version of this process going on toward the motor output end of the foreground. The flow starts in an area of the foreground that might be labeled the Make-It-So region, which is the primary source of support for the tree of atoms that eventually lead down to the specific muscle signals required to execute a movement.

#### 12.2.8.1 Molecular Framework: Summary

This short account of the molecular framework is intended only as a hint of what might be done to start a research program that is immune to the complex systems problem. The next stage would be to elaborate this account in enough detail to see how, in principle, various specific cognitive phenomena could be accounted for, then implement this as a software (and perhaps hardware) system designed to support a class of models consistent with the framework. After the software is built, large numbers of different candidate mechanisms need to be set up within the system and their properties explored under a variety of circumstances. An iterative process of exploration and analysis would then (hopefully) lead to a convergence on systems that are good approximations to the particular system that is used by the human brain. At that point, we might be able to say that we understand the neural-cognitive mapping in enough detail to know what kinds of nanotech interventions would have which effects on the system.

## 12.3 Conclusion

If complex systems are what they seem to be, then the universe contains some systems that are impenetrable to scientific analysis, in the sense that we can observe their behavior but cannot develop any kind of analytic proof that this behavior is the result of the underlying mechanisms. If this impenetrability sometimes occurs in systems that are partially complex, but also partially non-complex, we could find ourselves in a situation where we first explain some aspects of that system, but then convince ourselves that the rest of the system will eventually fall to the same scientific attack. Unfortunately, the stubbornly complex aspects of the system could resist attack for a very long time, because the mechanism behind those aspects might look utterly unreasonable—it might be the kind of mechanism we would never have guessed would be responsible.

When the ramifications of this idea are examined in depth, it appears that our approach to cognitive science—the entire methodology we use for unlocking the secrets of human cognition—might be in need of drastic revision.

The revision proposed in this paper is to find ways to build large sets of explanatory models, rather than just single models, and to insert these into simulations that can then be used to explore how all of these candidate models behave. In this way, we open the door to considering models that look unreasonable on the surface, but which may in fact be the only viable explanation for a given set of experimental data.

### *12.3.1 Postscript: The Urgency of New Software Tools*

If this new approach to cognitive science is to be implemented, one of the first prerequisites will be software tools capable of building models and organizing them into simulations. In the past, cognitive researchers have tended to build small pieces of software to implement their models, and this process has required them to be part-time software engineers as well as psychologists. The results have been mixed: such models are often very simple (even simplistic), and incapable of generalization. It would be impossible to expect the proposed new approach to cognition to be implemented unless researchers could be liberated from the burden of low-level programming.

Historically, the arrival of new tools has often been the vital catalyst that starts technological revolutions. A lack of the right tools can be seen as the single biggest factor that has caused the complex systems problem to go unrecognized for so long: with no way to do anything about it, there is little incentive to consider it. What is needed now is the kind of software that might trigger a new cognitive revolution

## References

Ackley, D.H., G.E. Hinton, and T.J. Sejnowski. 1985. A learning algorithm for Boltzmann machines. *Cognitive Science* 9: 147–169.

Bruce V., and A. Young. 1986. Understanding face recognition. *British Journal of Psychology* 77(Pt 3): 305–327.

Gardner, M. 1970. Mathematical games: The fantastic combinations of John Conway's new solitaire game 'life'. *Scientific American* 223(4): 120–123.

Horgan, J. 1995. From complexity to perplexity. *Scientific American* 272(6): 104–109.

Kuhn, T.S. 1962. *The structure of scientific revolutions*. Chicago: University of Chicago Press.
Loosemore, R.P.W., and T.A. Harley. 2010. Brains and minds: On the usefulness of localisation data to cognitive psychology. In *Foundational issues of neuroimaging*, ed. M. Bunzl and S.J. Hanson. Cambridge, MA: MIT Press.
Marslen-Wilson, W.D. 1990. Activation, competition, and frequency in lexical access. In *Cognitive models of speech processing: Psycholinguistics and computational perspectives*, ed. G.T.M. Altmann, 148–172. Cambridge, MA: MIT Press.
McClelland, J.L., and D.E. Rumelhart. 1981. An interactive activation model of context effects in letter perception: Part 1. An account of basic findings. *Psychological Review* 88: 375–407.
McClelland, J.L., D.E. Rumelhart, and The PDP Research Group. 1986a. *Parallel distributed processing: Explorations in the microstructure of cognition*, Psychological and biological models, vol. 2. Cambridge, MA: MIT Press.
McClelland, J.L., D.E. Rumelhart, and G.E. Hinton. 1986b. The appeal of parallel distributed processing. In *Parallel distributed processing: Explorations in the microstructure of cognition*, vol. 1, ed. D.E. Rumelhart, J.L. McClelland, and The PDP Research Group. Cambridge, MA: MIT Press.
Mitchell, M. 2008. *Complexity: A guided tour*. New York: Oxford University Press.
Rumelhart, David E., D.E. Rumelhart, G.E. Hinton, and R.J. Williams. 1986a. Learning representations by back-propagating errors. *Nature* 323: 533–536.
Rumelhart, D.E., J.L. McClelland, and The PDP Research Group. 1986b. *Parallel distributed processing: Explorations in the microstructure of cognition*, Foundations, vol. 1. Cambridge, MA: MIT Press.
Waldrop, M.M. 1992. *Complexity: The emerging science at the edge of order and chaos*. New York: Simon & Schuster.
Wolfram, S. 2002. *A new kind of science*, 737–750. Champaign: Wolfram Media.

# Chapter 13
# Narratives of Intelligence: The Sociotechnical Context of Cognitive Enhancement in American Political Culture

**Sean A. Hays**

## 13.1 Introduction

In American political culture, there are two competing narratives of intelligence. The first is predicated, in part, on the pioneering work of French psychologist Alfred Binet and the battery of tests he developed near the turn of the nineteenth century to help identify French school children in need of remedial instruction. Contrary to Binet's stated intentions and beliefs, when his methods were imported to America, they were used to construct the narrative of *IQ*, or the intelligence quotient. The bulk of this narrative was written by Charles Spearman, the creator of the statistical technique known as factor analysis as well as the theory of a general factor for intelligence, referred to as *g*. In this narrative, intelligence is a single property, measureable, heritable, comparable, and localized within the head. In both popular and scientific uses of this narrative, *g* has reached a degree of reification such that it is spoken of as if the skull could be split open, and *g* located in a specific place within the gray matter of the brain. Perhaps most important for American political culture, *g* is a largely fixed quantity, and is not particularly susceptible to environmental interventions, with the exception of physical or chemical damage to the brain itself. It plays a central role in the beliefs of many Americans with regard to social competition of all kinds, and, because of its history and significance, it has been a controversial narrative since its inception.

The second narrative of intelligence has been around as long as, if not longer than, the first. It is most closely associated with the recent and pioneering work of Howard Gardner, a Harvard developmental psychologist, who published his theory of multiple intelligences for the first time in 1983 (Gardner 1983). In this narrative of intelligence, cognition is fractured into pieces that are localized within different

S.A. Hays (✉)
The Centre for Nanotechnology in Society,
Arizona State University, Tempe, AZ, USA
e-mail: seanahays@gmail.com

S.A. Hays et al. (eds.), *Nanotechnology, the Brain, and the Future*,
Yearbook of Nanotechnology in Society 3, DOI 10.1007/978-94-007-1787-9_13,

parts of the brain. In some cases, they are closely interrelated, and in others, they bear no relationship to each other. Gardnerian psychology conceives of the various intelligences as being non-fungible, i.e. they cannot be exchanged for one another. Damage to, or an increase in, one kind of intelligence will not provide a linear decrease or increase in another kind of intelligence, and specific cognitive tasks cannot be made easier by increasing the capacity of unrelated parts of the brain. In a highly competitive society, the complexity of a theory such as Gardner's is less appealing in terms of a national narrative, but it plays an important role, nonetheless, in education theory and in the way schools are structured. Further, as I will demonstrate in the final portion of this paper, it is central to most of the scientific research and technological development being done in neuroscience today. Perhaps more interesting than the disjunction between the research being conducted and the dominance of a narrative that is anathema to that research, is that these two narratives often exist in, and are espoused by, the same person at different times. To be clear: The two theories are mutually exclusive, but many Americans fail to see the contradiction.

Part I of this paper describes the individual deep historical analysis of "intelligence" through a close engagement with the work of Stephen Jay Gould. As will be shown, the significance of intelligence is its central role in the desire for and fear of the development of cognitive enhancements and its relationship with pervasive and powerful social proclivities that might lend themselves to negative social outcomes if allowed to remain implicit and considered irrelevant. Further, as my analysis of Gould's work will show, the desire for—and perhaps more surprisingly opposition to—cognitive enhancement is actually predicated upon a series of common misunderstandings and logical errors with regard to the nature and reality of intelligence. Within the zeitgeist and certain branches of science, cognitive enhancement tends to refer to an increase in the overall ability to think. But the lesser known research science—as opposed to the more commonly referenced psychological, psychometric, and sociological research into intelligence—is considerably more incremental and recognizes the fictitious nature of the concept of a centralized, reified, and heritable intelligence (Gould 1996, 22). While Gould's work only marginally adheres to the historical method I am using in this analysis, it does allow me to specifically examine the intersection of science and desire in a way that makes clear the ongoing relationship between the two. It is a relationship that I argue is fueling an unfortunate social dynamic with regard to cognitive enhancement and that is acting to prepare the public to accept these new technologies in ways that are less than optimal from the perspective of social justice or economic efficiency.

Part II of the paper engages with the work of John Carson, and his institutional genealogical approach to recounting the history of "intelligence" as it developed more or less simultaneously in France and the United States (2007). I have included an analysis of Carson's work for two reasons. First, he not only offers us access to a meticulously assembled history of intelligence's development from a perspective distinct from, but sympathetic to, Gould's, but his work also serves as an excellent case study demonstrating the historical method I hope to employ with cognitive enhancement technologies in the longer term. Second, in general I would be remiss if in analyzing the interplay of *IQ* and the desire for human enhancement I did not include at least some engagement with Carson, as his work makes explicit the influence of publics and institutions on the development of the narrative of *IQ*. Generally, American society tends to see such development in terms of streetcars. I'm referring here to a theory in international relations—first articulated by Ned Lebow in *Contingency, Catalysts, and International System Change*—which argues that rather than institutions or structure being the driving forces behind political activity on the world stage, it is powerful individual actors or significant one-off events (2000, 591–593). Lebow refers to these as catalysts, but more memorable is his analogy to streetcars (2000, 601). Further, this theory holds that there is an inexorable progressive force behind international relations, and that if one charismatic actor doesn't come along, eventually another will, i.e., if Archduke Ferdinand hadn't come along to instigate The Great War, some other individual and his circumstances would have (Lebow 2000, 594). This emphasis on an inevitable, interchangeable, and essentially individual force behind world events—another streetcar will always be coming along shortly to move you from point A to point B—is similar to the common understanding of technological progress in America. Progress is seen as inevitable and inevitably upward but also driven and shaped largely by individual genius. It will become clear that my own take on technological progress is a hybrid of the institutional and streetcar views demonstrated by Carson and Gould, respectively.

I will conclude with an analysis of the significance of the deep historical analysis of *IQ* to the discussion of cognitive enhancement. This will establish a theoretical and historical context with which to examine the national survey data offered elsewhere in this text. Further, it offers an opportunity to discuss a significant problem of logic and epistemology inherent in the scientific analysis and explication of humans, the development of technology for their alteration or improvement, and the sociopolitical discussions and actions that shape and are shaped by those technological and scientific developments. I will argue that – absent what is likely a gross oversimplification of human cognition as represented by the unitary, biological, and heritable narrative of intelligence persistent in American culture today – the push to develop and use such enhancements would be greatly attenuated, or radically altered by a change in both expert and lay expectations for and fears of enhancement technology. Much of the opposition to the same research would be reshaped, and perhaps would assume a form more beneficial for successful governance of technological development. In essence, what I am arguing is that much of the research—and the demographically very specific open enthusiasm for enhancement—is a function of the belief that cognitive enhancement is both functionally possible and simple. The opposition to cognitive enhancement discovered by the Center for Nanotechnology in Society at ASU in its national survey described in the third chapter of this volume, and that has been prominently displayed in a variety of other publications,[1]

[1] For other examples, see *Beyond Therapy* (2003), the final report from the President's Council on Bioethics headed by Leon Kass, or the techno pessimism of Bill Joy's *Why The Future Doesn't Need Us*, for two of the most prominent examples.

is based on the belief that intelligence is a single attribute. Further, that it is one that is fundamental to the formation of identity, and critical to socioeconomic success in a competitive environment. If we were to change this single element—the belief in a heritable general intelligence—then the debate about cognitive enhancement would change radically. Given the complexity of the historical development of *IQ*, I will argue that it is actually unlikely that we will change it in any way that is significant to the public, and thus I will focus on recommendations for policy and research.

## 13.2 Defining IQ

The implicit understanding on the part of many—not all—researchers in psychology and other social sciences is that *IQ* is a fully fungible stock of ability within the brain and that enhancing any specific function will simply increase the total store of intelligence. The counter-argument—founded on the multiple intelligences championed by Howard Gardner, but with its roots firmly anchored in the now anachronistic science of phrenology, as Gould shows—often includes the belief that the development of various types of intelligence is a zero sum game. In a recent book on neural plasticity and the Internet, *The Shallows: What the Internet is Doing to Our Brains* by Nicholas Carr, the progression of neuroscience's understanding of the brain, its component functions, its ability to adapt, and the nature of the cognitive capacities therein is laid out in detail. Carr describes how, in virtually every instance, research into cognition is demonstrating that it is localized, task specific, and the relationship between areas and functions is non-linear, and largely non-fungible. Enhancing one area of human cognition must exact a price among other areas of intelligence, so the argument often goes. This understanding does not dominate—either explicitly or implicitly—the research being conducted either in cognitive science or the various fields producing technologies with the potential to enhance cognitive abilities. If it did, I contend, the very idea of enhancement would be far more controversial at an R & D and policy level than it is at this time. It is important to distinguish between R & D focused on science and technologies with enhancement potential, and scholarship focused more generally on the mind, humanity, and the implications of enhancement. I am arguing that the R & D is centered implicitly in the camp of multiple intelligences, while the narrative of unitary intelligence, or IQ drives much of the scholarship on this subject.

The specificity of potential enhancement technologies is best observed through a few examples of the technologies being developed or already in use. Perhaps the most famous brain implant is the BrainGate™ device to aid locked-in patients—who have full awareness but absolutely no fine or gross motor function—interact with the physical and social worlds (Hochberg 2008). The device links the motor cortex of the brain to a computer and interprets the neural signals generated when the user thinks about moving an arm or a cursor on a computer screen into action, either in a robot arm or on a computer monitor. A speech implant being developed by medical device start-up Neural Signals Inc., embeds a sensor in the brain of locked-in patients just above Broca's Area—long recognized as the language center of the brain—and translates neural signals generated when the patient attempts to pronounce a human phoneme into the corresponding sound through a computer speaker (Kennedy 2006). The scientists at Neural Signals report being able to train their patients to pronounce all 39 human phonemes—the basic units of human speech—through a computer (*Ibid*). Finally, in terms of implants, deep brain stimulation—which involves inserting electrodes into specific areas of the brain to deliver impulses that disrupt neural activity—has been shown to aid in controlling epilepsy and depression, both of which are believed to be conditions localized in certain regions of the brain.

Similar to deep brain stimulation, psycho pharmaceuticals for treating depression, epilepsy, Attention-Deficit Hyperactivity Disorder (ADHD), and other mental disorders, are localized in both their action and their intention. These drugs are designed to address a specific function of the brain. In the case of antidepressants, the goal is to control emotion—particularly chronic sadness and hopelessness—through preventing serotonin reuptake in particular areas of the brain. ADHD drugs are, typically, stimulants that increase concentration and executive function,[2] again acting primarily on the cortical regions of the brain. Epilepsy drugs target the temporal lobes of the brain in order to disrupt potentially fatal chains of misfiring neurons, and prevent or ameliorate seizures.

Each of the devices and drugs described above is specific in its intended function and its locus of activity. They are designed to modify or repair a specific brain function by addressing a failure in a particular physical region of the brain. Further, each of these drugs and devices has enhancement potential, though just as with its therapeutic value it is specific to the faculty it was originally designed for, and the only real difference is the state of the patient. A more systematic review of the scientific literature on neurological devices and drug would reveal a persistent relationship between cognitive enhancement R & D, and theories of multiple intelligences. Such theories hold that intelligence is application specific, localized within the brain, and can tell us little about the general competitive capacity of the individual (Carr 2010). While it is true that some cognitively specific enhancement might produce a general and persistent socio-economic advantage – e.g., an implant that allowed you to communicate faster and continuously with Wall Street trading computers – that could then be translated into political advantage, such a case says more about our economic and political systems than it does about the technologies or the underlying political culture.

While enhancement certainly would not be controversial for the same reasons as it is currently, if we were to change the public and scientific understanding of intelligence to something less rigid but still unitary, I predict it would remain controversial nonetheless. Perhaps less intuitively, I will argue that if intelligence were

[2] While, in some cases, "executive function" is used as a substitute for *g*, or IQ, it actually refers to a specific cognitive function, localized within the upper cortex, and dedicated to specific decision making processes, e.g. it works to assimilate specific sets of information and enable conscious decision making under certain circumstances. It is, however, not synonymous with intelligence or cognition, and cannot function without the other cognitive units within the brain.

understood to reside in various non-fungible, application-specific centers in the brain, then enhancement would actually become less controversial with the public, even as it became more controversial with researchers. I base this contention on the socio-political context in which Americans tend to encounter new technologies, one of supposedly free and fair competition. Choosing to trade excellence in one intellectual area in exchange for enhanced excellence in another would be far less controversial than the prospect of achieving an overall intellectual edge by increasing total intellectual capacity. This, I believe, should sound a strong cautionary note for anyone who is considering the long-term social and political implications of such technologies. I will conclude the chapter with a deeper discussion of the value of this case study—that of the historical and sociological composition of intelligence—to the broader discussion of human enhancement technology and to the specific case of cognitive enhancement.

The value of the case study contained in this chapter is twofold. First, it is an opportunity to demonstrate the theoretical and epistemological depth that historically rich qualitative analyses can add to efforts at Real Time Technology Assessment (RTTA) currently undertaken at the Center for Nanotechnology in Society and the Consortium for Science, Policy, and Outcomes (both at Arizona State University). RTTA is covered extensively elsewhere in this text, and all references here are from the seminal work by David Guston and Daniel Sarewitz, "Real Time Technology Assessment" (2002, 23–24). I believe that deep historical case studies,[3] as demonstrated here, can be one way to begin to fill in the historical and theoretical gaps extant in RTTA as it is being practiced today. I also believe that RTTA is the most promising method currently under development for establishing a regime of monitoring technological development, and constructing timely and effective policy responses. Thus, my primary goal actually lies beyond theoretical and ethical evaluations of human enhancement technologies.

The second point of value is to supply, through an examination of a very specific epistemological and logical problem inherent in discussions about cognitive enhancement, a framework for evaluating human enhancement generally. It should also be noted that, although I will only treat it superficially here, most scientific analyses of humans that stray beyond the confines of severely constrained biological perspectives could be subjected to the same type of critical analysis that I am performing here on human enhancement technology, and its attendant scientific researches. I will argue that much of the hope for, and fear of, cognitive enhancement stems from our misapprehending the nature of human intelligence,[4] and the tone and content of the national conversation about cognitive enhancement would change radically if

[3] I have referred to this kind of case study elsewhere in my work as "genealogy," as much of my method of historical analysis is derived from Nietzsche's ethic and method of history. However, the specific method of genealogy I am working to develop is still too poorly defined to function on its own, and thus, in this chapter I will refer, instead, to deep historical case studies.

[4] This misapprehension is founded generally on well-meaning, but misguided attempts to study the subject scientifically.

this common misunderstanding of intelligence were to change. Similarly, many of our fears and expectations of human enhancement technology stem from a common conception of human nature that I can generously say is based on scientific and cultural constructions which are far from logically or empirically rigorous. Just as variance in the seat and strength of intelligences radically changes the implications of cognitive enhancement for humanity and liberal democracy, the introduction of variability or scale to the concept of human nature changes the implications of human enhancement technology more broadly conceived.

## 13.3 The Web We Weave

In engaging with Stephen Gould's *The Mismeasure of Man*, I will begin to establish working definitions for both centralized and distributed intelligence in an effort to set the boundaries for my concluding discussion of the implications of these two definitions. Intelligence as a centralized, reified, heritable, unalterable, and variably distributed human trait occupies a central role in much of both popular and scholarly thinking on individual difference, equality, and performance in competitive environments. The various elements of the popular definition of intelligence above require a bit more explanation. In terms of the various logical errors in play with regard to intelligence—of which I will tell you more in a moment—centralization is perhaps the most contentious. Researchers interested in mental testing or cognitive development can be divided into two (asymmetrical) groups. The first group—and by far the largest for a variety of reasons, which will become clear as I go on—holds to the definition popularized by Charles Spearman—creator of Spearman's *g* and the use of factor analysis in analyzing the results of mental testing. This theory describes intelligence as a single entity within the head that acts as a sort of centralized bank filled with a more-or-less fungible mental ability we can draw upon for any of the wide variety of mental tasks we typically engage in on any given day (Gould 1996, 372–373). Alternatively, the second group of researchers is composed of adherents of the theory of multiple intelligences, the most prominent of which is Howard Gardener (2006). This theory holds that intelligence is divided into various specialized groups—verbal, mathematical, spatial, etc.—and the "stuff" fueling these various intelligences is non-fungible; it cannot be spent across intellectual boundaries. Gould notes that this theory of intelligence corresponds more or less to two schools of neurological research, the long since debunked phrenology and contemporary cognitive, behavioral, and developmental neuroscience (1996, 22).

Phrenology held that various types of mental ability were localized under the different prominences on the skull and by measuring or "reading" these bumps, one could tell something about the underlying personality and mental ability of the subject. While this may seem preposterous from our modern perspective, it was the leading scientific discipline investigating intellectual ability through the eighteenth and early nineteenth centuries. It is tremendously significant that as a society we have moved from a theory of multiple intelligences fueling the leading neurological science of the Victorian era, through the rise of centralized intelligence as embodied by Spearman's g, and back into applied research focusing on multiple intellectual centers in the brain. It is significant to the pursuit of human enhancement technology insofar as the leading theories underlying applied research often shape the research agenda and constrain the types of technologies considered feasible or useful to develop. Researchers focusing on mental testing and much of the public tend to conceive of intelligence in Spearman's centralized terms, but the researchers creating the technologies that may ultimately lead to practical enhancement appear to hold with Gardner and the phrenologists in conceiving of intelligence as being distributed to various centers within the brain. A theoretical frame that does not correspond with the frame driving the applied research is fueling our expectations, desires, and fears.

Often intelligence and cognition—one a supposed measure of overall mental ability and the other a more functionally conceived attribute separate from, say, memory—are conflated, and a measure like Spearman's g becomes a ready substitute for the ability to think. It is a part of American culture to assume that the categorization that follows the measurement of IQ is a function of clearly delineated analytical categories and coherent terms and principles. These assumptions are implicit in the commonplace way in which claims about IQ and stratification are made and accepted in popular culture and the scholarly literature (Herrnstein and Murray 1994; Jensen 1969, 1985; Pinker and Steven 2009). To be sure, there are many who continue to challenge the contemporary conceptualization of IQ and its place within the culture and politics of the United States, but, as will become apparent, the concept is still fundamental in forming American attitudes toward enhancement, particularly the role of IQ in competition. This conclusion is supported by our national survey data, as will be demonstrated elsewhere in this text.

Intelligence is afforded a special place in American society because it responds to a culturally constituted desire for objective, empirical, and most of all biological explanations in dealing with social problems (Gould 1996; Carson 2007). The rise of empirical psychology[5] began as a function of the French institutional system of education, and its desire to accurately channel the intellectually deficient into special schools and asylums, where they could receive the practical training most suited to them (Gould 1996, 222–225; Carson 2007, 113–136). Carson argues that it is one of the principal ironies of the rise of psychometrics in the United States that it was actually rejected in France in favor of the continued influence of panels of experts (2007, 113). It was psychometrics' concordance with certain extant American social desires and movements that lent it force in the US that it never was able to muster in the more homogeneous French society (Gould 1996, 27–33).

The essence of Gould's analysis is that the current form of intelligence and its implications for society are firmly rooted in the personal politics and ideologies of the prominent figures in the recent history of American psychology and mental testing.

[5] This combined elements of physiology with European laboratory psychology, and the continued influence of the American infatuation with various forms of craniometry to constitute the field of psychometrics.

A discussion of the form and veracity of intelligence cannot take place without reference to these figures, and this is one of the central problems with the most recent policy oriented piece of scholarship founded on IQ—Herrnstein and Murray's *The Bell Curve*—in that it takes for granted the authenticity and reality of IQ. It fails to acknowledge what Gould refers to as the soft social construction of the technology of intelligence and mental testing in an effort to more fully entrench the logical errors outlined above: reductionism, reification, dichotomization, and hierarchy.

While society and many scholars continue to take for granted the centralized, reified, heritable, and hierarchical nature of intelligence, the direction of enhancement research and scholarship as well as the public discussion about this research will continue to be dysfunctional. I will discuss the dysfunctions flowing from this understanding of intelligence in the final section of this chapter in detail. In brief, the bench and clinical science currently under way—very little of which is explicitly oriented toward enhancement but much of which has enhancement potential—is predicated upon a theory of intelligence which localizes mental process and potentials in different areas of the brain and has very little to say about any sort of overall mental ability. Conversely, the scholarship on mental testing, cognition, and enhancement takes as its foundation *The Bell Curve* version of intelligence. The result of this disjunction is that societal expectations—which appear to be fueled by psychological and sociological scholarship more than by familiarity with the actual research underway—do not accurately map onto the medical and technological research being conducted. The fear of the effects of cognitive enhancement is similarly founded on the theory of general intelligence. It is likely that if we could break the hold this definition of intelligence has on society and on significant portions of the academy, then the complexion of the discussion would change considerably as expectations and fears began to correspond more closely to the actual potential technologies under development. However, breaking this hold is significantly more difficult than it might appear upon reviewing only Gould's work. Repudiating the scientific and popular champions of general intelligence is simply not enough, as much of its development and popular entrenchment is grounded in systemic and historical features of American society. These systemic features have been rigorously treated by John Carson, whose work I will review in the next section.

## 13.4 You Would Be Right, if Only That was What Binet Actually Said

In John Carson's *The Measure of Merit: Talents, Intelligence, and Inequality in the French and American Republics, 1750–1940*, the long and sometimes sordid history of intelligence is told along with the history of the testing implements designed to measure it. The central argument in Carson's work is that the concept of centralized intelligence was quickly exported from the French to the American scientific establishment,[6] and then proceeded to mutate into something that was largely unrecognizable in France because of the very different social forces at play in America. The development of intelligence was simultaneous, but the character of the concept varied greatly based upon the underlying socio-political context in each country. This is in contrast to—or perhaps complements—Gould's position where the difference is driven by the politics and personality of the dominant researchers in the two countries, what I referred to previously as the "streetcar" theory of technological development.

Carson describes the initial conditions in France that prompted Alfred Binet, among others, to begin looking into mental testing for education purposes. The French system of education depended upon a tiered system of schools. These schools became progressively more difficult and acted to sort out the less promising students and ultimately funnel them back into industry and agriculture as the "brighter" students passed on to ever more rigorous levels of education. The social and economic fabric of France was—and largely still is—supported by this education system. The most promising graduates are groomed for government, civil service, science, and education, and institutionally the focus had been on using expertise to determine the students best equipped to enter these vital elements of a rigidly organized, top-down social and political structure.

In the late nineteenth and early twentieth centuries, the French government decided to attempt to alter the way in which this system of education operated. They were responding to two systemic problems perceived by both the public and the civil service. First, there was considerable popular fear of and expert frustration at the perceived negative influence of less-promising students on their peers. It was believed that such students were holding back their classmates through disruption and through the expenditure of limited classroom resources on dealing with their special needs. The second problem was one of bias and competence in the expert panels that had, up to this point, made the decisions about which students to advance and which needed remedial or special education. The government turned to scientists working in the burgeoning field of empirical psychology to provide them with a solution.

Alfred Binet—creator of what has become the modern IQ test—was asked to create a method of identifying struggling students so that they could be shifted to special schools and classrooms. In these special settings, they would receive instruction appropriate to what was seen as their inevitably limited economic role and where they would cease to have a negative impact on their unimpaired classmates. Binet constructed a battery of basic tasks that were believed to measure a representative cross section of mental abilities, the results of which could be used to identify any deficiencies relative to a normative baseline established through applying the same tests to students who had been identified as normal or exceptional. Binet saw no value in the scores of "normal" students beyond establishing a baseline against which to measure the results of struggling children (Carson 2007, 140–141).

[6] Carson and Gould both note that the idea had already existed popularly in both places that and this tended to fuel the direction that researchers chose to go in examining intelligence.

He believed that the tests told us nothing about intrinsic mental ability and was only valuable in identifying students in need of additional instruction.

Binet's test was ultimately scrapped in France in favor of the continued use of expert panels, in part because of the entrenched interests of the expert communities that composed the panels and set the rules. However, in establishing a "normal" baseline measurement Binet, along with his partner Theodore Simon, had created a tool with the potential to be applied to all students to establish general mental ability. Binet thought that this was preposterous, but his counterparts in the various fields of American psychology felt otherwise. The test was imported to America and in this different context began to transform into what I think is understand today as a tool for measuring general intellectual ability. The structural differences that led to this altered development were threefold. First, American society was far more heterogeneous than that in France and was rapidly acquiring ever greater diversity through a much higher rate of immigration. Second, the systems of education and government in America were, at least in theory, more egalitarian and less rigidly structured than in France. Finally, at the time the test was imported, American society was obsessed with discovering methods for biologically establishing racial difference and hierarchy in the wake of the Civil War and the introduction of former slaves into the political and free economic community.

Gould acknowledges these forces but focuses his analysis on the individuals who acted to bring about the symbiosis of science and social proclivity. These actors were shaped by social structure and they in turn acted to shape and reinforce that same sociopolitical environment through the introduction of scientific validation of already extant beliefs about the inherent inferiority of immigrants, women, and blacks. In France, no such justification was required in part because such conflicts were limited in a homogeneous society and because the far more rigid sociopolitical context precluded the possibility of resource and power capture by "out" groups without a need for scientific proof that they were inferior. In short, mental testing as developed in France and refined in America met an existing social need in the latter, and it is for this reason that it failed to die out as it did in France where it was not similarly necessary. Its persistence—coupled with the continuing power of those same sociopolitical structures that helped to shape it—can explain the continuing power of IQ in shaping the scientific and political agenda in America.

Cognitive enhancement should occupy a significant place in any discussion about technology's impact on competition and equality. In analyzing the complex history of the concept of intelligence in American society, Carson has demonstrated that its influence is far more pervasive than simply giving impetus to certain lines of scientific inquiry. It motivates a variety of public policy initiatives, explicitly or implicitly. Stephen Gould rightly noted that it was no coincidence that the publication of *The Bell Curve* coincided with the Republican takeover of the House of Representatives in 1994 (1996, 31). The desire to curtail or redirect social spending is one of the societal events that tend to reinvigorate deterministic, biological theories of intelligence that enable policy makers to attribute socioeconomic disparities to unalterable biological differences.[7] Lingering misconceptions about the ability to make group distinctions based on IQ shapes social attitudes with regard to the poor and to minority groups. As recently as the 1990s scholars have issued policy prescriptions for government intervention in minority communities based on group-level discrepancies in IQ, and this despite clear evidence that such distinctions lack logical or empirical support as demonstrated by Gould and Carson. Intelligence continues to wield significant moral and political force in the United States, and thus it is likely to shape the way we respond to and apportion cognitive enhancements.

Intelligence is significant in the context of enhancement and competition not merely because Americans largely remain convinced of its reality, its unity, and its heritability. It is significant because of the way in which it has historically been bound up with pre-existing ideological predispositions. The desire to draw clear social boundaries between racial and ethnic groups—particularly blacks and whites—is certainly less evident today than it has been in the past, but it is far from absent. In the Senate the segregationist proclivities of certain southern members of the Senate is once again under scrutiny, an issue many Americans hoped had died with Strom Thurmond and retired with Trent Lott. This desire to separate and distinguish between groups has long been served by the scientific community in the United States beginning with mid-nineteenth century psychologists turned craniometrists like Spearman, Simon, and Terman. Intelligence continues—however quietly—in service of such goals even today. Thus, in establishing policy with regard to cognitive enhancement we must beware of the undue influence of a concept like IQ.

It is easy enough to conceive of circumstances where the presumed inherited disparities in "native" intelligence were used not necessarily to justify the unequal distribution of cognitive enhancers but to justify their results. Unequal wealth distribution and access to the tools necessary for wealth creation, like education and high-paying jobs, have been justified in recent history by scholars like Herrnstein and Murray. They do so on the basis of racial differences in heritable intelligence, and this logic could easily be extended to quietly underwrite the widening of such disparities using cognitive enhancements (Herrnstein and Murray 1994). Alternatively, the same claims about racial differences could form the basis for government policies at various levels or, more likely, a social stigma that encourages the use of these products in minority communities. A technological

[7] Arthur Jensen opened his 1969 paper on racial differences in IQ, "How much can we boost IQ and scholastic achievement" with the line, "Compensatory education has been tried, and it apparently has failed." Jensen is here being explicit about the ideological motivations behind his study and the policy implications he believes it carries. Common resources—primarily tax receipts—according to Jensen are wasted when they are spent on "compensatory" education because the racial differences in innate *IQ* negate any perceived benefit in providing an equal and quality education to all American children. The paper is a precursor to *The Bell Curve*, which was far more direct in discussing the folly of compensatory education and the need to channel young people in educations and careers that are appropriate to their innate and unalterable intellectual ability.

fix for a social problem—particularly when it is supported by centuries of dubious science catering to racist ideology—is very often a seductive proposition and cognitive enhancing drugs could be seen to play such a role.

## 13.5 Conclusion

What is cognition's role in competition within a liberal democratic context?[8] A society arranged as a meritocracy and holding cognition to be a marginally "fixed" internal attribute that is essential in determining one's ability to compete effectively raises cognition to a far more central role in constructing hierarchies of power than is commonly acknowledged. It is far more common to hold that democratic institutions are designed to be blind to intellectual capacity and that neither rights nor the ability to equally and fairly participate in democratic governance are predicated upon what is commonly believed to be an innate characteristic. I could refer to this as the Bootstrap Doctrine. The commonly held cultural belief—particularly in the United States—that a desire to compete and better one's self is all that is really required to "pull oneself up by one's bootstraps." The myth of the self-made man is perhaps the most well known example of this trope.

Carson and Gould tell the same story in two different and equally significant ways. In relating the social history of IQ as a technology, they approached the subject from both the individual and the structural perspectives. Gould relates how the concept of intelligence developed in response to the desire for social control and scientific validation of extant racist tendencies on the part of a number of prominent scientists. Spearman, Goddard, Yerkes, and Burt each pursued the development and propagation of IQ as a way of lending the credibility and cache of empirical science to their preconceived notions about the inherent inferiority of immigrants, African Americans, and the socioeconomically disadvantaged. Further, they worked in service to a nation similarly committed. The zeal of any particular scientist notwithstanding, no scientific theory or the conclusions drawn from even the most meticulous data will take root if the social context is not already primed to some degree to accept it.

America from Reconstruction onward was primed to look for biological justifications for the retrenchment of many of the social injustices the nation had ostensibly just fought a bloody civil war to end. African Americans were seeking some measure of the freedom and equality they had been promised during and

[8] While there are similarities between competition in democratic societies and competition, in other socio-political contexts, the differences are profound enough to require a completely separate analysis and thus I will not be dealing with non-democratic societies. For the purposes of this paper, I will refer to competition generally but by that, I will mean the very specific circumstances of competition in western liberal democracies. Ideology is pervasive in the formation of institutions and the impact of classical liberal ideology on the formation western liberal democracies and market economies justifies a distinct analysis of the role of competition and technology.

immediately after the war. Conversely, white Americans were, on the one hand, suffering from something akin to buyers remorse and beginning to regret having created a whole new class of economic competitors and having destroyed a system that allowed them to maintain a safe distance from a people they considered wholly other. On the other hand, many Americans, particularly in the newly conquered southern states, had never accepted the prospect of African Americans being granted freedom, let alone equality. These two groups were thus seeking a basis for justifying the establishment of a system of legal and quasi-legal institutions to replicate the effects of slavery without its explicit reestablishment. For many of the same reasons they were seeking to restrain the flow of immigrants into the country with these very same tools in addition to a variety of anti-immigration legislation.

Gould notes that it was the socioeconomic circumstances that served to facilitate an already present desire to use science to classify and rank humanity. Carson's work allows us to understand the persistence of the technology created by the prominent psychometricians profiled by Gould. It is the structural elements of American society—particularly the way in which competition among social groups continues to be organized along racial and class lines, though this often goes unacknowledged in contemporary America—that maintain the popular perception of IQ as a centralized, inherent, real, and hierarchical thing in the head which can be used to segment society based on intrinsic ability without a need for vigorous government intervention. Essentially, IQ is psychology's contribution to the continuing myth of America as an open meritocracy. People are where they are because of a combination of drive and innate ability, and collective efforts to alter those socioeconomic strata are misguided at best.

Three elements combine to form the current context with respect to cognitive enhancement. The belief that America is an open society that rewards individuals in proportion to their merit, which is a combination of motivation and ability, is the first element. The second is the more factually accurate perception of America as a liberal democracy predicated upon free and fair economic and political competition. Finally, there is the belief that IQ is not only a real thing, but that it is also central to the formation of identity and that it confers a tremendous benefit in the two principal areas of American competition, economics and education. These three elements serve to produce a bifurcated effect in public perception of cognitive enhancement. In very specific interpersonal contexts—particularly where the their prosperity, or that of their offspring, is obviously at stake—Americans would appear to be willing to consider seeking cognitive enhancement for the personal benefits that they would gain in competition with their peers. On the other hand, they fear the widespread but uneven introduction of such technologies because of the possibility that they will not be included in the "in group," those that receive the benefits of cognitive enhancement.

I am arguing that a change in the popular perception of intelligence, away from IQ and toward the theory of multiple intelligences, would radically alter public perception of cognitive enhancement. The public would not fear cognitive enhancement's effects on competition in the same way as they do now if they understood intelligence to be something that is distributed to various functional areas of the brain, and variably distributed among the populace. The distribution being not in terms of greater overall quantities of mental ability, but in terms of greater or lesser aptitudes in these same functional areas. In other words, if the public believed that they and their children possessed varying degrees of mental abilities, in a variety of functional areas,[9] and that an increase in one area of intelligence was not equal to an overall increase in intelligence, then they would be more comfortable with the idea of cognitive enhancement. It would be possible for them to increase their child's artistic ability without negatively affecting their neighbor's ability to succeed as an engineer through unfairly skewing competition.

It is important to note that the changes I am hypothesizing would bring public expectations and concerns in line with the present state of science and research in the various fields of neuroscience. Contemporary neuroscience is focusing on examining and creating therapies for specific mental functions more-or-less discretely localized within the brain. It is likely that any enhancement technologies deriving from this research would be similarly specialized, i.e. enhancements focusing on memory, or increasing the speed of executive function, adapting verbal abilities, or changing perceptive abilities and processing. These technologies have very different implications for competition in liberal democracy than would a hypothetical technology designed to increase overall mental ability.

There would be other radical changes accompanying a shift away from general intelligence toward multiple intelligences. The use of standardized tests beyond primary and secondary schools would be problematized, as it became clear that they were measuring fiction and failing to capture more specific aptitudes which might be just as useful in determining success within discrete university programs. It would tend to push the American system toward a softer version of the French system of education and economic production. Individuals would be guided by information about their specific aptitudes toward pursuing advanced education or vocational training suited to those aptitudes rather than being more generically sorted. It is possible to envision increases in overall economic productivity as citizens who were previously prevented from reaching their full potential by the overly general way in which we sorted students by mental ability gave way to a system that focused on helping students find the career that maximized use of their dominant mental abilities. The American system as it stands not only produces unfair socioeconomic advantages both through structural defects and the uneven distribution of extant cognitive enhancers like early childhood nutrition and excellent education, but also denies to some opportunities for higher education through a focus on general mental ability rather than specific aptitude.

[9] e.g. some individuals with very high mathematical aptitude but low social intelligence and others with very high verbal intelligence but low spatial intelligence.

## References

Carr, Nicholas. 2010. *The shallows: What the Internet is doing to our brains*. New York: W. W. Norton & Company, Inc.

Carson, John. 2007. *The measure of merit: Talents, intelligence, and inequality in the French and American Republics, 1750–1940*. Princeton: Princeton University Press.

Gardener, Howard E. 2006. *Multiple intelligences: New horizons in theory and practice*. New York: Basic Books.

Gardner, Howard. 1983. *Frames of mind: The theory of multiple intelligences*. New York: Basic Books.

Gould, Stephen J. 1996. *The mismeasure of man*. New York: W.W. Norton & Company.

Guston, David H., and Daniel Sarewitz. 2002. Real-time technology assessment. *Technology in Society* 24: 93–109.

Herrnstein, Richard J., and Charles Murray. 1994. *The bell curve: Intelligence and class structure in American life*. New York: The Free Press.

Hochberg, Leigh R. 2008. Turning thought into action. *The New England Journal of Medicine* 359(11): 1175–1177.

Jensen, Arthur R. 1969. How much can we boost IQ and scholastic achievement? *Harvard Educational Review* 39: 1–123.

Jensen, Arthur R. 1985. The nature of the black-white difference on various psychometric tests: Spearman's hypothesis. *The Behavioral and Brain Sciences* 8: 193–258.

Joy, William. 2000. Why the future doesn't need us. *Wired* magazine 8.04, April.

Kass, Leon. 2003. *Beyond therapy: Biotechnology and the pursuit of happiness* . New York: Harper Perennial.

Kennedy, Philip R. 2006. The future of brain computer interfacing in a brave new world. *BioSystems Reviews*, February.

Lebow, Richard N. 2000. Contingency, catalysts, and international system change. *Political Science Quarterly* 115(4): 591–616.

Pinker, Steven. 2009. My genome, myself. *The New York Times*, 11 Jan, MM24+.

# Chapter 14
# Towards responsible use of cognitive-enhancing drugs by the healthy*

**Henry Greely, Barbara Sahakian, John Harris, Ronald C. Kessler, Michael Gazzaniga, Philip Campbell, and Martha J. Farah**

Society must respond to the growing demand for cognitive enhancement. That response must start by rejecting the idea that 'enhancement' is a dirty word, argue **Henry Greely and colleagues**.

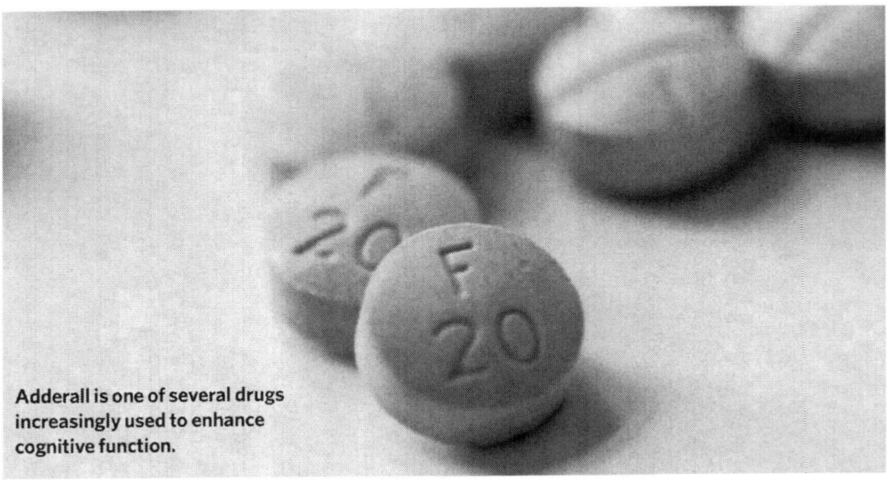


**Adderall is one of several drugs increasingly used to enhance cognitive function.**

---

*Springer Science+Business Media Dordrecht/Nature, 456 (No. 7223), 2008, p. 702-705, Towards responsible use of cognitive-enhancing drugs by the healthy, Greely H, Sahakian B, Harris J, Kessler RC, Gazzaniga M, Campbell P & Farah MJ, with kind permission from Springer Science+Business Media Dordrecht 2012.

H. Greely (✉)
Stanford Law School, Crown Quadrangle, 559 Nathan Abbott Way, Stanford, California 94305-8610, USA
e-mail: hgreely@stanford.edu

S.A. Hays et al. (eds.), *Nanotechnology, the Brain, and the Future*,
Yearbook of Nanotechnology in Society 3, DOI 10.1007/978-94-007-1787-9_14,

Today, on university campuses around the world, students are striking deals to buy and sell prescription drugs such as Adderall and Ritalin — not to get high, but to get higher grades, to provide an edge over their fellow students or to increase in some measurable way their capacity for learning. These transactions are crimes in the United States, punishable by prison.

Many people see such penalties as appropriate, and consider the use of such drugs to be cheating, unnatural or dangerous. Yet one survey (McCabe et al. 2005) estimated that almost 7% of students in US universities have used prescription stimulants in this way, and that on some campuses, up to 25% of students had used them in the past year. These students are early adopters of a trend that is likely to grow, and indications suggest that they're not alone (Maher 2008).

In this article, we propose actions that will help society accept the benefits of enhancement, given appropriate research and evolved regulation. Prescription drugs are regulated as such not for their enhancing properties but primarily for considerations of safety and potential abuse. Still, cognitive enhancement has much to offer individuals and society, and a proper societal response will involve making enhancements available while managing their risks.

B. Sahakian
Department of Psychiatry, University of Cambridge, MRC/Wellcome Trust Behavioural and Clinical, Neuroscience Institute, Cambridge, UK
e-mail: jenny.hall@cpft.nhs.uk

J. Harris
Institute for Science, Ethics and Innovation, and Wellcome Strategic Programme in The Human Body, its Scope, Limits and Future, University of Manchester, Oxford Road, Manchester M13 9PL, UK
e-mail: john.harris@manchester.ac.uk

R.C. Kessler
Harvard Medical School, Department of Health Care Policy, 180 Longwood Avenue, Boston, Massachusetts 02115-5899, USA
e-mail: kessler@hcp.med.harvard.edu

M. Gazzaniga
Sage Center for the Study of Mind, University of California, Santa Barbara, California 93106-9660, USA
e-mail: m.gazzaniga@psych.ucsb.edu

P. Campbell
Nature, 4 Crinan St, London N1 9XW, UK
e-mail: nature@nature.com

M.J. Farah
Center for Cognitive Neuroscience, University of Pennsylvania, 3720 Walnut Street, Room B51, Philadelphia, Pennsylvania 19104-6241, USA
e-mail: mfarah@psych.upenn.edu

## Paths to enhancement

Many of the medications used to treat psychiatric and neurological conditions also improve the performance of the healthy. The drugs most commonly used for cognitive enhancement at present are stimulants, namely Ritalin (methyphenidate) and Adderall (mixed amphetamine salts), and are prescribed mainly for the treatment of attention deficit hyperactivity disorder (ADHD). Because of their effects on the catecholamine system, these drugs increase executive functions in patients and most healthy normal people, improving their abilities to focus their attention, manipulate information in working memory and flexibly control their responses (Sahakian and Morein-Zamir 2007). These drugs are widely used therapeutically. With rates of ADHD in the range of 4–7% among US college students using DSM criteria (Weyandt and DuPaul 2006), and stimulant medication the standard therapy, there are plenty of these drugs on campus to divert to enhancement use.

A newer drug, modafinil (Provigil), has also shown enhancement potential. Modafinil is approved for the treatment of fatigue caused by narcolepsy, sleep apnoea and shift-work sleep disorder. It is currently prescribed off label for a wide range of neuropsychiatric and other medical conditions involving fatigue (Minzenberg and Carter 2008) as well as for healthy people who need to stay alert and awake when sleep deprived, such as physicians on night call (Vastage 2004). In addition, laboratory studies have shown that modafinil enhances aspects of executive function in rested healthy adults, particularly inhibitory control (Turner 2003). Unlike Adderall and Ritalin, however, modafinil prescriptions are not common, and the drug is consequently rare on the college black market. But anecdotal evidence and a readers' survey both suggest that adults sometimes obtain modafinil from their physicians or online for enhancement purposes (Maher 2008).

A modest degree of memory enhancement is possible with the ADHD medications just mentioned as well as with medications developed for the treatment of Alzheimer's disease such as Aricept (donepezil), which raise levels of acetylcholine in the brain (Grön, et al. 2005). Several other compounds with different pharmacological actions are in early clinical trials, having shown positive effects on memory in healthy research subjects (see, for example, Lynch and Gall 2006). It is too early to know whether any of these new drugs will be proven safe and effective, but if one is it will surely be sought by healthy middle-aged and elderly people contending with normal age- related memory decline, as well as by people of all ages preparing for academic or licensure examinations.

## Favouring innovation

Human ingenuity has given us means of enhancing our brains through inventions such as writ- ten language, printing and the Internet. Most authors of this Commentary are teachers and strive to enhance the minds of their students, both by adding substantive information and by showing them new and better ways to process that information. And we are all aware of the abilities to enhance our brains with adequate exercise, nutrition and sleep. The drugs just reviewed, along with newer technologies such as brain stimulation and prosthetic brain chips, should be viewed in the same general category as education, good health habits, and information technology — ways that our uniquely innovative species tries to improve itself.

Of course, no two enhancements are equivalent in every way, and some of the differences have moral relevance. For example, the benefits of education require some effort at self- improvement whereas the benefits of sleep do not. Enhancing by nutrition involves changing what we ingest and is therefore invasive in a way that reading is not. The opportunity to benefit from Internet access is less equitably distributed than the opportunity to benefit from exercise. Cognitive-enhancing drugs require relatively little effort, are invasive and for the time being are not equitably distributed, but none of these provides reasonable grounds for prohibition. Drugs may seem distinctive among enhancements in that they bring about their effects by altering brain function, but in reality so does any intervention that enhances cognition. Recent research has identified beneficial neural changes engendered by exercise (Hillman, et al. 2008), nutrition (Almeida, et al. 2002) and sleep (Boonstra, et al. 2007), as well as instruction (Draganski 2004) and reading (Schlaggar and McCandliss 2007). In short, cognitive-enhancing drugs seem morally equivalent to other, more familiar, enhancements.

Many people have doubts about the moral status of enhancement drugs for reasons ranging from the pragmatic to the philosophical, including concerns about short-circuiting personal agency and undermining the value of human effort (Farah 2004). Kass et al. (2003), for example, has written of the subtle but, in his view, important differences between human enhancement through biotechnology and through more traditional means. Such arguments have been persuasively rejected (for example, Harris 2007). Three arguments against the use of cognitive enhancement by the healthy quickly bubble to the surface in most discussions: that it is cheating, that it is unnatural and that it amounts to drug abuse.

In the context of sports, pharmacological performance enhancement is indeed cheating. But, of course, it is cheating because it is against the rules. Any good set of rules would need to distinguish today's allowed cognitive enhancements, from private tutors to double espressos, from the newer methods, if they are to be banned.

"We should welcome new methods of improving our brain function."

As for an appeal to the "natural", the lives of almost all living humans are deeply unnatural; our homes, our clothes and our food — to say nothing of the medical care we enjoy — bear little relation to our species' "natural" state. Given the many cognitive-enhancing tools we accept already, from writing to laptop computers, why draw the line here and say, thus far but no further?

As for enhancers' status as drugs, drug abuse is a major social ill, and both medicinal and recreational drugs are regulated because of possible harms to the individual and society. But drugs are regulated on a scale that subjectively judges the potential for harm from the very dangerous (heroin) to the relatively harm- less (caffeine). Given such regulation, the mere fact that cognitive enhancers are drugs is no reason to outlaw them.

***Based on our considerations, we call for a presumption that mentally competent adults should be able to engage in cognitive enhancement using drugs.***

## Substantive concerns and policy goals

All technologies have risks as well as benefits. Although we reject the arguments against enhancement just reviewed, we recognize at least three substantive ethical concerns.

The first concern is safety. Cognitive enhancements affect the most complex and important human organ, and the risk of unintended side effects is therefore both high and consequential. Although regulations governing medicinal drugs ensure that they are safe and effective for their therapeutic indications, there is no equivalent vetting for unregulated "off label" uses, including enhancement uses. Furthermore, acceptable safety in this context depends on the potential benefit. For example, a drug that restored good cognitive functioning to people with severe dementia but caused serious adverse medical events might be deemed safe enough to prescribe, but these risks would be unacceptable for healthy individuals seeking enhancement.

Enhancement in children raises additional issues related to the long-term effects on the developing brain. Moreover, the possibility of raising cognitive abilities beyond their species-typical upper bound may engender new classes of side effects. Persistence of unwanted recollections, for example, has clearly negative effects on the psyche (Schacter 2002).

An evidence-based approach is required to evaluate the risks and benefits of cognitive enhancement. At a minimum, an adequate policy should include mechanisms for the assessment of both risks and benefits for enhancement uses of drugs and devices, with special attention to long-term effects on development and to the possibility of new types of side effects unique to enhancement. But such considerations should not lead to an insistence on higher thresholds than those applied to medications.

> ***We call for an evidence-based approach to the evaluation of the risks and benefits of cognitive enhancement.***

The second concern is freedom, specifically freedom from coercion to enhance. Forcible medication is generally reserved for rare cases in which individuals are deemed threats to themselves or others. In contrast, cognitive enhancement in the form of education is required for almost all children at some substantial cost to their liberty, and employers are generally free to require employees to have certain educational credentials or to obtain them. Should schools and employers be allowed to require pharmaceutical enhancement as well? And if we answer "no" to this question, could coercion occur indirectly, by the need to compete with enhanced classmates and colleagues?

Questions of coercion and autonomy are particularly acute for military personnel and for children. Soldiers in the United States and elsewhere have long been offered stimulant medications including amphetamine and modafinil to enhance alertness, and in the United States are legally required to take medications if ordered to for the sake of their military performance (Moreno 2006). For similar reasons, namely the safety of the individual in question and others who depend on that individual in dangerous situations, one could imagine other occupations for which enhancement might be justifiably required. A hypothetical example is an extremely safe drug that enabled surgeons to save more patients. Would it be wrong to require this drug for risky operations?

Appropriate policy should prohibit coercion except in specific circumstances for specific occupations, justified by substantial gains in safety. It should also discourage indirect coercion. Employers, schools or governments should not generally require the use of cognitive enhancements. If particular enhancements are shown to be sufficiently safe and effective, this position might be revisited for those interventions.

Children once again represent a special case as they cannot make their own decisions. Comparisons between estimates of ADHD prevalence and prescription numbers have led some to suspect that children in certain school districts are taking enhancing drugs at the behest of achievement-oriented parents, or teachers seeking more orderly classrooms (Diller 1996). Governments may be willing to let competent adults take certain risks for the sake of enhancement while restricting the ability to take such risky decisions on behalf of children.

The third concern is fairness. Consider an examination that only a certain percentage can pass. It would seem unfair to allow some, but not all, students to use cognitive enhancements, akin to allowing some students taking a maths test to use a calculator while others must go without. (Mitigating such unfairness may raise issues of indirect coercion, as discussed above.) Of course, in some ways, this kind of unfairness already exists. Differences in education, including private tutoring, preparatory courses and other enriching experiences give some students an advantage over others.

Whether the cognitive enhancement is substantially unfair may depend on its availability, and on the nature of its effects. Does it actually improve learning or does it just temporarily boost exam performance? In the latter case it would prevent a valid measure of the competency of the examinee and would therefore be unfair. But if it were to enhance long-term learning, we may be more willing to accept enhancement. After all, unlike athletic competitions, in many cases cognitive enhancements are not zero-sum games. Cognitive enhancement, unlike enhancement for sports competitions, could lead to substantive improvements in the world.

Fairness in cognitive enhancements has a dimension beyond the individual. If cognitive enhancements are costly, they may become the province of the rich, adding to the educational advantages they already enjoy. One could mitigate this inequity by giving every exam-taker free access to cognitive enhancements, as some schools provide computers during exam week to all students. This would help level the playing field.

Policy governing the use of cognitive enhancement in competitive situations should avoid exacerbating socioeconomic inequalities, and should take into account the validity of enhanced test performance. In developing policy for this purpose, problems of enforcement must also be considered. In spite of stringent regulation, athletes continue to use, and be caught using, banned performance-enhancing drugs.

***We call for enforceable policies concerning the use of cognitive-enhancing drugs to support fairness, protect individuals from coercion and minimize enhancement-related socioeconomic disparities.***

## Maximum benefit, minimum harm

The new methods of cognitive enhancement are "disruptive technologies" that could have a profound effect on human life in the twenty-first century. A laissez-faire approach to these methods will leave us at the mercy of powerful market forces that are bound to be unleashed by the promise of increased productivity and competitive advantage. The concerns about safety, freedom and fairness, just reviewed, may well seemless important than the attractions of enhancement, for sellers and users alike.

Motivated by some of the same considerations, Fukuyama (2002) has proposed the formation of new laws and regulatory structures to protect against the harms of unrestrained biotechnological enhancement. In contrast, we suggest a policy that is neither laissez-faire nor primarily legislative. We propose to use a variety of scientific, professional, educational and social resources, in addition to legislation, to shape a rational, evidence-based policy informed by a wide array of relevant experts and stakeholders. Specifically, we propose four types of policy mechanism.

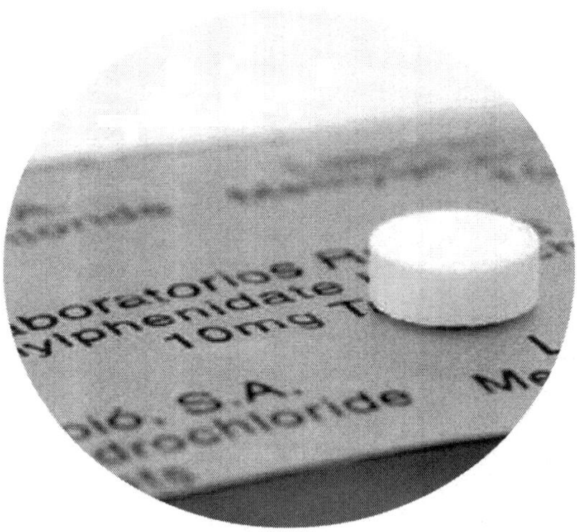

The prescription drug Ritalin is illegally traded among students

The first mechanism is an accelerated programme of research to build a knowledge base concerning the usage, benefits and associated risks of cognitive enhancements. Good policy is based on good information, and there is currently much we do not know about the short- and long-term benefits and risks of the cognitive-enhancement drugs currently being used, and about who is using them and why. For example, what are the patterns of use outside of the United States and outside of college communities? What are the risks of dependence when used for cognitive enhancement? What special risks arise with the enhancement of children's cognition? How big are the effects of currently available enhancers? Do they change "cognitive style", as well as increasing how quickly and accurately we think? And given that most research so far has focused on simple laboratory tasks, how do they affect cognition in the real world? Do they increase the total knowledge and understanding that students take with them from a course? How do they affect various aspects of occupational performance?

***We call for a programme of research into the use and impacts of cognitive-enhancing drugs by healthy individuals.***

The second mechanism is the participation of relevant professional organizations in formulating guidelines for their members in relation to cognitive enhancement. Many different professions have a role in dispensing, using or working with people who use cognitive enhancers. By creating policy at the level of professional societies, it will be informed by the expertise of these professionals, and their commitment to the goals of their profession.

One group to which this recommendation applies is physicians, particularly in primary care, paediatrics and psychiatry, who are most likely to be asked for cognitive enhancers. These physicians are sometimes asked to prescribe for enhancement by patients who exaggerate or fabricate symptoms of ADHD, but they also receive frank requests, as when a patient says "I know I don't meet diagnostic criteria for ADHD, but I sometimes have trouble concentrating and staying organized, and it would help me to have some Ritalin on hand for days when I really need to be on top of things at work." Physicians who view medicine as devoted to healing will view such prescribing as inappropriate, whereas those who view medicine more broadly as helping patients live better or achieve their goals would be open to considering such a request (Chatterjee 2004). There is certainly a precedent for this broader view in certain branches of medicine, including plastic surgery, dermatology, sports medicine and fertility medicine.

Because physicians are the gatekeepers to medications discussed here, society looks to them for guidance on the use of these medications and devices, and guidelines from other professional groups will need to take into account the gatekeepers' policies. For this reason, the responsibilities that physicians bear for the consequences of their decisions are particularly sensitive, being effectively decisions for all of us. It would therefore be helpful if physicians as a profession gave serious consideration to the ethics of appropriate prescribing of cognitive enhancers, and consulted widely as to how to strike the balance of limits for patient benefit and protection in a liberal democracy. Examples of such limits in other areas of enhancement medicine include the psychological screening of candidates for cosmetic surgery or tubal ligation, and upper bounds on maternal age or number of embryos transferred in fertility treatments. These examples of limits may not be specified by law, but rather by professional standards.

Other professional groups to which this recommendation applies include educators and human-resource professionals. In different ways, each of these professions has responsibility for fostering and evaluating cognitive performance and for advising individuals who are seeking to improve their performance, and some responsibility also for protecting the interests of those in their charge. In contrast to physicians, these professionals have direct conflicts of interest that must be addressed in whatever guidelines they recommend: liberal use of cognitive enhancers would be expected to encourage classroom order and raise standardized measures of student achievement, both of which are in the interests of schools; it would also be expected to promote workplace productivity, which is in the interests of employers.

Educators, academic admissions officers and credentials evaluators are normally responsible for ensuring the validity and integrity of their examinations, and should be tasked with formulating policies concerning enhancement by test-takers. Laws pertaining to testing accommodations for people with disabilities provide a starting point for discussion of some of the key issues, such as how and when enhancements undermine the validity of a test result and the conditions under which enhancement should be disclosed by a test-taker.

The labour and professional organizations of individuals who are candidates for on-the- job cognitive enhancement make up our final category of organization that should formulate enhancement policy. From assembly line workers to surgeons, many different kinds of employee may benefit from enhancement and want access to it, yet they may also need protection from the pressure to enhance.

***We call for physicians, educators, regulators and others to collaborate in developing policies that address the use of cognitive-enhancing drugs by healthy individuals.***

The third mechanism is education to increase public understanding of cognitive enhancement. This would be provided by physicians, teachers, college health centres and employers, similar to the way that information about nutrition, recreational drugs and other public-health information is now disseminated. Ideally it would also involve discussions of different ways of enhancing cognition, including through adequate sleep, exercise and education, and an examination of the social values and pressures that make cognitive enhancement so attractive and even, seemingly, necessary.

***We call for information to be broadly disseminated concerning the risks, benefits and alternatives to pharmaceutical cognitive enhancement.***

The fourth mechanism is legislative. Fundamentally new laws or regulatory agencies are not needed. Instead, existing law should be brought into line with emerging social norms and information about safety. Drug law is one of the most controversial areas of law, and it would be naive to expect rapid or revolutionary change in the laws governing the use of controlled substances. Nevertheless, these laws should be adjusted to avoid making felons out of those who seek to use safe cognitive enhancements. And regulatory agencies should allow pharmaceutical companiesto market cognitive-enhancing drugs to healthy adults provided they have supplied the necessary regulatory data for safety and efficacy.

***We call for careful and limited legislative action to channel cognitive-enhancement technologies into useful paths.***

## Conclusion

Like all new technologies, cognitive enhancement can be used well or poorly. We should welcome new methods of improving our brain function. In a world in which human workspans and lifespans are increasing, cognitive enhancement tools — including the pharmacological — will be increasingly useful for improved quality of life and extended work productivity, as well as to stave off normal and pathological age- related cognitive declines (Beddington 2008). Safe and effective cognitive enhancers will benefit both the individual and society.

But it would also be foolish to ignore problems that such use of drugs could create or exacerbate. With this, as with other technologies, we need to think and work hard to maximize its benefits and minimize its harms.

"Many kinds of employee may benefit from enhancement."

**Acknowledgement** This article is the result of a seminar held by the authors at Rockefeller University. Funds for the seminar were provided by Rockefeller University and *Nature*.

**Competing interests** B.S. consults for a number of pharmaceutical companies and Cambridge Cognition, and holds shares in CeNeS. R.C.K. consults for and has received grants from a number of pharmaceutical companies.

## References


1. McCabe, S. E., Knight, J. R., Teter, C. J. & Wechsler, H. *Addiction* 100, 96–106 (2005).
2. Maher, B. *Nature* 452, 674–675 (2008).
3. Sahakian, B. & Morein-Zamir, S. *Nature* 450, 1157–1159 (2007).
4. Weyandt, L. L & DuPaul, G. *J. Atten. Disord.* 10, 9–19 (2006).
5. Minzenberg, M. J. & Carter, C. S. *Neuropsychopharmacology* 33, 1477–1502 (2008).
6. Vastag, B. *J. Am. Med. Assoc.* 291, 167–170 (2004).
7. Turner, D. C. *et al. Psychopharmacology* 165, 260–269 (2003).
8. Grön, G., Kirstein, M., Thielscher, A., Riepe, M. W. & Spitzer, M. *Psychopharmacology* 182, 170–179 (2005).
9. Lynch, G. & Gall C. M. *Trends Neurosci.* 29, 554–562 (2006).
10. Hillman, C. H., Erikson, K. I. & Kramer, A. F. *Nature Rev. Neurosci.* 9, 58-65 (2008).
11. Almeida, S. S. *et al. Nutr. Neurosci.* 5, 311–320 (2002).
12. Boonstra, T. W., Stins, J. F., Daffertshofer, A. & Beek, P. J. *Cell. Mol. Life Sci.* 64, 934–946 (2007).
13. Draganski, B. *et al. Nature* 427, 311–312 (2004).
14. Schlaggar, B. L. & McCandliss, B. D. *Annu. Rev. Neurosci.* 30, 475–503 (2007).
15. Farah, M. J. *et al. Nature Rev. Neurosci.* 5, 421–425 (2004).
16. Kass, L. R. *et al. Beyond Therapy: Biotechnology and the Pursuit of Happiness* (President's Council on Bioethics, 2003); available at www.bioethics.gov/reports/beyondtherapy
17. Harris, J. *Enhancing Evolution: The Ethical Case for Making Better People* (Princeton Univ. Press, 2007).
18. Schacter, D. L. *Seven Sins of Memory* (Houghton Mifflin, 2002).
19. Moreno, J. D. *Mind Wars: Brain Research and National Defense* (Dana Press, 2006).
20. Diller, L. H. *Hastings Cent. Rep.* 26, 12–18 (1996).
21. Fukuyama, F. *Our Posthuman Future: Consequences of the Biotechnology Revolution* (Farrar, Straus and Giroux, 2002).
22. Chatterjee, A. *Neurology* 63, 968–974 (2004).
23. Beddington, J. *et al. Nature* 455, 1057–1060 (2008).

# Chapter 15
# The Opposite of Human Enhancement: Nanotechnology and the Blind Chicken Problem*

**Paul B. Thompson**

## 15.1 Introduction

Like genetic manipulation, but perhaps with more realistic possibilities, nanotechnology is linked to a variety of post-human futures, where consciousness can be "downloaded" onto electronic media, where human sensory apparatus will be linked to spatially dispersed information gathering devices, where intelligence will be distributed amongst various brains and computing capabilities and where the vagaries of the human body will be bolstered by devices that increase its physical power and resistance to external threats. This paper will not engage the ethical and ontological issues of the distant post-human future directly. Instead, I will probe its opposite: the disenhancement of non-human animals' capabilities in the present and near term, a set of technological possibilities exemplified by the blind chicken problem, discussed below. Here we have a set of ethical quandaries that have already been widely discussed, and yet, I will argue, little progress has been made in articulating exactly what the ethical issue actually is. I make no overt claims about the ethics of human enhancement, though my suspicion is that similarly inchoate concerns pervade this area as well.

A surprisingly large literature has developed in response to proposals for relieving distress that animals experience in certain food commodity production environments by means of technological alteration of animals' ability to experience distress.

---

*Springer Science+Business Media Dordrecht/Nanoethics, 2, 2008, p. 305–316, The Opposite of Human Enhancement: Nanotechnology and the Blind Chicken Problem, Paul B. Thompson, Received: 29 October 2008/Accepted: 6 November 2008/Published online: 22 November 2008, with kind permission from Springer Science+Business Media Dordrecht 2012.

P.B. Thompson (✉)
Department of Philosophy, Michigan State University,
503 South Kedzie Hall, East Lansing, MI 48824, USA
e-mail:thomp649@msu.edu

S.A. Hays et al. (eds.), *Nanotechnology, the Brain, and the Future*,
Yearbook of Nanotechnology in Society 3, DOI 10.1007/978-94-007-1787-9_15,

Blind chickens, who suffer less in crowded conditions than sighted birds, are emblematic of these proposals. They are discussed in the opening section of the paper. Although these proposals typically provoke powerful and highly negative moral responses, the animal welfare arguments in support of them should be taken seriously, as argued in the second section. The next three sections review the literature on animal disenhancement starting with the work of Bernard Rollin and moving on to his critics. The first group of critics argues against all genetic engineering of animals. They may be less relevant to the topic at hand than the second group, who argue that disenhancement itself is problematic, rather than the technical means for accomplishing it. Nevertheless, the larger point in reviewing this debate in the context of nanotechnology enabled human enhancement is to illustrate its inconclusiveness, so it is useful to touch on the full range of views. The two sections on Rollin's critics note a pattern of part/whole and type/token fallacies. In the concluding section, I argue that none of the philosophical attempts to resolve the contradiction between intuition and reasoned argument have been successful, and speculate that at least some responses to nanotechnology enabled enhancement of human cognitive capabilities will mimic the tensions and contradictions exhibited by this literature on the blind chicken problem.

## 15.2 Of Blind Hens

Viewed philosophically, the blind chicken problem dates to a 1999 paper written by a group of Danish researchers led by philosopher Peter Sandøe (1999). The paper discussed ethical issues raised by research on a strain of congenitally blind chickens that were less likely to exhibit signs of stress or agitation under crowded conditions. This suggested that blind chickens might be a response to some of the animal welfare problems in poultry production, notably the aggressive behavior of hens crowded together in the battery cage system of egg production that is, at this writing, still the most widely used approach in North America.

The range of possible reactions was exhibited following a 2001 National Public Radio broadcast of the Morning Edition program focused on animal biotechnology where I said the following:

> There's a strain of chickens that are blind, and this was not produced through biotechnology. It was actually an accident that got developed into a particular strain of chickens. Now blind chickens, it turns out, don't mind being crowded together so much as normal chickens do. And so one suggestion is that, 'Well, we ought to shift over to all blind chickens as a solution to our animal welfare problems that are associated with crowding in the poultry industry.' Is this permissible on animal welfare grounds?
>
> Here, we have what I think is a real philosophical conundrum. If you think that it's the welfare of the individual animal that really matters here, how the animals are doing, then it would be more humane to have these blind chickens. On the other hand, almost everybody that you ask thinks that this is an absolutely horrendous thing to do (Kastenbaum 2001).

Not only did I hear from acquaintances I had not seen in 20 years, I was subjected to numerous inquiries from strangers and a few angry phone calls from the U. S.

poultry industry. Poultry producers challenged the suggestion that blind chickens were even being studied (they were wrong) and were irate at the suggestion that they would actually use them, but they were not the only ones who were hostile. For a time, an animal protection group posted a website claiming that I had advocated blinding chickens, urging their membership to write NPR in protest. One can still find a number of disapproving websites where I am described as a "philosopher" in quotation marks.

My critics seemed to think that I was actually promoting the use of blind chickens, though the original context of the David Kastenbaum story makes it even clearer than the quotation above that I was calling attention to the ethical reaction that most people experience when they hear this kind of experiment described. Perhaps they were objecting to the fact that I described it as a "conundrum." How could he think blind chickens could be more humane under any circumstances? But seeing that the blind chicken problem really is a conundrum is critical to the relevance that this problem has for nanotechnology and the ethics of enhancement. Blind chickens are not products of nanotechnology, but we can expect to see an ever-lengthening list of converging technologies that mimic the ethical tensions of the blind chicken problem. I submit that a closer examination of blind chickens reveals a philosophical problem that will not easily be solved, and that will be the source of much confusion and possible mischief in future discussions of human enhancement. By framing the problem in connection to blind chickens instead of human enhancement, we may see that at least some dimensions of the philosophical problem can be generalized beyond ethical intuitions that we associate specifically with the human species. The fact that blind chickens currently exist and have been created using classical techniques of animal breeding indicates both that the kind of problem I have in mind is not "science fiction," and also that it is not uniquely tied to our ability to manipulate matter at the nano scale.

In the following section I will discuss some of the technological strategies for going beyond blind hens in very general terms. Understood as the convergence of molecular biotechnologies, information technology and materials science, nanotechnology will almost certainly be implicated in any actual attempt to pursue such strategies, if only because this convergence can be expected to play some role in the development of almost any technology that depends on manipulations at the cellular level. As Patrick Lin and Fritz Allhoff write, "With nanotechnology, so much is unknown that scientists are really not in a position to accurately forecast what is likely or not and by when," (Lin and Allhoff 2007, p. 12). As such, it is prudent to cast the net of nanoethics widely, and to discuss possible scenarios well in advance of our ability to identify specific technological applications (much less specific nanotechnological applications) that might lead to their realization.

Thus though the main point of this article does not depend upon the existence of nanotechnologies to accomplish ends analogous to congenitally blind hens, it is actually quite plausible to expect a connection between human enhancement nanotechnology and animal disenhancement, at least when considered in terms of technical capability. Some human enhancement technologies envision an interface with neurological activity. There is every reason to expect that the development of an interface technology for "nanoaugmentation" could also be utilized to selectively disrupt specific neural activities, including pain receptors or even sight. What is more, the initial development of such an interface will certainly be done on animal, rather than human brains. Any number of medical technologies, from vaccination to embryo transfer, have been developed and deployed in veterinary contexts well before their use in the human species. This is not to say that we should expect short-term application of neural interface technologies in agricultural settings. Costs may well be prohibitive over the short run, and might remain so. What is more, the conundrum noted above itself might dissuade food animal producers from adopting nanodisenhancement, should it become technically and economically feasible, just as they have thus far resisted blind hens.

## 15.3 Why Change Animal Nature?

In fact, the conundrum is both practical and philosophical. The practical dimensions may indeed be less relevant to nanotechnology and enhancement than the philosophical ones, but it may still be useful to develop the context in which technologies that seek the opposite of enhancement might realistically be deployed. Blind chickens are emblematic of a potentially large class of animals that are modified in response to so-called "production disease." Production diseases are animal pathologies that occur as a result of or in association with livestock production practices. Hens confined at high density in battery cages are prone to feather pecking and cannibalism, aggressive behaviors that may have a defensive, territorial function in the wild. In egg production systems, such behavior is harmful to other hens and leads to injuries that impose cost on the producer in the form of reduced production and increased veterinary care. Beak-trimming is one response to this "production disease," but trimming each individual hen's beak to limit pecking is itself harmful and costly. Other animals and other production practices lead to different production diseases. Large-breasted broiler chickens are susceptible to leg and muscle problems. High producing dairy cows are susceptible to mastitis. Many animals kept in confined settings exhibit obsessive, repetitive movements called stereotipies. In the wake of a global movement toward more humane animal production, researchers are involved in a constant search for responses to these problems. Currently researchers are applying the techniques of genetic engineering, cloning, and cellular manipulation in search of ways to reduce both the suffering and economic cost associated with production disease, (see USDA [United States Department of Agriculture] 2006), and surgical techniques, such as beak trimming, are widely used in industry. Nano-enabled devices or methods for disrupting an animal's ability to experience pain or distress would certainly be adopted if they were available on a cost-effective basis. Or would they?

It is useful to distinguish two conceptually different but equally radical routes to the technological solution to production disease. One might be called the Dumb Down approach. Here researchers identify the genetic or neurological basis for certain characteristics or abilities (such as sight), and produce animals that lack them by removing or otherwise disabling them either genetically or through a nano-mechanical intervention in cellular or neurological processes. The end result of a genetic process might be the headless commodity-producing organism described as "football birds" by Fred Gifford (2002), though I believe that this is exceedingly unlikely. Surgeries that disrupt neurological processes, on the other hand, might well be feasible, and the question is whether some form of nano- or converging technology would make them cost effective. The alternative might be called the Build Up approach. Here, researchers work with cells in vitro, designing scaffolding and other mechanisms that might be produced according to instructions encoded in DNA, to wind up with an organism that yields the animal products (meat, milk and eggs) currently produced using pigs, cows and chickens, (Edelman et al. 2005). This approach might truly yield a quasi-living system that might even involve some elements of animate neural control of organ functions or muscle tension, but without a central nervous system or brain. The practical conundrum is that the need for a response to production disease suggests that agricultural researchers and the animal products industry should be pursuing both Build Up and Dumb Down research streams, though the potential for "yuck factor" responses on the part of the public suggest that perhaps they should not.

There is a philosophical conundrum here because our leading theories of animal ethics tell us that this would be a good thing to do, but our moral intuitions tell us that it is an absolutely horrendous thing to do. Philosophers have used the word "intuition" in many ways, but here I refer to a large class of seemingly immediate and involuntary cognitive experiences. Perceptual intuitions are raw sensations, like the cylindrical white shape I now see against a dull gray background. The white shape is my coffee cup and the background is my desk. However, seeing them as my coffee cup and desk may involve additional processing that I could defer at will, but it is hard to imagine how I could not see these shapes, so long as I can see at all. Linguistic intuitions are the "sense" that we make of words and sentences when they are spoken or visually presented to us. Here, too, there is an involuntariness, a compulsive character that cannot be resisted. If someone says "Move over, loser," I can pretend that I have not heard, but I cannot actually choose whether or not I want to understand, (though I must, of course, understand English idioms to have this linguistic intuition). Moral intuitions are similar in that they are immediate, seemingly involuntary, and do not involve any conscious or thoughtful judgment. When confronted with a given situation (either in practice or, as above, by description), we just react to it as "wrong." It is quite possible that, as in the case of language, we are culturally educated into our moral intuitions, but this does not alter the fact that we seem unable to choose whether or not we will have them.

To use the term "intuition" in this sense does not imply a commitment to intuitionism or any other moral theory holding that intuitions are morally authoritative. Indeed, the normal case is that intuitions blend seamlessly into more carefully considered judgments. In the case of moral intuitions, we typically experience no dissonance between our immediate reaction and the judgment we reach when we thoughtfully review a situation in light of moral principles. But despite their immediacy, intuitions are not always reliable. Sometimes we realize that what we thought we saw or heard was not in fact what was there, or what was actually said was not at all what we thought we heard. The same is true for moral intuitions. In many cases where our first reaction is to think that something is morally wrong, we may be brought around to the idea that it is not wrong after all by reasoning carefully about the situation and considering all of the relevant details. But some moral intuitions are quite robust, and our sense of rightness or wrongness about them may remain even when thinking more carefully about them fails to support the initial reaction. Such intuitions produce conundrums.

The thought of blind chickens producing our table eggs is repulsive; it just strikes us as wrong. But leading theories of animal ethics do not support this judgment. Peter Singer's approach to animal welfare, for example, tells us that we should give equal consideration to interests, without regard to the animal that has these interests. We should take the suffering of animals into account in making our decisions and should not favor choices that produce trivial human benefits simply because the harm or suffering these choices cause happen to occur in nonhuman animals (Singer 2002). Relevant in the present case are interests in avoiding the suffering that is associated with production disease. Conventional animals have these interests, and experience the suffering. Modified animals lack the interests and do not experience the suffering. If our goal is to minimize the unnecessary suffering in the world, as utilitarian philosophers have advocated for over 200 years, the choice seems direct. Organisms that lack the capacity to suffer cannot be harmed, so taking steps to create such organisms seems to be what a utilitarian would have us do.

Perhaps, one might think, a stronger animal rights view would not support this. The position advocated by Tom Regan, for example, would not support the use of blind chickens, for example, because even blind chickens still have an internal life experience, a sense of present and past, and a capacity to live their lives in a manner conducive to their own individual proclivities and interests. They are, as Regan would have it, subjects-of-a-life, and it would be wrong to treat them solely as for our own purposes (Regan 1983, 2003). Gifford's football bird, however, eliminates the capability of experiencing an internal life experience altogether. By Regan's own reasoning, animals (such as insects or protozoa) that lack any conscious capability altogether are not subjects-of-a-life. If we can develop an animal that produces meat, milk or eggs and is not a subject-of-a-life, there is nothing or no one to be harmed by doing so. Further, if doing that is a step toward removing ordinary pigs, cattle and chickens from the production circumstances where their rights are, in Regan's view, currently being violated, it would seem that his ethic of "empty cages" weighs in on the side of developing such literally mindless animals. Thus, to repeat, at least some versions of the blind chicken strategy seem to be supported by animal ethics, but almost everyone thinks that this would be an absolutely horrendous thing to do.

Is it possible that at least some of the moral revulsion expressed by authors who write about human enhancement is similar to the repugnance associated with blind chickens? I am tempted to think that it is, and that high-minded rhetoric stressing the uniqueness of human beings actually obfuscates whatever it is that gives rise to these moral intuitions. However, I will not probe the relationship between human enhancement and animal disenhancement further. Instead I will recount some of the efforts that philosophers have made to pinpoint the trouble that gives rise to an intuition that blind chicken strategies are wrong. My aim will be to show that these efforts are unsatisfactory to the point of being obfuscatory. Here I hint that debates over human enhancement have the potential to follow a similar trajectory, though, of course, that remains to be seen.

## 15.4 Animal Biotechnology and Animal Ethics

The basic conceptual elements of the Dumb Down approach were described over 20 years ago by Bernard Rollin (1986), who gave an extended discussion of them in his 1995 book, The Frankenstein Syndrome. Rollin was thinking primarily of laboratory animals that would be genetically engineered to exhibit particularly devastating forms of human disease for the purpose of biomedical research. In Rollin's view, there are compelling ethical reasons to conduct research on these diseases, and animal models of genetically-based disease would be especially useful in developing therapies. However, the suffering that such animals would endure is tremendous. Because the entire point of creating these animals is to exhibit the disease, there would be no escape. How can this kind of genetic engineering for medical research be ethically justified, Rollin asked?

In partial answer to his own question, Rollin speculated that it might be possible to perform additional genetic engineering as a palliative to the suffering that animals created to model disease might endure. It might, for example, be possible to genetically modify the animal's pain receptors, so that the animal would not experience the torturous and continual pain associated with the disease process. He wrote that it might even be possible to create totally decerebrate animals, animals that experience no conscious life at all. Such animals would have brain functions necessary to maintain breathing, blood circulation and other automatic life support processes, but would lack any capacity for conscious sensory stimulation or awareness (Rollin 1995, 1998).

Rollin originally raised the possibility of animals that cannot suffer in the context of querying the circumstances under which it would be permissible to change an animal's telos, a term that he coined to indicate the genetically based needs, drives and behaviors characteristic of species, subspecies and breeds. His general answer to this question was the Principle of Welfare Conservation: technologically modified animals should not have worse welfare (susceptibility to disease and experience of pain or frustration) than unmodified animals of the same species or breed. Applying the Principle of Welfare Conservation to genetic engineering, Rollin argued that there is nothing intrinsically wrong with changing the genetic make-up of animals, so long as this change did not create animals that were more likely to experience pain, suffering or other deprivations of welfare as a result. Following the same animal ethics logic sketched above, Rollin presumed that organisms biologically incapable of conscious experience or awareness of pain cannot have compromised welfare. This logic was implicit in the first edition of Peter Singer's Animal Liberation (Singer 1975) and it was made explicit by many animal advocates explaining why their concern for animals did not also extend to plants. Plants do not have a "welfare" in the relevant sense, though clearly they can be made better or worse off (Varner 1990).

In short, Rollin concluded that (a) genetic engineering is acceptable as long as the transgenic animal is not made worse off than comparable non-transgenic animals. But the compelling human needs addressed by transgenic animals developed to study disease presented him with a challenge. How can attempts to alleviate the horrible suffering of human victims of genetic disease be denied? So he further concluded that (b) it is not only acceptable but desirable to render animals that would suffer under these conditions genetically incapable of experiencing suffering, that is to do genetic engineering that places them into something very much like a persistent vegetative state. If vegetables do not have a welfare that can be harmed, vegetative animals do not either. To put this as succinctly as possible, if we are to choose between two possibilities, one of which involves experimental animals enduring constant pain and suffering and the latter of which involves creating animals incapable of suffering (hence enduring none), the latter is, on Rollin's logic, obviously the preferable course of action. In his most recent book, Rollin has argued that animal models of human disease are permissible only if the researcher can describe some protocol for alleviating animal pain, though he has abandoned the view that decerebrate animals will ever be produced through genetic engineering (Rollin 2006). Although he does not discuss nanotechnology in this connection, his view would appear to endorse any feasible nano-enabled device for achieving anesthesia or analgesia in animals of the sort he describes.

Rollin's writings have spawned a number of critics, who can be classified into three groups. Most have focused on element (a), his claim that genetic engineering is acceptable in cases where the welfare of animals is not compromised. For this first group of critics any kind of genetic engineering is said to violate "species integrity" or the "dignity of the creature" (Balzer et al. 2000; Colwell 1989; Sapontzis 1991). Some of these critics (the second group) have followed Rollin's use of the term telos to describe the genetic basis of species-characteristic proclivities and drives, but have argued that it would be wrong to alter telos (Fox 1990; Mauron 1989). Thus the key philosophical problem for this group of critics has been to express the basis for their objection in persuasive terms, and various linguistic innovations characterize their attempts. Of course it would be possible to argue that the "yuck factor" response is itself a morally sound argument, and this strategy represents the second and smallest group of respondents (Ortiz 2004). Critics in a final group (the third group) do not object to genetic engineering as such, but have questioned whether Rollin has characterized the sense in which an animal can be made worse off too narrowly (Appleby 1999; Bovenkirk et al. 2001; Holland 1995; Thompson 1997), and blind chickens might be a case in point. Are blind chickens "worse off" than sighted ones? We can consider each type of criticism in turn to see whether the conundrum is addressed.

## 15.5 Rollin's Critics: Against Genetically Engineered Animals

Critics in the first group argue that genetic engineering of animals is intrinsically wrong and have sought some form of argument that at least mimics a Kantian categorical imperative. In this respect, they are making a philosophical argument that parallels the animal rights approach in animal ethics, but the authors I have cited above all appear to recognize that it is not an individual creature that is harmed by blind chicken strategies or modifications of telos, and that to the contrary, the individual animals are better off than they otherwise might be. As such, the "dignity of the creature" noted by Balzer, Rippe and Schaber is dignity in an abstract sense (see also Heeger 2000), or perhaps it is the integrity of the animal that is being affected adversely (Rutgers and Heeger 1999; Warkentin 2006). Rob De Vries (2006) has undertaken a careful analysis of the way that the term 'animal integrity' has been applied to the evaluation of genetic engineering. His analysis shows that for authors who use these terms, "dignity", "integrity" or "telos" must be regarded as something characteristic of species or kinds, perhaps as articulated in the genome, and understood as capable of being harmed or disrespected even in cases where the individual is benefited.

In examining whether the tests De Vries notes can be met in a coherent manner it is worth following out the animal rights logic in a bit more detail. As noted already, it is unlikely that blind hens will win any endorsements from Tom Regan, yet the problem lies not with the fact that they less capable than sighted chickens. Any chicken kept in a cage will violate Regan's ethic of animal rights. What would a rights theorist say about the comparison between blind and sighted hens in egg production? Both cases violate rights, but isn't it a worse offense to inflict suffering on top of that? One general problem in applying rights theory to production disease is that the theoretical commitments of the rights view are so firmly opposed to the very idea of animal production that they seem wholly inapplicable to the ethics of making the best of a bad situation. But the blind chicken problem is not simply eliciting the intuition that keeping chickens in crowded environments is wrong. It is the further intuition that making the best of a bad situation (from the standpoint of the animal's subjective experience) is actually the wrong thing to do. This means that considerations relevant to species or kinds would override the rights of actual animals (i.e., the individuals who instantiate those species or kinds), and this is something that Regan has argued against time and time again (Regan 1983, 1995).

It is possible that something like violation of dignity or integrity captures the essence of a widely felt aversion to biotechnology (see McNaughton 2004). At the least, the language of telos and species integrity gives people something to latch on to. Furthermore the terminology of integrity permits a public airing of the issues, and Bernice Bovenkirk, Frans Brom, and Babs van den Bergh (Bovenkirk et al. 2001) have argued that this is itself the primary argument for adopting this kind of language. But these responses are still opposed by several key points that remain in Rollin's favor. One is that biotechnology can be used to help people and animals, to better their lives. Appeals to integrity and dignity can become pompous when thrown in the face of creatures (of whatever species) who are actively enduring suffering right now.

Second, the claim that it is wrong to violate species integrity, the dignity of the creature or telos seems to overstate the case. At a minimum, critics would need to explain whether such arguments would also forbid routine forms of animal breeding (Sandøe et al. 1996). Bovenkirk, Brom and van den Bergh admit this point, describing the concept of integrity as "flawed but workable," (Bovenkirk et al. 2001, p. 20).

Whatever disclaimers are issued by these authors, the rhetoric of dignity and integrity makes it sound as if actual animals are harmed by manipulations of DNA in a Petri dish. The arguments of Bovenkirk, Brom and van den Bergh seem especially problematic, as they explicate the concept of integrity in connection to human beings and environments, illustrating that it is possible to damage the integrity of either while nevertheless doing things that might be thought good for them from a utilitarian perspective. But in both cases, it is actual people or actual ecosystems that bear the brunt of such an affront. As Rollin himself has argued vehemently, species integrity, dignity and telos are abstractions intended to describe animals as types; they do not describe actual, living and breathing animals at all (Rollin 1998). The point here is that all these objections to Rollin border on, if not actually committing, something between a division fallacy and a type-token fallacy, a confusion between the description or conceptualization of certain interests exhibited by a class of individuals and the actual interests of individuals so classified. Even if it makes sense to say that there is something to debate with respect to species integrity or dignity, it is a confusion to presume that this has anything at all to do with the integrity or dignity of individual animals.

It is also worth noting that Rollin's arguments on transgenic animal disease models differ from the blind chicken problem in important respects. First, there are no human patients anxiously awaiting a cure that provide the ethically compelling force that motivates his entire argument. Second, Rollin's mouse models for human disease were being made worse off through genetic engineering, and "decerebration" was offered as a compensating response. In the case of production disease, there are alternative ways of improving animal welfare, namely, improve the environment. Thus these cases are not strictly comparable. On the one hand, blind chickens or football birds seem to more problematic than decerebrate mice because there is no reason to think that creating them is the only way to address a compelling need. On the other hand, it is doubtful that either the Dumb Down or Build Up strategies violate Rollin's criterion of the Conservation of Welfare. These animals have improved welfare relative to the normal animals in the comparison class, albeit because they are incapable of having their welfare compromised in the normal way. The welfare of a football bird or a neurologically nano-disabled pig may be zero, but zero is better than less than zero.

## 15.6 Rollin's Critics: The Trouble with Disenhancement

This brings us to the second and third groups of critics, all of whom are focused specifically on livestock rather than disease model biotechnology. These critics are not explicitly attempting to develop arguments that would oppose all forms of genetic engineering applied to animals, though is it not entirely clear what forms of modification would be acceptable. Importantly for the present case, the focus of the argument is on modifications that can be understood as disenhancements. Here, the argument is that we would quite reasonably think that a blind chicken is worse off than a sighted one, that strategies to reduce the mental or experiential capacity of animals actually violate Rollin's principle of conservation of welfare.

As noted above, one possible response is simply to accept the disquieting intuitions as definitive: we feel like this is wrong, so it is. This form of argument was put forward by Leon Kass in a widely read response to animal cloning entitled "The Wisdom of Repugnance" (Kass 1997) and was reiterated as a general indictment of biotechnology in the domain of foods (and especially animal foods) by Mary Midgley (2000). Neither Kass nor Midgley was responding directly to Rollin, however. A paper by Sara Ortiz (2004) provides a careful discussion of the literature generated in response to Rollin's argument, concluding simply that the thrust of these critiques is to take the objection to biotechnology "beyond welfare," as if that were enough. However, Ortiz's analysis is useful here in part because in admitting (or in the spirit of her analysis it might be more appropriate to use the word "recognizing") that these arguments not only fail to engage animal welfare concerns but may in fact run counter to them, she also recognizes that there is a conundrum here.

In simply siding with our intuitions, however, authors like Kass, Midgley and Ortiz appear to accept an analysis which admits that that there are no operative reasons at work and no real moral argument that can be deployed to support the conclusion they wish to endorse. Indeed, Rollin himself seems to agree with this judgment when he, in virtually every paper cited above, expresses doubt that such applications of biotechnology will ever come into being because the public's aesthetic response will make the products unmarketable. But in calling this an aesthetic response, Rollin is also stating that it has no moral force. It is, in fact, a form of aesthetic revulsion that, however decisive it may be in determining the economic fate of eggs from blind chickens, appears to morally unjustified in the face of practical opportunities to alleviate the distress of farm animals suffering from production disease. Thus although this second group of critics is making a different type of argument than the first, they, too, seem all too willing to condone the continuation of real harm to animals.

Critics in third group want to claim that disenhancements are, in fact forms of harm to actual animals. Bovenkirk, Brom and van den Bergh's discussion of animal integrity (discussed above) is, in fact, intended to make such a claim, rather than being put forward as a catch-all objection to genetic engineering. They, like Allan Holland (1995) and Mike Appleby (1999), frame the critique in terms of the need to respect what it is animals, by nature, typically are. Appleby cites the Brambell Committee's injunction that animals should be able to engage in natural behaviors (Brambell 1969). For Appleby, the argument begins with a now fairly standard characterization of animal welfare in terms of animal bodies (e.g., physiological and veterinary health), animal natures (behavioral drives) and animal minds (subjective experiences such as pain or satisfaction). Appleby's claim is that animal natures are defined by species typical norms. Thus, blind chickens are worse off than sighted chickens simply because they are blind.

The counter argument to Appleby holds that behavioral drives should be recognized as relevant to the welfare of animals that actually have these drives. Though animal welfare scientists often focus on species-typical data to identify behavioral drives, the justification for doing this is methodological, rather than ontological. In fact, the presumption is that these drives are genetically based; hence changed genetics should change drives. What is more, some behavioral drives are clearly related to conditioning rather than species-typical norms. An animal that has been conditioned to fear human beings will have its welfare compromised on simply seeing one. It is in that animal's nature to fear the presence of humans. This kind of reaction may not be "species typical," but that fact provides no basis for claiming that these conditioned responses are irrelevant to a given animal's welfare. Appleby is thus just wrong to equate animal natures with species-typical norms. What matters is whether a given animal's nature is compromised, and this may have nothing to do with what is typical for the species.

Holland's quasi-Kantian argument suggests that we are disrespecting the animal itself when we undertake measures that alleviate suffering so that we may continue in what is, at bottom, an exploitative relationship. Thus, blind chicken strategies are like offering assembly line workers an aspirin in lieu of better working conditions. Both are responses that ameliorate distress, but do so in a way that is an affront to the dignity of the distressed individual. I am myself the third critic in this second group, and I offered two criticisms in a 1997 paper (Thompson 1997) that also includes much more detailed discussion of some reasons to think that those who have chosen to articulate the problem in terms of integrity or dignity have gotten it wrong. In that paper I argued that Rollin's Principle of Conservation of Welfare could as easily be applied to behavioral or surgical interventions as to genetic research, but that here we would be reluctant to think that eliminating a capacity or felt need makes an animal no worse off than it was before. I also offered the following claim:

> Clearly the telos that is characteristic of any species (including humans) is instantiated only in the individuals of the species. If we recognize immorality in acts that would modify a human genome to the point that the resulting individual would no longer be characteristic of the human species, why is it not also immoral to modify the genome of other animals so that the resulting individuals are uncharacteristic of their species? Until someone can offer a non-arbitrary reason for making this distinction, radical forms of transgenesis for animals should be regarded as morally problematic (Thompson 1997, p. 20).

Today, I am not confident that any of these objections do more philosophical work than the suggestion that we err because we violate the integrity or dignity of animals. I would submit that the approaches of Holland, Appleby and myself all have the virtue of avoiding the most egregious logical fallacies and the misleading tone of the other critics, but I am not sure that anyone has done anything much more that paste a philosophical label on the "yuck factor" intuition that blind chicken problem provokes. My own treatment is particularly vulnerable to this criticism.

It is one thing to talk about natural behavior or an animal's nature as constructs that are intended to call attention to an animal's entire life, or to the way that it fits within its farm environment. This kind of talk may be useful in calling our attention to how farm animals actually fare. It is precisely this point that Bovenkirk, Brom and van den Bergh have in mind when they defend the term "integrity." It is something else again to say that such talk provides a unilateral argument against adjusting the fit between animal and environment by adjusting the animal, rather than the environment. "Integrity" or "animal nature" may give us terms on which to hang our considered moral intuition that there is something wrong with blind chickens, football birds and the Dumb Down strategy in general, but it is a response that invites us to conflate actions that actually cause harm to real, live farm animals with actions that actually relieve harm, when compared to the alternative that would be most likely to prevail. It is only when our understanding of actual welfare associated with possible alternative courses of action is in view that the considerations of animal ethics have force. All of the options thus far considered for explaining why blind chickens and the Dumb Down approach might be morally wrong do so by taking our attention away from the conditions in which animals actually live.

## 15.7 Resolving the Conundrum or Admitting Defeat

Is it possible to resolve the blind chicken conundrum? In fact, I think not, but there are a number of lingering points that must be addressed before admitting defeat. First, we must recognize that a driving factor behind the persistence of our intuition that there is something wrong here may well be the presumption that there are other, more straightforward ways to address livestock production disease. Why not give the chickens more room? In fact, the answer to a question like this is very similar to the answer that factory owners might give to someone who takes them to task for offering aspirin rather than improving working conditions: it's easier said than done. In fact, chickens in non-cage systems also experience stress associated with visual stimulation, though in their case it may have more to do with large group size than crowding. In any case, beak trimming is believed to be necessary in virtually all egg production systems that operate at a commercially-viable scale (Savory 1995; Tausin 2002).

A thorough discussion of animal production disease would take the present inquiry into the details of diverse agricultural commodity markets and animal production systems, trying the patience of any reader interested in the ethics of nanotechnology or human enhancement. It must suffice to say that technological responses to the animal welfare problems associated with livestock production disease become ethically attractive when seemingly more desirable correctives become difficult if not impossible to implement in the face of political and economic circumstances. Of course the very repugnance these responses provoke can also change those circumstances. The practical, agricultural ethics of production disease will depend very much on where one sits and the choices that one has. To the extent that political and economic realities remain what they are, it is important for anyone involved in livestock production to at least consider the argument for blind chickens.

Thus if one possibility is simply to go with the "wisdom of repugnance," another possibility is to accept the possibility that what ethical theory tells us is right, and that our intuitions are simply mistaken in this family of cases. In this connection, it is worth noting that the Build Up and Dumb Down strategies actually engage our intuitions in somewhat different, if also overlapping, ways. The reaction to Build Up may simply be, "Yuck. I don't want to eat that." Manipulating cells in a Petri dish does not seem on the face of it to involve moral concerns, and the organisms that will result from this line of research seem distant enough from actual chickens for one to plausibly claim, "No animals were harmed in producing this meal." Although built up organisms might still face the aesthetic revulsion that Rollin sees as the primary barrier to football birds, I, at least, would be inclined to say that willfully choosing animal suffering over this approach is morally problematic in a fairly straightforward way. In contrast, the intuitive rejection of Dumb Down strategies puts morality into play in a way that (prima facie, at least) counters the arguments from animal welfare: it's not only disgusting and distasteful, it's morally wrong.

The persistence of this intuition suggests that the blind chicken problem is a problem for human conduct in itself and not in regard to its impact on animals. It is our attitude to animal natures that troubles Appleby and Holland, but this attitude does no harm to the organism that emerges at the convergence point between Dumb Down and Build Up. That organism lacks the capacity to be, in Tom Regan's terms, the subject of a life. There is no subject there to which we can show disrespect. Were we to come upon such an organism existing in nature, would using it agriculturally provoke any distasteful intuitions at all? It seems unlikely. The problems seem to arise in connection with setting out to displace familiar animals in favor of this mindless blob. Contrary to the quasi-Kantian elements in Holland's analysis, it is not so much that doing so shows disrespect to any actual sheep, pig, cow or chicken. Indeed, perhaps we will continue to associate with much smaller numbers of them in mutually satisfactory settings. The problem seems to be that the entire project exhibits the vices of pride, of arrogance, of coldness and of calculating venality. The suspicion that a more radical restructuring of industrial animal production might alleviate the need for blind chickens reinforces this presumption (see Davis 1996).

To put the point another way, it is not the disrespect that animals suffer that is focus of what is wrong with blind chicken strategies. It is disrespectfulness as a pattern of behavior or a character trait on the part of the agent that is at the heart of the issue. In this connection, Holland's analysis can be seen to trade upon an ambiguity associated with the word "respect". In its usual Kantian connotation, respect is owed to others, and we say that they are harmed when they do not get it. I take it that this is exactly what Holland intends when he says that blind chicken strategies disrespect animals even while alleviating their distress. But respectfulness is a virtue that can describe the character of a person even when their actions in particular circumstances achieve less than they might like. To revisit one last time the analogy to assembly line workers whose dignity is offended by the offer of aspirin in lieu of better working conditions, we might perceive a factory owner up against the wall of economic competition far more favorably than the market leader who sets the terms of competition. It is not that workers in the latter's factory are harmed in a manner that workers in the former's are not. It is the relative virtue or character of each owner that is at issue in marking the moral difference.

Can we then say that our intuitions about the wrongness of blind chickens are captured when we articulate this as a problem in human virtue, a problem with the kind of moral character that people who would do such a thing might have? If so, does this answer carry over to the qualms that may be felt about human enhancement? I am not sanguine with a positive response to either question, though at present I have nothing more insightful to offer. The suggestion that we can draw upon virtue ethics resituates the philosophical problem by shifting our attention away from a better account of harm to animals and toward those practices and traditions we associate with good and bad moral character. But it is hardly clear that resituating the argument this way makes it any more convincing. We are still left with the practical problem of suffering from production disease and thus we are still left with a conundrum.

The more important point to draw is that the inchoate nature of concerns about blind chickens has spawned a minor industry of linguistic innovation, where various forms of intrinsic value, species or animal integrity, dignity and telos have been proffered as the thing being harmed, compromised or offended. Despite Bovenkirk et al. (2001) suggestion that such innovations at least give us a way to express the fact that many of us think that this is an absolutely horrendous thing to do, these terms do little or nothing to articulate why it is wrong. It is not altogether clear that resituating the moral issue as one focused on the virtue of the agents is an improvement, either. My conjecture is that the coming debate over nano-enabled human enhancement will find itself facing analogous conundrums, allegedly resolved by analogous terminological innovations. As noted at the outset, we should expect that any interface for neural enhancements will also enable disenhancements, as well. But more generally, the gap between persistent intuitions and explicitly articulated moral concerns looms large for the debate over human enhancement itself. Too often linguistic innovations that name our inchoate concerns also allow us to neglect our responsibility to probe the basis and legitimacy of our moral intuitions. It may even encourage us to think that we have resolved a conundrum when we have, in fact, done nothing more than conceal it.

## References

Appleby, M.C. 1999. *What should we do about animal welfare?* Oxford: Blackwell Science.

Balzer, P., K.P. Rippe, and P. Schaber. 2000. Two concepts of dignity for humans and non-human organisms in the context of genetic engineering. *Journal of Agriculture and Environmental Ethics* 13: 7–27.

Bovenkirk, B., F.W.A. Brom, and B.J. van den Bergh. 2001. Brave new birds: The use of integrity in animal ethics. *The Hastings Center Report* 32(1): 16–22. doi:10.2307/3528292.

Brambell, F.W. 1969. *Report of the technical committee to enquire into the welfare of animals kept under intensive livestock husbandry systems*. London: Her Majesty's Stationery Office.

Colwell, R.K. 1989. Natural and unnatural history: Biological diversity and genetic engineering. In *Scientists and their responsibilities*, ed. W.R. Shea and B. Sitter, 1–40. Canton: Watson Publishing International.

Davis, K. 1996. The ethics of genetic engineering and the futuristic fate of domestic fowl. United Poultry Concerns Website. Available at http://www.upc-online.org/genetic. html. Accessed 13 Jan 2006.

de Vries, R. 2006. Genetic engineering and the integrity of animals. *Journal of Agriculture and Environmental Ethics* 19: 469–493. doi:10.1007/ DOI:dx.doi.org s10806-006-9004-y DOI:dx.doi.org.

Edelman, P.D., D.C. McFarland, V.A. Mironov, and J.G. Matheny. 2005. In vitro-cultured meat production. *Tissue Engineering* 11: 659–662. doi:10.1089/ten.2005.11.659.

Fox, M.W. 1990. Transgenic animals: Ethical and animal welfare concerns. In *The bio-revolution: Cornucopia or Pandora's box*, ed. P. Wheale and P. McNally, 31–54. London: Pluto.

Gifford, F. 2002. Biotechnology. In *Life science ethics*, ed. G. Comstock, 191–224. Ames: Iowa State Press.

Heeger, R. 2000. Genetic engineering and the dignity of creatures. *Journal of Agriculture and Environmental Ethics* 13: 43–51.

Holland, A. 1995. Artificial lives: Philosophical dimensions of farm animal biotechnology. In *Issues in agricultural bioethics*, ed. T.B. Mepham, G.A. Tucker, and J. Wiseman, 293–306. Nottingham: University of Nottingham.

Kass, L. 1997. The wisdom of repugnance. *The New Republic* 216: 17–26. June 2.

Kastenbaum, D. 2001. Analysis: Debate over genetically altered fish and meat. Morning Edition (December 4, 2001). Transcript available online at http://www.npr.org. http://www.npr.org/templates/story/story.php?storyId=1134248. Accessed 25 June 2008.

Lin, P., and F. Allhoff. 2007. Nanoscience and nanoethics: Defining the disciplines. In *Nanoethics: The ethical and social implications of nanotechnology*, ed. F. Allhoff, P. Lin, J. Moor, and J. Weckert, 3–16. Hoboken: Wiley-Interscience.

Mauron, A. 1989. Ethics and the ordinary molecular biologist. In *Scientists and their responsibilities*, ed. W.R. Shea and B. Sitter, 249–265. Canton: Watson Publishing International.

McNaughton, P. 2004. Animals in their nature: A case study on public attitudes to animals genetic modification and 'nature'. *Sociology* 38: 533–551. doi:10.1177/0038038504043217.

Midgley, M. 2000. Biotechnology and monstrosity. *Hastings Center Report* 30(5): 7–15. doi:10.2307/3527881.

Ortiz, S.E.G. 2004. Beyond welfare: Animal integrity, animal dignity and genetic engineering. *Ethics and the Environment* 9: 94–120. doi:10.2979/ETE.2004.9.1.94.

Regan, T. 1983. *The case for animal rights*. Berkeley: University of California Press.

Regan, T. 1995. Are zoos morally defensible? In *Ethics on the ark*, ed. B.G. Norton, M. Hutchins, E.F. Stevens, and T. Maple, 38–51. Washington, DC: Smithsonian Institution.

Regan, T. 2003. *Animal rights, human wrongs: An introduction to moral philosophy*. Lanham: Rowman & Littlefield.

Rollin, B. 1986. The Frankenstein thing. In *Genetic engineering of animals: An agricultural perspective*, ed. J.W. Evans and A. Hollaender, 285–298. New York: Plenum Press.

Rollin, B. 1995. *The Frankenstein syndrome: Ethical and social issues in the genetic engineering of animals*. New York: Cambridge University Press.

Rollin, B. 1998. On telos and genetic engineering. In *Animal biotechnology and ethics*, ed. A. Holland and A. Johnson, 156–187. London: Chapman & Hall.

Rollin, B. 2006. *Science and ethics*. New York: Cambridge University Press.

Rutgers, B., and R. Heeger. 1999. Inherent worth and respect for animal integrity. In *Recognizing the intrinsic value of nature*, ed. M. Dol, M. Fentener van Vlissingen, S. Kasanmoentalib, T. Visser, and H. Zwart, 41–53. Assen: Van Corcum.

Sandøe, P., N. Holtung, and H.B. Simonsen. 1996. Ethical limits to domestication. *Journal of Agriculture and Environmental Ethics* 9: 114–122.

Sandøe, P.B., L. Nielsen, L.G. Christensen, and P. Sørensen. 1999. Staying good while playing God—The ethics of breeding farm animals. *Animal Welfare* 8: 313–328.

Sapontzis, S.F. 1991. We should not manipulate the genome of domestic hogs. *Journal of Agriculture and Environmental Ethics* 4: 177–185. doi:10.1007/BF01980315.

Savory, C.J. 1995. Feather pecking and cannibalism. *World's Poultry Science Journal* 51: 215–219. doi:10.1079/WPS19950016.

Singer, P. 1975. *Animal liberation*. New York: Avon Books.

Singer, P. 2002. *Animal liberation*, revisedth ed. New York: Harper- Collins.

Tausin, R. 2002. Furnished cages and aviaries: Production and health. *World's Poultry Science Journal* 58: 49–63. doi:10.1079/ DOI:dx.doi.org WPS20020007 DOI:dx.doi.org.

Thompson, P.B. 1997. Ethics and the genetic engineering of food animals. *Journal of Agriculture and Environmental Ethics* 10: 1–23. doi:10.1023/ DOI:dx.doi.org A:1007758700818 DOI:dx.doi.org.

USDA (United States Department of Agriculture). 2006. ARS Project: Identification and manipulation of genetic factors to enhance disease resistance in cattle. Available at http://www.ars.usda.gov/research/projects/projects.htm? ACCN_NO=405817&showpars=true&fy=2003. Accessed 13 Jan 2006. Page last modified 12 Jan 2006.

Varner, G. 1990. Biological functions and biological interests. *Southern Journal of Philosophy* 27: 251–270.

Warkentin, T. 2006. Dis/integrating animals: Ethical dimensions of the genetic engineering of animals for human consumption. *AI & Society* 20: 82–102. doi:10.1007/s00146-005- DOI:dx.doi.org0009-2 DOI:dx.doi.org.

# Chapter 16
# National Citizens' Technology Forum: Nanotechnologies and Human Enhancement

**Patrick Hamlett, Michael D. Cobb, and David H. Guston**

## 16.1 Executive Summary

Many observers believe that the "converging technologies" of nanotechnology, biotechnology, information technologies, and cognitive science (NBIC) could lead to radical and pervasive enhancements of human abilities. Both supporters and critics of NBIC technologies acknowledge that their continued development and deployment portend dramatic social and cultural challenges. Stakeholders see a need for *informed citizen input* early in the process of developing such technologies. Indeed, the legislation that authorizes the US National Nanotechnology Initiative (P.L. 108–93) speaks to the importance of public input in decision-making about such research and development.

This report discusses the results of one major effort at public input. In March 2008, the Center for Nanotechnology in Society at Arizona State University (CNS-ASU) and its collaborators at North Carolina State University held the nation's first "National Citizens' Technology Forum" (NCTF), on the topic of nanotechnology and human enhancement. Organizers selected from a broad pool of applicants a diverse and roughly representative group of 74 citizens to participate at six

P. Hamlett
Department of Interdisciplinary Studies, College of Humanities and Social Sciences, 106D 1911 Building, Campus Box 7107, Raleigh, NC 27695-7107
e-mail: phamlett@ncsu.edu

M.D. Cobb (✉)
Department of Political Science, North Carolina State University, 223 Caldwell, Campus Box 8102 Raleigh, NC 27695
e-mail: mike_cobb@ncsu.edu

D.H. Guston
The Center for Nanotechnology in Society, Arizona State University, P.O. Box 875603, 85287-5603 Tempe, AZ, USA
e-mail: david.guston@asu.edu

S.A. Hays et al. (eds.), *Nanotechnology, the Brain, and the Future*, Yearbook of Nanotechnology in Society 3, DOI 10.1007/978-94-007-1787-9_16,

geographically distinct sites across the country. Participants received a 61-page background document – vetted by experts – to read before deliberating. They also completed a pre-test questionnaire to record their initial attitudes and understandings of the topic. They deliberated face-to-face in their respective geographic groups for one weekend at the beginning of the month, and they deliberated electronically across their geographic groups in nine, 2-h sessions during the rest of the month. Electronic deliberations included question-and-answer sessions with a diverse group of topical experts. The NCTF concluded with a second face-to-face deliberation at each site. Participants drafted reports that represented the consensus of their local groups, and they completed a post-test questionnaire to record their perspectives on the NCTF and any changes in their attitudes and understandings.

### *16.1.1 Findings from the Reports Include the Following*

**Unanimous support**. The reports from all six sites highlighted concern over the effectiveness of regulations for NBIC technologies and the need to provide public information, including more public deliberative activities and K-12 education, about NBIC technologies.

**Near-unanimous support**. Five of six site reports expressed concern about the equitable distribution of new enhancement technologies. The also determined therapeutic technologies to be of greater importance than enhancement research. The sites emphasized the important role that stakeholders might play in setting that research agenda. They felt the need for careful monitoring of such technologies, and the development of international safety standards for them. Finally, they desired the development of such technologies to maximize their benefits with both public and private investment.

**Majority support**. Four of six site reports identified as important the formal inclusion of ethicists and ethical considerations into decision-making for NBIC technologies, the careful protection of individual privacy in the development and deployment of these technologies, and the potentially problematic role of health insurance in limiting access to new enhancement technologies.

**Split support**. Three of six site reports noted concerns that NBIC technologies might fall into the hands of terrorists or have other unanticipated military applications, concerns about potential environmental consequences, and ensuring the protection of civil liberties and free choice – particularly the choice to refuse enhancements.

Findings about the participants' views on human enhancement technologies from the pre- and post-test questionnaires include: (1) reduced certainty about the benefits of human enhancement technologies; (2) increased worry about the affordability of NBIC enhancements and overwhelming support for the government to guarantee access to them if they prove too expensive for the average American; (3) reduced, but still strong, support for publicly funded research for developing human enhancement technologies; (4) conflicting emotions – continued, extensive hope and increased worry – about NBIC developments; and (5) opposition to many particular kinds of hypothetic human enhancements as described in the background literature.

Findings about the participants' experience on the NCTF process from the pre- and post-test questionnaires include: (1) significant increases in the percentage of participants who hold opinions about NBIC technologies; (2) significant substantive learning by participants about the details of nanotechnology and human enhancement technologies; (3) very high levels of individual support for the conclusions of the respective geographic groups; (4) increased feelings of efficacy and trust as a result of participants' role in the NCTF; and (5) changes in preferences from on-line mediated deliberations to face-to-face deliberations.

We conclude that average citizens want to be involved in the technological decisions that might end up shaping their lives. Citizens remain supportive of research that might lead even to transformational technologies, if reliable information about and attentive and trustworthy oversight of their development exists. Such information and oversight should not be restricted to environmental health and safety but should include social risks such as equity, access, and civil rights. With the appropriate information and access to experts, citizens are capable of generating thoughtful, informed, and deliberative analyses that deserve the attention of decision makers.

## 16.2 Introduction

A new area of technological change has been emerging onto the agendas of decision makers around the globe: the "converging technologies" of nanotechnology, biotechnology, information technologies, and cognitive science (NBIC). Many observers believe that these new technologies could lead to radical and pervasive enhancements of human abilities. Some visionaries expect NBIC technologies to dramatically enhance strength and endurance, alleviate or eliminate pain, improve or restore sight and hearing, enhance memory, speed information processing, spark artistic expression, and extend life.

Some anticipate, however, significant social change as these technologies move into widespread use, and many are concerned about public reactions them: What would relationships between "enhanced" and "un-enhanced" people in society be like? What does fairness mean when previously immutable aspects of a person's abilities are alterable? What would significantly increased life expectancy do to families, to work, to cultural continuity and innovation, and to society more generally?

These concerns have led to a flurry of interest among scholars, policymakers, and interest groups both in the United States and Europe. A number of committees, conferences, and scholars have generated in-depth reports about human enhancement technologies in general and NBIC implications in particular (see p. 284 for Selected Further Readings). Despite their disagreements on the prospective value of these new technologies, both supporters and critics of NBIC acknowledge that their continued development and deployment portend dramatic and powerful social and cultural challenges.

Such promises and challenges raise the stakes for the development and introduction of NBIC technologies, and many people across government, business, academe, and public interest and advocacy groups see a great need for *informed citizen input* early in the process of developing such potentially revolutionary technologies. With numerous examples of major technologies having become entangled in divisive political conflict and legal action—e.g., nuclear energy and genetically modified foods—decision makers are often eager to find ways to elicit the values and concerns of ordinary people and incorporate them into the process of developing these technologies. Indeed, the federal legislation that authorizes much of the US National Nanotechnology Initiative (Public Law 108–93) speaks to importance of public input in decision-making about nanotechnology research and development.

## 16.3 Consensus Conferences and Citizens' Technology Forums

In recent decades, new techniques for eliciting informed, deliberative public opinion have been developed and used in several countries. These techniques are often more helpful than traditional public opinion polls when the topics of concern are those, like emerging technologies, about which the public has initially very modest levels of information.

One of these practices, developed in Denmark and known as a "Consensus Conference," involves recruiting ordinary, non-expert citizens, providing them with background information and access to experts on the particular topic, and assisting them as they deliberate toward a set of agreed-to recommendations. The Danish Parliament's Board of Technology, which organizes the consensus conferences, helps communicate the recommendations to the parliament, the press, and the public.

Over the past 10 years, a technique based on the Danish consensus conferences – called the "Citizens' Technology Forum" (CTF) – has been developed by scholars at North Carolina State University for use in the American context. To the original Danish model, the CTF adds the Internet as a mode of interaction, in addition to face-to-face interactions, among the citizen participants. On-line communication allows deliberations involving multiple groups of citizens in multiple geographic locations – a crucial innovation if such a process is to take root across a country that spans a continent and has multiple population centers, compared to one roughly twice the size of Massachusetts with one central city.

The CTFs conducted in the US, which have examined topics including genetically modified foods, climate change, and nanotechnology, have usually been run in university contexts as research and demonstration projects and have not been part of official policy making bodies. As part of this research orientation, many CTFs have included questionnaires administered to the participants before and after their participation, allowing researchers to collect significant amounts of data about processes of learning, attitude changes, and personal interactions in which citizens engage.

## 16.4 The National Citizens' Technology Forum

In March 2008, the first National Citizens' Technology Forum (NCTF) took place. It employed the basic CTF process, but this time involved six locations across the country, and the participation of 74 individuals.

The NCTF was organized under the auspices of the Center for Nanotechnology in Society at Arizona State University (CNS-ASU), which is funded by the National Science Foundation to perform research, training and outreach on the societal aspects of nanotechnology. The six sites participating in the NCTF, representing six distinct regions of the country, were:

- the University of New Hampshire (Durham), in the North East;
- Georgia Institute of Technology (Atlanta), in the South;
- the University of Wisconsin (Madison), in the Upper Midwest;
- the Colorado School of Mines (Golden), in the Mountain region;
- Arizona State University (Tempe), in the South West; and
- the University of California (Berkeley), on the West Coast.

The results of the study, therefore, are not limited to one section of the country, but reflect a truly national, informed, deliberative public assessment of NBIC potentials.

Each campus formed a facilitation team including a faculty leader and other assisting faculty and students. A complete list of facilitation team members can be found in Appendix A. Dr. Patrick Hamlett (North Carolina State University, NCSU), coordinated the overall project, including the on-line components, and Dr. Michael Cobb (NCSU) oversaw the data gathering and analysis elements. Drs. Hamlett and Cobb both have experience in running the earlier CTFs and, under subcontract from CNS-ASU, coordinated many of the operational aspects of the NCTF as well.

### *16.4.1 Panelists*

Each geographic site recruited its own panelists using newspaper and Internet advertising. While some sites attracted large numbers of volunteers and other sites attracted fewer – possibly due to the exotic and unfamiliar nature of the technologies in question – each site endeavored to create panels that were broadly representative of the communities from which they were drawn. Prospective panelists also answered a questionnaire to elicit demographic information and discover any possible conflicts of interest. Efforts at matching local and, in aggregate, national demographics were largely successful (see Appendix B), although both applicants and participants were somewhat more liberal and educated than the population as a whole. A small number of potential panelists were excluded for reasons of conflict.

Panelists were required to have Internet access in order to participate, although sites also arranged for the use of local libraries or other accessible venues if that became a hardship for participants. Because of the intensive nature of the NCTF and the considerable time commitment involved– two full weekends of face-to-face (F2F) meetings and 18 h of Internet, or keyboard-to-keyboard (K2K) communication – organizers paid the participants $500 upon completion of their duties.

### 16.4.2 Background Materials

The organizers prepared a 61-page background document and delivered it to each panelist before the first F2F meeting. The document, describing the emergence of NBIC technologies and the debates about their anticipated social impacts, was drafted and edited by many researchers across CNS-ASU.

Following the Danish pattern, an Oversight Committee reviewed drafts of the document to help ensure that the materials were accurate, balanced, and accessible. The Oversight Committee consisted of Ida-Elisabeth Andersen, a project manager for the Danish Board of Technology in Copenhagen, and David Rejeski, the director of the Project on Emerging Nanotechnologies of the Woodrow Wilson International Center for Scholars in Washington, DC. The background document is available at http://www4.ncsu.edu/~pwhmds/.

### 16.4.3 Pre-and Post-tests

A pre- and post-test questionnaire was developed and administered to all panelists. The questionnaires assess several possible impacts of participation by the citizens, including factual learning and shifts in attitudes about NBIC technologies, as well as qualities of the deliberative process itself, including the presence and strength of cognitive and affective pathologies of deliberation and the level of consensus among the participants.

### 16.4.4 First F2F Weekend

During the first weekend of the NCTF, citizens gathered for face-to-face discussions that were led by facilitators from each of the campuses. The panelists discussed the background materials, the structure and goals of the project, and began to raise whatever concerns or issues they found significant. While the background document provided substantial information and framed the inquiry, the panelists had significant control over what specific issues or concerns should be addressed.

### 16.4.5 Internet Elements

The panelists from all six sites joined for nine, 2-h synchronous online discussion sessions. During these Internet sessions, panelists from all the sites were exposed to the concerns, interests, values, and perspectives of panelists at all the other sites. During some of these sessions, content experts joined the discussions to respond to follow-on questions developed by the panelists and to fill in any gaps in the background materials. The content experts included technical specialists, a philosopher, and a specialist in regulatory processes. The content experts who participated were:

- Dr. Roberta M. Berry, the Georgia Institute of Technology (a specialist on the legal, ethical, and policy implications of life sciences research and biotechnologies);
- Dr. Steven Helms Tillery, Arizona State University (a specialist on cortical neuroprosthetics);
- Dr. Maxwell J. Mehlman, Case Western Reserve University (a specialist in the federal regulation of medical technology);
- Dr. Kristin Kulinowski, Rice University (Executive Director of the Center for Biological and Environmental Nanotechnology); and
- Dr. Jason Scott Robert, Arizona State University (a philosopher of science and the bioethicist).

### *16.4.6 Final F2F Weekend*

The panelists gathered for a second and final F2F weekend during which they reconsidered the issues, problems, and concerns they had expressed during the first weekend in light of the additional information and discussions provided by the Internet sessions. Working with a facilitator, they then deliberated toward a set of policy recommendations that all panelists could endorse. The panelists themselves then compiled these recommendations into each site's Final Report.

### *16.4.7 Final Reports*

After each panel reached a consensus among its members about what recommendations to advance, the panelists at each site wrote a Final Report representing that consensus. Each site's Final Report (available at http://www4.ncsu.edu/~pwhmds/final_reports.html) contains the specific recommendations that each panel endorsed, along with a discussion of issues, concerns, and values the panelists believe should be important in the management of NBIC technologies.

While the panelists at each site had been exposed to the concerns and issues panelists at the other sites thought were important, there was no effort to reach a single consensus involving all six sites; thus, each Final Report represents concerns and issues specific to that site. Nevertheless, when we compare the Final Reports, we find significant overlap among all six sites.

### 16.4.8 Findings from the Reports

An examination of the recommendations contained in the Final Reports illustrates the areas of concern expressed by panelists from all areas of the country about NBIC technologies of human enhancement.

**Regulatory adequacy**. All sites (6 of 6) expressed significant concern about effective regulation of new NBIC technologies. Some sites recommended creating a new regulatory agency specifically charged with managing these technologies, while others recommended strengthening the Food and Drug Administration (FDA).

**Public information.** All sites (six of six) strongly endorsed programs intended to keep the public informed about developments in human enhancement technologies, including conducting more deliberative panels and including discussions in high school and K-12 education.

**Access & equity**. Nearly all the sites (5 of 6) included recommendations that emerging enhancement technologies be made available on an equitable basis to those who need them most.

**Funding accountability**. Nearly all the sites (5 of 6) recommended that funding be prioritized for the treatment of disease before enhancements and that stakeholders should have a say in research decisions.

**Safety**. Nearly all the sites (5 of 6) included recommendations for the careful monitoring of enhancement technologies, including the development of international safety standards for them.

**Entrepreneurship & development**. Nearly all the sites (5 of 6) included recommendations that the development of these technologies should maximize their benefit, and that both public and private investment in these technologies is critical.

**Ethical consideration.** A majority of the sites (4 of 6) recommended that ethicists and ethical considerations should be a formal part of decision-making about these technologies.

**Privacy**. A majority of the sites (4 of 6) recommended that individual privacy be carefully protected in the development and deployment of enhancement technologies.

**Health insurance**. A majority of the sites (4 of 6) were concerned about whether health insurance providers should be required to cover the costs of enhancements, and how health providers can provide adequate information about enhancements and other alternatives in health care.

**Military uses**. Half of the sites (3 of 6) worried that NBIC enhancements might fall into the hands of terrorists or have other unanticipated military applications.

**Environmental impacts**. Half of the sites (3 of 6) expressed concerns about possible environmental degradation from the use of NBIC technologies, especially in areas of waste management and toxicity.

**Rights**. Half of the sites (3 of 6) wanted to ensure that individuals retain the right to refuse enhancements and that civil liberties and free choice be protected if these NBIC technologies are deployed.

### *16.4.9 Findings from the Pre- and Post-tests*

Examination of responses to the pre- and post-tests provides statistically reliable data about the panelists' attitudes toward NBIC and human enhancement technologies. The data also provide insights into the quality of the deliberation in the NCTF.

Deliberation resulted in reduced certainty among the participants about the benefits of enhancing human capabilities. Pre-deliberation, 82% were at least somewhat certain the benefits would exceed the risks; post-deliberation the percentage of these respondents dropped to 66%. Conversely, deliberation slightly strengthened participants' perception that most scientists were confident the benefits would exceed the risks (92% pre-deliberation and 96% post-deliberation).

Despite concerns about risks, participants overwhelmingly favored government's guaranteeing access to human enhancements if they proved to be too costly for the average American. Before deliberation, 57% held that government should provide such guarantees; after deliberation, 63% said it should. On the other hand, deliberation resulted in a significant increase in the belief that individuals should have to pay out of pocket for most kinds of enhancement. Before deliberation, 74% thought insures should pay for most kinds of enhancements; after deliberation that percentage had shrunk to 55%.

Deliberations increased general concern on the part of participants about NBIC developments but not at the expense of optimism. The percentage of those who expressed worries about NBIC technologies increased from 65% pre-deliberation to 80% post-deliberation, while the percentage of those who were not worried at all decreased from 35% pre-deliberation to 21% post-deliberation. Despite the shifts, the percentage of those who describe themselves as "hopeful" about NBIC technologies was 98% pre-deliberation and 98% post-deliberation.

Deliberation increased specific worry about affording enhancements. Before deliberation, 63% were at least somewhat worried that the average family would not be able to afford enhancements; after deliberation, that percentage increased to 76%. Similarly, before deliberating, 48% of participants were at least somewhat worried that their own family would not be able to afford enhancements; after deliberating, that percentage increased to 60%.

Despite increased concerns about costs, the panelists increased their support for individual responsibility for meeting the costs of enhancements. Those who believe that individuals, not insurance companies, should pay for enhancements shifted from 14% pre-deliberation to 32% post-deliberation. Those who thought that we should avoid technologies that interfere with natural human development increased from 39% (29% strongly) pre-deliberation to 53% (41% strongly) post-deliberation.

Deliberation reduced support for government spending for research on human enhancements. Before deliberating, participants' average score was 7.3 on an 11-point scale, where "11" meant they favored dramatically increased government spending and "1" meant dramatically decreased government spending. After deliberating, the average score fell to 6.3, which was the sharpest decline among five stimuli (health services, new energy sources, space exploration and weapons for defense). This finding is supported by another question that forced participants to decide between spending on enhancements versus space exploration; preference for spending on enhancements over space remained high, but it declined from 90% to 81%.

Deliberation resulted in opposition to most kinds of hypothetical human enhancements that they were described in the Background Materials. Participants were asked to express their support or opposition to five kinds of enhancements on a five-point scale. After deliberating, participants opposed all enhancements except for "implants to catch diseases before they became dangerous." Before deliberating, participants also supported "bionic eyes" and were neutral about using nano-wires and implants to communicate with other people or computers. Respondents remained opposed to "administering drugs to prisoners to prevent escapes."

Some scholars who study small group deliberations – like those that go on in the NCTF – worry that such groups too easily fall prey to dynamics that can distort their results. Among these pathologies are what are known as "reputational cascades" and "social effects" which, they fear, induce members of deliberating groups to endorse statements of the group that, in fact, they reject personally. Thus, in order not to stand out from an apparent majority position, isolated individuals may agree to provisions that they actually object to. The pre- and post-test questionnaires attempted to assess the presence of such processes within the NCTF. The results strongly suggest that such pathologies were not present in these deliberations, and that panelists, in fact, deliberated.

Given the highly speculative nature of the NBIC technologies, and the general lack of public knowledge about their development and implications, the panelists showed significant firming of opinions about them. Comparing pre-deliberation and post-deliberation results, the percentages who believed that the risks of NBIC technologies exceed the benefits increased from 6% to 28%, the percentage who believed that the risks equaled the benefits increased from 16% to 23%, and the percentage who thought that the benefits would exceed the risks also increased from 23% to 46%. Overall, the percentage pre-deliberation that had no opinion about the relative risks and benefits decreased from 55% to just 3% post-deliberation.

The panelists showed significant increases in their substantive knowledge of nanotechnology and human enhancements. The pre- and post-tests assessed substantive learning by asking a set of factual questions and companion questions about the level of certainty of the panelists' answers to those factual questions. Deliberation increased panelists' knowledge on the factual question from an average of four correct responses of six to an average of 5.3 correct responses. When the panelists' level of certainty was included in the analysis – by having panelists say whether they were certain or were guessing and, e.g., rewarding correct and certain answers more highly than correct guesses – panelists' knowledge improved from 3.7 to 9.0 (on a scale from −6 to +12).

The panelists demonstrated high levels of support for the specific provisions of each group's final report and high levels of congruence between their individual preferences and the contents of those reports. Overall, 89.9% of participants agreed or strongly agreed that their group's consensus report accurately reflected their individual preferences. Similarly, 81.2% said that they personally endorse almost every major point in their group's Final Report, while an additional 15.9% said that they personally objected to a few of the major points, and only 2.9% personally objected to many of the major points in the Final Report.

The panelists' sense of internal efficacy – that is, their feeling of being competent to discuss issues like those raised in the deliberations as measured across several questions in the pre- and post-test – increased significantly. Similarly, their sense of trust – that is, their notion that other people will not attempt to take advantage of them – increased. However, their feelings of external efficacy – that is, their belief that their opinions or actions can actually affect political outcomes – decreased after the deliberations.

The panelists found face-to-face deliberations to be significantly preferable to on-line only or to mixed formats. Those who preferred online communication shifted from 18% pre-deliberation to 3% post-deliberation. Those who favored face-to-face communication shifted from 33% pre-deliberation to 70% post-deliberation. In addition, those who favored online and face-to-face communications equally shifted from 49% pre-deliberation to 27% post-deliberation.

## 16.5 Conclusions

We offer five conclusions from this national scale study.

First, average citizens very much want to be involved in the decisions that shape technologies that, in turn, shape their lives. Given good information, access to experts, and the time to discuss their concerns with other citizens, average people are able to learn the important details of even very complex issues, and to generate *thoughtful, informed, deliberative* recommendations. They also fully expect governmental and private sector decision-makers to listen to their ideas.

Second, although average people sometimes express reservations and concerns about new technologies, they remain supportive of scientific and technical creativity and innovation. What they desire, however, is effective, trustworthy, and attentive monitoring of those new technologies. They believe that there have been too many episodes of highly touted new technologies that generated unexpected dangers for them to passively accept whatever technologies the market may generate.

Third, average citizens insist that they have continuous access to reliable, nonpartisan information about new technologies, and that they have frequent and repeated opportunities to express their concerns about how new technologies are managed.

Fourth, in addition to concerns about individual and environmental health and safety, average citizens express concerns for a wider array of social risks that they think are important in the development of new technologies. For instance, issues of economics, equal access, and equity are important, as are technological impact on personal freedom, civil rights, and political rights. Ordinary people have a fairly nuanced and sophisticated view of the role of new technologies in their everyday lives and in society.

Fifth, decision-makers in the government and in the private sector should pay careful attention to the concerns and issues expressed in this study. These panelists spent several weeks studying the issues involved in NBIC technologies, proposed trenchant questions to content experts, and engaged each other – both in their local panels and with the panelists from across the country – in clarifying, exploring, impressing political, cultural, moral, and economic values that they think will be affected by these technologies. These were thoughtful, committed, and well-informed panelists, not misinformed, hysterical, individuals being manipulated by outside groups.

## Appendix A: List of Facilitation Teams at Participating Universities

Arizona State University

David Guston, Professor of Political Science and Director, CNS-ASU
Cynthia Selin, Assistant Research Professor, CNS-ASU
Roxanne Wheelock, Graduate Assistant, CNS-ASU

Colorado School of Mines

Carl Mitcham, Professor, Director, Hennebach Program in the Humanities
Jennifer Schneider, Assistant Professor of Liberal Arts & International Studies

Georgia Institute of Technology

Susan Cozzens, Associate Dean of Research, Ivan Allen College
Ravtosh Bal, Graduate Assistant, School of Public Policy/Georgia State University

University of California, Berkeley

David Winickoff, Assistant Professor of Bioethics and Society
Mark Philbrick, Graduate Assistant, Department of Environment, and Management
Javiera Barandiaran, Graduate Assistant, Goldman School of Public Policy

University of New Hampshire

Tom Kelly, Professor, Director, University Office of Sustainability
Elisabeth Farrell, Program Coordinator, Culture & Sustainability, Food, & Society Initiatives

University of Wisconsin, Madison

Daniel Kleinman, Professor of Rural Sociology
Jason Delborne, Post-doctoral Research Associate, Holz Center for Science and Technology Studies

## Appendix B: Summary Demographic Statistics

| | Applicant | Panelists | National |
|---|---|---|---|
| Sex | 42% male | 50% male | 49% male |
| | 58% female | 50% female | 51% female |
| Education | 25% some college | 29% some college | 50% some college |
| | 33% college degree | 31% college degree | or degree |
| | 33% grad school | 31% grad school | 9% grad school |
| Party ID | 48% democrat | 44% democrat | 36% democrat |
| | 11% republican | 9% republican | 27% republican |
| | 30% independent | 36% independent | 37% independent |
| Political | 48% liberal | 41% liberal | 25% liberal |
| Ideology | 14% conservative | 14% conservative | 36% conservative |
| | 28% moderate | 27% moderate | 35% moderate |
| Race | 71% White | 65% White | 66% White |
| | 16% Black | 15% Black | 12% Black |
| | 5% Asian | 6% Asian | 4% Asian |
| | 5% Hispanic | 7% Hispanic | 15% Hispanic |
| | <1% Native Amer | 2% Native Amer | |
| Household | 9%<$15K | 9%<$15K | |
| Income | 16%>$15K<$35K | 21%>$15K<$$35K | |
| | 21%>$35K<$50K | 16%>$35K<$50K | Median household |
| | 23% .$50K<$75K | 20%>$50K ,$75K | income=$46K |
| | 15%>$75K<$100K | 16%>$75K<$100K | |
| | 16%>$199K | 17%>$100K | |
| Median age | 37 years old | 39 years old | 37 years old |

## Appendix C: Future Scenes of Nanotechnology and Human Enhancement Included in Background Materials

Included in the background material: "The following fictional scenes are extrapolations from current nanoscale research; they have been vetted for their technical plausibility by scientists currently working in nanoscale research. We hope these scenes will stimulate you to reflect upon the meanings, potentials, and problems surrounding nanotechnology. The goal is to cultivate our collective ability to govern the implications of our technological ingenuity."[1]

[1] Technical background on the generation of the scenes may be found in C. Selin (2011). "Negotiating Plausibility: Intervening in the Future of Nanotechnology." *Science and Engineering Ethics* 17(4):723–37.

## C.1 Engineered Tissues

*What are your thoughts on synthetically grown tissues and organs? Using tissue-printing technology, this system is able to build tissues with a vascular structure enabling the building of new organs.*

Newly developed artificial tissues have been approved for use in wound healing as well as for skin grafts. These artificial tissues are made by "seeding" cells into a bioengineered scaffold where upon they reorganize it into a material suitable for use as an artificial tissue. In the process of tissue engineering, the cell makes use of the scaffold components as nutrients. The starting scaffold is usually three-dimensional Jell-O like material called a collagen gel. Made up mostly of water, sugars, and carbohydrates the gel also contains fibrous proteins like collagen, fibrin, and fibronectin, which allow the cells to interact with the scaffold. The fibrous proteins are large and tend to form bundles of fibers, or fibrils. After some time the cells use up the scaffold materials reorganizing some of them into an artificial tissue that can then be used for surgical procedures.

Because the tissue is grown from the patient's own cells there is almost never any rejection of the transplant. In some cases such as cancerous tissues, this is not possible. However, using compatible cells from an appropriate donor gives a high success rate with no risk to the cell donor. Further developments of tissue engineers have made it possible to replace not only tissues, but also organs. One such technology is tissue printing which would allow one to produce whole organs from gel scaffolding and cells in an ingenious way.

This advanced technique allows cells to be arranged within the scaffold in order to shape the tissue into larger structures. Cells are arranged by inserting them into a device analogous to an inkjet printer where cells are ink. The cells are then printed in a two dimensional pattern such as a circle. After a circle of cells is laid down on top of a sheet of scaffold, another layer of scaffold is placed on top, followed by yet another circle of cells and another sheet of scaffold. Several circles placed in this way will reorganize the scaffold to form a tubular tissue, thus creating a tissue with a vascular system. This is one of the biggest breakthroughs in tissue engineering, because it allows blood and nutrients to flow through the artificial tissue. Tissue printing thus allows us to develop microstructures. These developments have lead to externally grown tissues that can replace vital organs, as well as more general tissues like skin, bone, muscles, and arteries. The lack of transplant materials is no longer a problem.

## C.2 Living with a Brain Chip

*What are your thoughts on using cranial chips to enhance cognition? This cranial chip features a data feed that puts information into the brain while the user is resting.*

The next generation of cranial chip implants enables data transmission directly to the brain during rest without interfering with sleep. This data feed feature dramatically decreases the amount of time needed to assimilate new data each day, in fact the chipped person will just wake up knowing what was streamed into their head the previous night. The presence of the chip interferes with REM sleep, but the new data feed does not actually disrupt or alter in any way the sleep of the person with the implant.

The new disruptor cage is constructed out of more advanced materials that are lighter and more comfortable for the wearer. No longer is it necessary to lock head, neck, and torso in to a rigid structure, the new generation of disruptor cages need only to lock to the head and upper vertebrae of the neck. This new format still provides the same protection against magnetic damage to the brain, advances in real time processing now allow for emergency shut off if the magnetic pulses are not directed exactly at the chip. The use of rare earth magnets in a wider net around the cranium makes for a more thorough disruption of the chip (even while undergoing data feed). This improves sleep by removing annoying dream sequences, restlessness, or need for sedatives previously common in past cranial chip implants.

These advances in cranial chip disruptors will work with all cranial chips. However, those with the newer (Gen. 3.4 or higher) cranial chips will see the most improvements and those who receive the soon to be released Gen. 4.0 will be able to take advantage of many new options. The 4.0 chips, like those before it, are a sandwich of carbon nanotubes, and gate molecules that are covered in neural growth promoters. The 4.0 chip features advances in neuron-to-chip interface, allowing more neurons to contact the chip in ways that are more functional. This in turn increases the rate of information in and out of the chip, further increasing cognitive ability.

With this increase in connectivity of brain to chip and chip to brain comes increased assimilation and learning time. After implantation (still an outpatient procedure), it will take 30–90 days of neuron growth around the chip for it and the brain to become fully integrated. Upon chip integration, the newly chipped person will need to attend 9 months of intensive classroom based courses, where they are taught new ways to think, process thoughts, and to categorize memories and data.

It is during this time, as the chip becomes enabled, that they will begin to feel the effects of the continuously running chip. As the brain becomes dependant on the chip, the implantee will find it difficult to sleep. The first effects will be tossing and turning at night, followed by repetitive dreams, and finally inability to sleep. It is at this point that the cranial chip disruptor is needed and technicians will work with the chip-implanted person (and spouse if necessary) insuring proper technique in fitting the disruptor, allowing the user to have the best nights sleep ever.

## C.3 Automated Sewer Surveillance

*What are your thoughts on tracking individuals using their genetic material? Ultra fast sequencing technology is used to analyze the DNA in harvested wastewater, thus screening large populations.*

Capitalizing on recent advances in very fast genome sequencing technologies, Sentinel Genetics is pleased to offer its new real-time in-stream wastewater sequencing system. Genetic material is randomly harvested from the waste-stream, usually at the sewage treatment facility. The automated system then prepares the DNA for sequencing and individual samples can be sequenced to the extent necessary to compare it to the National Registry in less than 1 s. A small bank of sequencers can process tens of thousands of samples each hour.

Sentinel Genetics developed the single strand sequencing technology, which works by quickly pulling strands of DNA through tiny nanoscale pores. Breakthroughs in micro and nanoscale mechanical devices that are small enough to automate preparations with the very small DNA strands have allowed for sequencing prices as low as pennies per thousands. Due to the large amount of non-human DNA that is in a wastewater stream, it was only through this high speed processing of samples at low price that large scale screening of municipal populations could become cost beneficial.

The database of America's genetic information has been available to law enforcement agencies since the inception of the United States Genomic Registry, but only in the last several years has it been complete enough to look for individuals. The Sentinel Genetics Sequencer data processing system is fully compatible with the Registry and provides advanced algorithms for comparing genomic and partial genomic material against the data in the Registry. By combining the massive throughput of the treatment-facility-based sequencer bank with portable units for signal triangulation through upstream testing, it is possible to track the location of individuals in metropolitan areas.

## C.4 Disease Detector

*What are your thoughts on diagnosing disease before you are ill? Doc in the Box is a device that tracks an individuals protein levels to monitor changes that imply early stage illness or disease before symptoms emerge.*

BioMarker Detector created Doc in a Box with the ability to track a person's health status on a day-to-day basis from the comfort of their home. Doc in a Box is able to detect and record the health level of an individual by examining multiple proteins that are present in their blood, which are collected through a nearly invisible needle causing no detectable pain. The proteins present in the blood will fluctuate, either up or down, as the body changes. These changes can be due to many different naturally occurring events such as puberty, pregnancy, or menopause, along with more unfortunate changes such as getting cancer, flu, or Alzheimer's disease. Doc in a Box is able to measure the amounts of specific proteins, or biomarkers, which are correlated to particular diseases, infections, or changes in the human body. These biomarkers are measured and recorded over time as health markers and tracked to develop a particular pattern specific for each individual called a biosignature. When there is a change in the body, there is an immediate change in the biomarkers outside the range of the biosignature and detected by Doc in a Box.

Since the Doc in a Box is detecting markers on the molecular level, users will be informed of a cold or flu before a sore throat or cough ever occur. With the ability of Doc in the Box to detect diseases pre-symptomatically, people will be able to get treatment before they feel the illness and far before it is too late to treat the disease. For cancer patients, there will be biological implications of cancer before a tumor develops and before the cancer has time to spread. For Alzheimer's patients, early detection of biomarker changes will enable more effective treatment options, possibly before any memory loss.

## *C.5 Barless Prison*

*What are your thoughts on a barless prison? NanoCage has developed a caged drug that is injected into prisoners that becomes activated by radio control if prisons cross-designated boundaries.*

Ever since the first true nanomedicine product came on the market, a caged cancer drug that releases once bound to the cancer cell, researchers have been working towards utilizing these technologies for control purposes. This week it was announced that NanoCage, in collaboration with United Penitentiary Systems, have developed the first barless prison. Upon entry, inmates are injected with a cocktail of caged drugs that have a variety of effects when released via radio control. The base technology utilizes focused radio waves to target deep tissue tumors in places such as the abdominal cavity.

The basis for security is a net of radio transmitters that surrounds the facility. As a prisoner crosses the perimeter threshold, the radio signals will cause the release of one type of caged drug. For instance, if the prisoner crosses an inner 'warning' perimeter, a drug will be released that causes extreme vertigo and mild nausea. If the prisoner continues, the next perimeter will signal the release of incapacitating sedatives, and if the next signal is reached, it will trigger a fatal dose of narcotics. These perimeters are spaced far enough apart to prevent unintentional crossing of more than the first.

The caged drug is connected to an antenna that upon receipt of a specific radio signal causes the physical break down of the carbon-nanotube-based cage. The package including the antenna is roughly half the size of a red blood cell. A coating of biocompatible molecules minimizes the physiological side effects from the caged drugs. There are, on very rare occasions, mild inflammatory responses that can be treated with over the counter anti-inflammatory drugs. Because some degradation of the caged drugs occurs naturally in the body, supplemental injections are advised every 6 weeks and always after drugs have been released.

Guards in barless facilities will be equipped with radio transmitters that can be aimed at individual inmates or larger areas to quell local unrest. The transmitters used by the guards will be unable to access the frequencies that trigger the fatal dosages.

NanoCage and United Penitentiary Systems claim this is the new model for working prisons, where inmate labor is unencumbered by restraints or monitoring devices and physical investment costs are not much more than traditional factories. The perimeter of these facilities need only be physically secured to keep people from trespassing on the grounds.

## C.6 Bionic Eyes

*What are your thoughts on visual enhancement? Opti-scan is an optical implant that looks and functions like a normal eye, yet has new enhancements enabling magnification, visualizing infrared, and night vision.*

Penetrode Inc. presents the Opti-scan visual enhancement system, the latest in ocular prosthetics. Opti-scan is capable not only of restoring sight to the blind but also of providing them with additional capabilities beyond those of the normally sighted. The housing of the implant is designed to mimic the external appearance of the eye and comes with an iris capable of changing colors to suit the daily tastes of our customers. A series of small motors implanted within the eye socket will provide human like eye movements while allowing for much greater tracking speeds than is possible with normal muscle.

The heart of the technology is a thin film photosensitive ceramic panel that is located in the back of the eye. These panels take light signals and transduce them into electrical signals that stimulate the ganglial cells. The stimulated ganglial cells allow the signal to be processed along the optical nerve to the visual cortex. If there is extensive damage to the ganglial cells or the optical nerve then the signal can be routed directly to the lateral genicuate nucleus, which is where the optic nerve connects to the visual cortex.

A massive zoom/magnification function will allow for telescopic sight similar to that of a high-grade set of binoculars and the ability to greatly magnify nearby objects achieving magnification power similar to that of many laboratory microscopes. Opti-scan uses digital magnification features similar to those found in most digital cameras to achieve this additional functionality. Opti-scan is also available with night vision, thermal imaging, and high definition video and still photo capture. Images captured through the Opti-scan can be downloaded via Bluetooth and Wi-Fi to any personal computing device. Depending upon the condition of your optic nerve, Opti-scan can be implanted through outpatient surgery and after a brief, 2-week course of training and therapy, you, and your new eyes will be fully functional.

# Selected Further Readings

Berloznik, R., R. Casert, C. Enzing, M. van Lieshout, and A. Versleijen. 2006. *Technology assessment on converging technologies: Literature study and vision assessment* [Background document for the STOA Workshop].Brussels: European Parliament.

Bostrom, N. 2003. Human genetic enhancements: A transhumanist perspective. *Journal of Value Inquiry* 37(4): 493–506.

ETC Group. 2006. NanotechRX: Medical applications of nano-scale technologies: What impact on marginalized communities? (www.etcgroup.org).

Fukuyama, F. 2004. The world's most dangerous idea: Transhumanism. *Foreign Policy* (Sept/Oct):42–43.

Lee, P., and M. Robra. 2005. *Science, faith and new technologies: Transforming life.* Vol. 1, *Convergent technologies*. Geneva: World Council of Churches.

Nanotechnology Task Force. 2007. *Nanotechnology: A report of the US food and drug administration*. Washington, DC: US Government Printing Office.

President's Council on Bioethics. 2003. *Beyond therapy: Biotechnology and the pursuit of happiness*. Washington, DC: U.S. Government Printing Office.

Roco, M., and W. Bainbridge, eds. 2002. *Converging technologies for improving human performance: Nanotechnology, biotechnology, information technology and cognitive science*. Arlington: National Science Foundation and Department of Commerce report; published by Springer in 2003, http://wtec.org/ConvergingTechnologies/1/NBIC_report.pdf.

Taylor, M.R. 2006. *Regulating the products of nanotechnology: Does FDA have the tools it needs? Project on emerging nanotechnologies*. Washington, DC: Woodrow Wilson International Center.

Zonnefeld, L., H. Dijstelbloem, and D. Ringoir, eds. 2008. *Reshaping the human condition: Exploring human enhancement*. The Hague: Rathenau Institute, in collaboration with the British Embassy, Science and Innovation Network, and the Parliamentary Office of Science and Technology.

# Chapter 17
# Panelists' Reports by State: Arizona, California, Colorado, Georgia, New Hampshire, and Wisconsin (a–f)

## 17.1 Chapter 17a 2008 National Citizens' Technology Forum on Human Enhancement, Identity, and Biology

### *17.1.1 Arizona Panelists' Final Report*

Editor's Note: These reports have been republished in approximately their original form, with only modest reformatting to preserve the visual continuity of this volume. The reports are the product of the participants in the National Citizen's Technology Forum, and we wanted to preserve their authentic voice. No one specific view or conclusion can be attributed to any particular author.

Authors: Bohnke, Heather; Brandtfox, Trisha; Burns, James; Cartwright, Charles; Corso, Matt; Gulcelik, Guliz; Hull, Bart; Johnson, Darlene; Manriquez, Santiago; Romanick, Tamera; Ryan, Terry; Scott, Stuart; Thompson, Kirk; Zeise, Lynda

#### 17.1.1.1 Introduction

It is the belief of this group that NBIC technologies present important challenges and opportunities that we must face; we have the utmost confidence that facing them will lead to a better future. The multidisciplinary spectrum of these technologies is so vast that special attention should be paid to the implications, benefits, and risks of human enhancement as a rising field of research and development. The ethical, social, economic, and political consequences of NBIC technologies will be present in everyday life. Special care should be taken to avoid excessive prudence or reckless adoption. Our decisions now will affect both the present and the future of humanity and life.

S.A. Hays et al. (eds.), *Nanotechnology, the Brain, and the Future*, Yearbook of Nanotechnology in Society 3, DOI 10.1007/978-94-007-1787-9_17,

We are excited, as well as concerned, about improved quality of life for all people. Because we are at the beginnings of these converging technologies, it is important that we are proactive in our asking of questions. The answers will most likely lead to a bright future!

#### 17.1.1.2 Socio-Economic

The socio-economic repercussions of the adoption of NBIC technologies are several. Some of them we can envision; many more we cannot. These repercussions can range from the aggravation of existing racial, social and economic divides, even to the creation of new ones. It carries the possibility of eliminating some current societal and economic problems to the creation of new ones. However, in presenting and adopting these new developments we should aim to maintain the ideals that allow the individual to become who she/he strives to be, and safeguard the values of liberty and free will, and the pursuit of happiness.

We believe that all people regardless of race, creed, color, or economic status should have equitable access to the benefits of these technologies. The information about these technologies must be presented to the public in terms that will allow understanding of the concepts, benefits, and risks, so as to lead to informed decisions.

We are concerned that under-represented and/or minority groups (e.g., Native Americans) will not have a say in the decision-making processes of these technologies, and that their voices of concern will be ignored.

Discrimination towards an individual, race, class, faith, etc., should be prevented regardless of the degree to which they have adopted NBIC technologies.

The future of NBIC technologies should rest with the needs of advancing humanity past the scourges of the human condition. These include poverty, disease, and manual labor. The direction should be bound by the concerns of the public.

Finally, citizens of the world want more information shared by the media, government, and industry to advance awareness.

#### 17.1.1.3 Safety

We encourage the development of international safeguards and standards for all human enhancement technologies for both public and private sectors. We are concerned about who can be trusted, where reliable information can be found, and who is going to assure human and environmental safety now and in the future. We are also concerned that the "government" regulates itself and there is no oversight of that process. For example, atomic energy sites and old military bases are now "Superfund" sites because of the environmental cleanup costs – the government created a problem and is now responsible for toxic environmental cleanup.

There is a need for the creation of a regulatory body composed of experts from a variety of fields, e.g. ethicists, chemists, etc. Its responsibilities would be the oversight of NBIC safety and efficacy. Its accountability would lie within itself as a regulatory body, the government, and the scientific community. However, regulations from this body or the government should refrain from instilling fear or instituting regulations out of fear.

#### 17.1.1.4 Human Identity

We believe in an overriding sense of both individuality and personal identity, and an environment that nurtures free will, in which each person has the right to use or refuse enhancement. It is important to safeguard the ideal that every individual is in fact a unique and sovereign entity in his or her own right. We should also strive to protect and respect the sanctity of the idea of an individual, unique, soul. Mortality is important to the human identity, yet the desire to improve the quality and length of life is also part of the human identity.

Regarding the debate of restoration vs. enhancement, we are divided in opinion. Some of us would like to see an adjusted bell curve implemented as a standard for normalcy. The adaptation of the bell curve would work not as a definitive statement of identity but as an establishment of when a person is either submitting to a restorative treatment or an enhancement procedure.

Others of us fear the use of an adjusted bell curve skewed in favor of enhancement as the new standard of normalcy. Implementation of this adjusted bell curve can lead to a new human identity that is fabricated by a societal obsession with enhancements.

Some of us are concerned that the personality, spirit, and identity of the individual will become altered and/or distorted.

#### 17.1.1.5 Government

We, citizens of the world, are inherently responsible for the roles of our governments in participating with the global NBIC market. Active awareness of the decision-making process must be made transparent and accountable to the global community. We are concerned that the United States citizens are not involved in or truthfully informed of the appropriation of federal funds used for research and development of NBIC technologies.

#### 17.1.1.6 Environmental

NBIC technologies should be used to find new environmental solutions to both new and old environmental problems, e.g., medical and biological waste. These externalities should not adversely affect the sanctity of life. In addition, we are concerned that we do not have adequate knowledge about, or means of disposal of, "waste" produced by NBIC technologies. For example, nano waste particles are so small they can easily contaminate the whole environment without anyone's knowledge. Another example is recycled drinking water causing negative effects on the human system.

We are concerned that these types of technologies will upset the natural order of the planet, people, plants, and animals, including life cycles and food chains.

#### 17.1.1.7 Development Issues

Regarding development of NBIC technologies, we believe in the following:

- We should depart from a tendency and custom to legislate, enforce, and regulate out of fear. The evolving technologies should be allowed to grow and not be choked by regulatory concerns.
- Funding should remain both public and private to avoid monopolies.
- Human enhancement technologies need to be more understandable to non-scientists.
- We are concerned that the accessibility of this technology will be easily available to terrorists, black market labs, and/or other individuals with the intention to harm others.
- We are concerned that only a select few will benefit from NBIC technologies due to how accessible they will or will not be.

#### 17.1.1.8 Health

We are on the cutting edge of an exciting and wonderful health revolution for the advancement of humankind. The goals of these technologies should be prevention, treatment, and cures over enhancement, and prioritizing humanitarian gain over special interest gain.

However, we still have some concerns. They are:

- prolonged exposure to NBIC in the body;
- risks of dehumanization as a result of NBIC;
- people living longer will further stress current planetary resources;
- a society filled with artificially enhanced individuals may become dependent on the medical profession;
- Immortality should not be a goal of NBIC health advancements lest it eliminate the meaning of living.

#### 17.1.1.9 Regulatory Challenges

We are concerned about who makes the rules, the qualifications of the rule-makers, and their accountability to the public.

What makes NBIC technologies challenging from a regulatory standpoint:

- Permanence and dependability
- Enforcement and oversight
- Unforeseeable repercussions

- Limited understandings
- Widespread effects
- Generational effects
- Endangerment of organisms
- Maintaining safety and effectiveness without stifling progress and innovation

We are concerned about secret government infiltration and the invasion of privacy with these technologies. For example, (1) injecting nano tracking devices into humans to track their every move; (2) big brother watching and controlling the masses; (3) government directing where money goes against public interest.

#### 17.1.1.10 Conclusions

The new and unique challenges presented by nanotechnology creates a need for new and unique safeguards and procedures to both protect and preserve a new and better quality of life for generations to come. We recommend that governments cooperate on a global scale. We would like to suggest the creation of a new regulatory and investigative bi-partisan committee that includes citizens, natural and social scientists, ethicists, philosophers of science, and governmental officials. It should be the responsibility of the government and individuals to be informed of the current state of affairs of NBIC technologies, hold industry accountable, and promote active decision-making and participation in the advancement of NBIC technologies.

## 17.2 Chapter 17b 2008 National Citizens' Technology Forum on Human Enhancement, Identity, and Biology

### *17.2.1 California Panelists' Final Report*

Atwood, Christina; Bautista, Teresita; Chu, Angela; deJesus, Mary; fleming craig; Heath, Alan; Ho, Carmen; Lewis, Vanessa; Moses, Daniel; Prescott, Charles; Willis, Janine

Editor's Note: These reports have been republished in approximately their original form, with only modest reformatting to preserve the visual continuity of this volume. The reports are the product of the participants in the National Citizen's Technology Forum, and we wanted to preserve their authentic voice. No one specific view or conclusion can be attributed to any particular author.

This report was produced by a group of citizens from Northern California, as part of a nationwide public deliberation project. Participants were selected from a pool of volunteers, with the aim of constituting a panel that reflects the diversity of California's population in terms of ethnicity, income, and gender. The group received and reviewed an extensive set of background materials concerning the convergence of nanotechnologies, biotechnologies, information technologies, and cognitive science (collectively NBIC), and their possible applications in the area of human enhancement. They gathered in person for an initial weekend of consultations, and participated in multiple online sessions together with the individuals from the other five sites nationwide. These online sessions included Q&A sessions with a number of experts in related disciplines, as well as an exchange of views among locations. The process culminated in a final weekend of meetings, resulting in the following consensus report. The opinions and words expressed here are those of the participants.

March 31, 2008

#### 17.2.1.1 Introduction

The goal of this report is to present a protocol for the testing and development of human enhancement products that will ensure the physical, cultural, social and political safety of human beings and protect our global environment, while simultaneously encouraging the innovative, aggressive and steadfast development of these new technologies. The convergence of NBIC technologies presents a tremendous set of potential benefits and risks. We want to ensure equitable access to the benefits, and minimize the public's exposure to the risks.

Thus, the federal government should assume a broad proactive approach towards approving the development and use of these technologies including thorough, unbiased testing and the strict disclosure of all information. This requires coordination and cooperation among multiple government agencies, with adequate funding and authority to carry out their missions, without detracting from their existing responsibilities. Additionally, collaboration between the public and private sectors is an important element of an overall governance strategy. This includes identifying funding mechanisms that allow private organizations to contribute to the public good.

Currently, NBIC technologies cut across multiple industries and areas of application, and are characterized by a great deal of uncertainty. We are concerned by the apparent lack of a comprehensive, cohesive set of policies concerning the following areas:

- Allocation of funding
- Enforcement of regulations
- Disclosure of potential risks and benefits
- Testing and approval of new products using converging technologies
- Public education

We recognize that overregulation could stifle productive innovation, especially at such an early stage of deployment. We encourage the development of beneficial applications, but believe that public safety, individual rights and privacy should be a higher priority than profitability. We also encourage the United States government to continue its efforts at international collaboration and exchange in these areas.

Finally, we endorse participatory processes such as this National Citizens Technology Forum, and urge that similar opportunities for public input be ongoing.

#### 17.2.1.2 Specific Recommendations

Within each priority area, recommendations are numbered in accordance with the following categorization:

- The policy-making and priority-setting processes for NBIC
- Environmental Concerns
- Privacy
- Public Welfare and Safety
- Alternatives (to NBIC-based human enhancement) and Prevention

Allocation of funding

*Policy Process*

With public funds in short supply and competition between agencies for these monies, we need to establish a system to prioritize the allocation of funds.

Agencies and projects requesting public funding for nanotech should clearly demonstrate that the monies would be used first for treating, preventing or curing disease or other human suffering; and, second for human enhancement beyond "normal" capabilities. Military applications should be the third level of priority, unless shown to be necessary for national security.

We recommend that government introduce methods for increasing stakeholders' ability to have a say in how funds for non-military research are allocated. By stakeholders, we refer to the public, NGOs, and others that represent the public interest. Some methods for achieving this may include congressional commissions that bring together scientists, consumer groups, and others without a stake in the outcome of funding decisions, citizen institutional review boards, and others. Academics in this field are discussing alternative methods that the government should consider.

Regarding access to information, while recognizing some information needs to be classified and that much information is already available, we recommend that greater efforts be made to make the details of products being produced with government funding as available as possible.

There should be funding dedicated specifically to monitoring, testing and ensuring the public and workplace safety and the environment. This includes funding for inspectors and adequate agency staffing to carry out these tasks effectively.

The government should not allow religious values to affect public or private funding for emerging human enhancement technologies.

### *Environmental Concerns*

Incentives should be used to encourage companies to develop NBIC-based solutions to clean up pollution resulting from human enhancement activities.

### *Alternatives and Prevention*

For every dollar of public money invested in NBIC technologies for disease remediation, a proportionate amount must be allocated towards research in, the promotion of, or increasing the accessibility of preventative medicine.

Public research funds should target disease prevention, particularly AIDS, hypertension, diabetes, heart disease, cancer, etc., along with repair and replacement of body parts.

We recommend that, for each family of enhancement applications, we assess the availability of lower-risk and/or more cost-effective alternatives to NBIC technologies before allocating significant funding.

## Enforcement of Regulations

### *Policy Process*

We recommend the formation of a new oversight body explicitly focused on NBIC technologies, comprised of representatives from existing government agencies, including EPA, FDA, Homeland Security, HHS, OSHA, etc., in order to implement the policy recommendations made in this report effectively. It is important that the individual members have the necessary expertise and time to dedicate to these issues.

The federal government should continue to seek international cooperation with regard to developing and implementing policies to manage the risks and benefits of NBIC technologies.

### *Environmental Concerns*

Severe civil and criminal penalties should be levied against companies that develop or use NBIC technologies that damage the environment. It is important that these penalties are not simply seen as a routine cost of doing business, but are substantial enough to prevent such actions.

### *Privacy*

Medical information must be kept private and confidential. If a medical procedure can or will jeopardize a patient's privacy, the patient must be able to make an informed decision about whether or not to proceed. Existing regulations guaranteeing privacy should be extended as appropriate to cover new privacy risks arising from NBIC-based applications.

Health insurers should be prohibited from discriminating against individuals based on genetic testing or new methods for early disease detection, whether in group or individual policies. This includes both the denial of policies and coverage for specific therapies.

Employers should be prohibited from discriminating against individuals for employment and workplace opportunities based on NBIC-derived medical information or treatments.

Legislation is needed to guarantee that the military and other security-related organizations, including the CIA, NSA, FBI, Homeland Security, and federal, state and local law enforcement, cannot use these technologies to conduct surveillance on people residing in the U.S. without due process. Because NBIC-based technologies pose a serious risk of abuse of privacy, these rights must be protected by the Constitution. To this end, it is necessary to review whether they are adequately covered in the current Constitution.

*Public welfare and safety*

All military personnel must be given full disclosure of any risks to personal health and safety derived from the use of NBIC-based applications for military purposes and must be allowed to consent or not, without retribution or coercion. Furthermore, the military must be responsible for the effects of any implants or personal deployment of NBIC-based technologies and actively assist with re-integration into civilian life.

### Disclosure

*Public welfare and safety*

All test results affecting public safety and welfare must be fully disclosed in a timely manner upon discovery. It is in the public's interest to have all information concerning health and safety readily available so that an informed decision can be made by each individual as well as by society as a whole. Knowledge is power.

All consumer products containing nanomaterials or produced using nanotech must be clearly labeled as such.

All worksites where workers are handling or exposed to nanomaterials must clearly post notices of the potential human risks of these materials, as well as verbally inform workers of these risks.

### Testing and Approval

*Environmental Concerns*

All NBIC-based technologies should go through vigorous testing regarding the effect of the specific nanomaterials on the environment.

When testing human enhancement products, we must keep in mind the integral relationship between the earth and humans. We are completely dependent on the earth and it is our responsibility to take every action to protect it. Recycling and waste strategies should be tested ahead of times. Private and public developers should pay for their own waste management and clean up. Those that do not comply with regulation should be penalized. We should ensure that no irreparable harm would come to the physical environment of our earth or surrounding atmosphere.

Non-biased, neutral experts should complete testing. Studies by industry or private business interests alone will not be considered sufficient evidence for the approval of new technologies. Follow-up testing must be conducted on a regular basis.

Producers of nanomaterials or nano-based consumer products need to be held responsible for the environmental impact of their products for their entire lifespan: from the extraction or production of raw materials to the conditions under which it is produced – including worker safety – to the proper disposal and/or recycling of the product itself and wastes/byproducts of production.

*Public welfare and safety*

Where feasible, testing should be done on artificial/virtual subjects before testing on humans and animals.

Before any human enhancement technology is approved for use on the market, thorough cost-benefit analyses should be conducted to compare these with any existing alternative therapies.

In regards to human testing, testing should only occur with willing participants. Testing should not target certain ethnic and prisoner communities. Testing should only occur after testers have actively provided testees with as much information as possible about the materials, procedures, side effects, potential harms, physiological/emotional changes, and long-term effects.

The communities surrounding test facilities should be made aware of the testing procedures, possibility of dangerous outcomes, waste management procedures, and all changes in the environment before testing. Neighborhoods and towns should have the authority to say whether they approve the testing. If the environment is damaged, the company should pay for clean up and compensate the community appropriately.

Under no circumstances is it ok to release dangerous, toxic, or untested particles/substances into the environments of communities and countries that do not have the privilege of financial or regulatory protection.

All neighborhoods and cities should be equally protected from adverse consequences of testing despite the economic advantages or disadvantages of the community.

Public Education

*Public welfare and safety*

The public must be educated particularly regarding the potential benefits and harms involved in employing NBIC-based technologies for human enhancement, such as misuse, contamination, etc. This can be accomplished via public service announcements, public school education, neighborhood workshops, press releases, talk shows, mass emails (e-blasts), white papers, FAQs, and others.

Patients seeking or eligible for NBIC-based treatments and human enhancement options should be informed by their physicians of alternatives. Complete information should be available and accessible to the public in both printed and electronic form.

The results of clinical trials of these technologies should be disseminated to workers/unions, consumers, educators, NGOs, and academics via the methods listed above.

## 17.3 Chapter 17c 2008 National Citizens' Technology Forum on Human Enhancement, Identity, & Biology

### *17.3.1 Colorado Panelists' Final Report*

Eric Brown, Teri Burgess, Nichole Carter, Abraham Eng, Starlyn First, Brett Kuenne, An Light, Ricky Lott, Patrick Mingus, Rose Murray, Alex Ramirez, Eldrine Richardson, Ariel Thomas, Tara Van Bommel

Editor's Note: These reports have been republished in approximately their original form, with only modest reformatting to preserve the visual continuity of this volume. The reports are the product of the participants in the National Citizen's Technology Forum, and we wanted to preserve their authentic voice. No one specific view or conclusion can be attributed to any particular author.

Open Letter to Honorable Senators Wayne Allard and Ken Salazar and Representatives Degette, Lamborn, Musgrave, Perlmutter, Salazar, Tancredo, and Udall

#### 17.3.1.1 Introduction

In March 2008, small groups of volunteer citizens gathered for two weekends to consider what guidelines might best steer the development of some very powerful, new technologies. Of specific concern were nanoscience, nanotechnology, and the ways these are merging with biotechnology, information technology, and cognitive science. The four together often are referred to as nano-bio-info-cogno (NBIC) science and technology, which may provide powerful ways to enhance human behavior and experience.

The 14 of us were one of six groups, and we met at the Colorado School of Mines in Golden, Colorado. Along with two in-person weekends, we got together via computer conference for nine 2-h sessions with members of other groups and NBIC experts. (More information about this project is available at the National Citizens Technology Forum web site, http://cns.asu.edu/nctf).

The activity as a whole was sponsored by the National Science Foundation (NSF) as part of its effort to ensure that NBIC development takes into account a broad spectrum of perspectives from all citizens. The idea behind the forum is that it is important for citizens to consider how NBIC technologies should be developed *before* they are actually implemented. The hope is that concerned citizens will be able to provide decision makers in government, business, and society with the *informed, deliberative opinions* of ordinary people who have taken the time to study these issues with some care.

#### 17.3.1.2 Enhancement

Because of this process, we have formulated some recommendations for developing, educating the public about, and regulating nanotechnology (nanotech). Our comments are directed primarily toward the implementation of nanotech for human enhancement. Enhancement is defined as the improvement of human and cognitive abilities. These abilities are said to expand the knowledge of how the human brain works, and are leading researchers to explore ways to modify its processes.

Although some may argue cognitive and human enhancement is comparable to earlier inventions such as modern electricity and computers, in fact nanotech is vastly different in that the broad scale of biological, cognitive, and informational applications is unlike anything seen before. Our recommendation is that nanotech be utilized for remediation to serve the goal of helping humans gain access to equality across the board as regards quality of life. Therefore, we suggest prioritization of funding should be given for issues of remediation.

Ideally, this type of technology should be available to those who need it the most no matter their income level. We strongly recommend that legislative action ensure that private insurers cover these needs and, failing that, government will step in to subsidize costs. Everyone, regardless of socioeconomic or cultural status, deserves equal access.

#### 17.3.1.3 Education

We discovered that nanotech is a broad field encompassing a diverse array of scientific and technological developments, yet the public remains mostly unaware of these developments and their far-reaching implications. We therefore felt compelled to make sure the general public becomes educated on the nature of these technologies, from a thorough, accurate, and easily accessible source.

One reason for such strong feelings of unease stems from the possible effects of nanotech on humanity, society, and the ecosystem. The developments in the field of nanotech are revolutionary. For example, currently in clinical trials are Brain Machine Interface (BMI) chips, which when implanted in the deep tissue of the brain allow a person to communicate with a computer via their brain signals.

This is only the beginning for nanotech, since it is in its infancy. Promising developments range from bionic eyes to nanoparticles that detect chemical and hormonal changes early, thus eliminating disease before symptoms emerge. For many, the possibility of the elimination of disease and the ability to attain a greater quality of life is a bright prospect, yet there are many possible adverse outcomes.

The dark side of nanotech was a ubiquitous concern in our group discussions. For many the maintenance of privacy and personal identity are problematic. Nanotech could make possible nanochips that allow us to communicate directly with computers or even link to the Internet. As citizens, we need to consider who should have access to our information, and for what uses.

To ensure the effective dissemination of information, with the goal of creating greater public awareness we propose the following policy recommendations:

1. Continued citizens' forums, funded annually by the National Science Foundation, to re-evaluate nanotech issues and update policy recommendations based on changes in emerging research and public opinion.
2. Create a federally managed online clearinghouse that consolidates all current resources and information on nanotech. These resources should be advertised in a variety of popular media.
3. The development of nanotech science exhibits explaining the technology, its relevance, its implications for the near and distant future, and nanotech products such as sunscreen, beauty products, food products.
4. Grants from the federal government to fund curriculum in public schools.
5. Convene an international nanotech summit involving government agencies, non-governmental agencies, industry leaders, and citizens. The goal of the summit is to engage in international dialogue on the development of nanotech, promote the exchange of ideas, and ultimately draft an international treaty of nanotech, which would establish appropriate regulations. The treaty should, at the least, restrict the use of nanotechnology in ways which might contaminate the human race or the environment, as well as in certain military applications in addition to avoiding empowering extremist groups by giving them access to nanotech, prohibit the exploitation of under privileged groups in relation to testing and implementation of the technology, and promote the open exchange of ideas among nations.

#### 17.3.1.4 Regulation

It is our position that these new advancing technologies will reach into areas that are not overseen by current regulatory bodies, namely the FDA and the EPA. Therefore, it is our desire to see a new regulatory body established to both extend regulatory oversight over human enhancing technologies, and to alleviate the burden on current regulatory agencies.

This new Human Enhancement Regulatory Agency (HERA) would not only be responsible for the extensive testing of these types of products and enhancements, but also would be the United States' point of contact with the rest of the world.

It is imperative that the global community reaches consensus on how these technologies will be governed. Because these human-enhancing technologies are inseparable from their hosts who are free to travel across international borders, it would be in the best interests of all to find common ground with respect to regulation and implementation of said technologies.

In addition to the creation of a new regulatory agency and a commitment to the international community, we would like to voice our concern about the potential use of these technologies for coercive behavioral modification, such as the use of implants to control prisoners.

It is also foreseeable that the application of non-reversible enhancing technologies in a military context would be the first step toward an arms race that would have the inevitable result of the complete dehumanization of the future soldier. Such forced implementation of these technologies should never be allowed in a free society, and therefore should be banned.

Nanotech is going to revolutionize the world. We believe that an informed public can alter the course of this technology, so as to avoid the possible disastrous outcomes of a technology which runs rampant without proper regulation, and to ensure that nanotech is used for the greatest good for the greatest number.

Finally, we completely acknowledge and support the ability of our representatives to be flexible in accommodating these technologies as they become available. However, no matter how far this technology advances it is never acceptable for our government to use such advances to usurp civil liberties and freedoms that are guaranteed to U.S. citizens under our Constitution.

## 17.4 Chapter 17d 2008 National Citizens' Technology Forum on Human Enhancement, Identity, & Biology

### *17.4.1 Georgia Panelists' Final Report*

Georgia Panelists: Adair, Allison; Alistairre, Rexxor; Bagheri, Johan; Curtis, Jennifer Leah; FitzHugh Foster; Goedeker, Michelle; Hairston, Timothy; Iglesias, Diana; Johnson Katherine; Johnson, Ashley; Naranjo, Juan; Ravi, Kokila; Reed, Jonathan; Shepherd, Carolyn; Singletary, Richard

Editor's Note: These reports have been republished in approximately their original form, with only modest reformatting to preserve the visual continuity of this volume. The reports are the product of the participants in the National Citizen's Technology Forum, and we wanted to preserve their authentic voice. No one specific view or conclusion can be attributed to any particular author.

The 2008 Atlanta National Citizens' Technology Forum included 13 participants drawn from a diverse range of ages, educational levels, ethnic backgrounds, and professions. The following report reflects the deliberations and consensus of our group.

We are enthusiastic about using nanotechnologies for human repair and regeneration. For example, grafts made from our own skin, regenerated limbs, sensors to release insulin automatically for diabetics, precisely targeted treatments for cancer, chips that can restore brain functions for people with Alzheimer's and

Parkinson's diseases – all these things are exciting developments. We hope to see them move forward.

We have more mixed views, however, on the possibility of using nanotechnology for human enhancement. We agree that individuals should be able to choose enhancements if they want them, but we also picture that some limits will be necessary. For example, we approve of the use of nanotechnologies in the military where they are used to prevent loss of life, particularly through robotics, but we are worried about effects on the human body of applications like a biotechnologically enhanced soldier who can stay up for days.

We have several major concerns about the applications of nanotechnology in biological sciences. We are concerned (1) that there is currently no agency capable of regulating the technologies, leading to a situation in which development may be driven primarily by greed and not by improvements for humanity; (2) that some of technologies could be dangerous if they fell into the wrong hands; (3) that the technologies could have long term effects on human health and the environment; (4) that the high cost of the technologies will lead to unequal access, which will lead to greater gaps between the haves and the have-nots; and (5) that the public will not receive complete information on the benefits and risks.

#### 17.4.1.1 Top Ten Questions

A number of major questions need to be answered as this set of technologies moves forward.

1. How will these emerging technologies benefit humanity as a whole – who decides who gets what, for what purpose, and why?
2. How do we ensure that nanotechnologies do not fall into the hands of those who want to control or cause harm?
3. Where is the funding coming from and does the funding, give certain rights to the technologies for the funders?
4. How do we ensure that there is a careful analysis of the long-term side effects (i.e. on people, plants, animals and the environment) of these emerging technologies?
5. How will the maintenance of these technologies be developed and deployed?
6. Given the critical nature of regulating these emerging technologies, how do we ensure that a separate governing body with adequate resources and relevant competencies will be established and deployed to implement appropriate policies, guidelines, rules, and laws?
7. How do you control the applications of nanotechnologies?
8. What are the marketing strategies for these emerging technologies?
9. Will there be an advisory panel to decide ethical questions and if so who?
10. How can we ensure that the public will receive balanced information on the benefits and risks?

#### 17.4.1.2 Recommendations

Regulations

Given the critical nature of emerging nanotechnologies, and given the fact that existing agencies are not capable of regulating these technologies, this committee recommends that a new regulatory agency be established on a national basis with an independent civilian board (see next section). This agency should be adequately staffed.

- The agency will have a director who is or has a scientific/technical background with a Ph.D. or M.D. education and who has practiced or researched in a field of Nanotechnology.
- The agency will be staffed with individuals who have appropriate backgrounds in the natural and social sciences, technology, philosophy, law and the humanities.

The agency must be charged with establishing rules, regulations, protocols, and laws for:

- Research & Development
- Commercialization and maintenance
- Privacy
- Sectors including but not limited to medical/health, general public, industrial, and national security

The agency should develop and implement standards for legal ramifications including but not limited to:

- Ownership
- Liability
- Limitations
- Consequences
- Enforcement

The agency should work to develop international relationships that have a focus on synergistic efforts for the betterment of humanity, e.g., the International Space Station. It should also develop a set of guidelines to help insure an open line of communication between the agency and military applications. This new organization should collaborate with existing agencies to develop and implement security policies and laws that ensure that the safety and welfare of all humanity are protected and safeguarded so that these technologies are not used in a detrimental manner. Although funding will provide certain intellectual property rights to NBIC technologies sponsors, these rights should be subject to national security concerns.

Policing

We recommend that an independent Civilian Board monitor the regulatory agency just described. The Board would consist of well-informed lay people with various backgrounds who will serve for limited terms. The Board would analyze materials on issues such as privacy and safeguards and make recommendations on the use and development of NBIC technologies.

### Long Term Effects

As we explore these new technologies, we recommend that those working in the field be cognizant of the long-term effects that these technologies pose for quality of life. Since we do not know these effects, we recommend that, after the devices have been approved for use, there be a study that tracks a real world sample of users. The study should carefully follow medical progress or regression, and provide data on broader changes in the lives of the people involved. This study also would carry over into the fields of agriculture and animal life. This data is very important and should be accurate and unbiased in order to show the various improvements that need to be made.

To avoid over-reliance on these technologies, we also recommend that governing bodies and the healthcare industry get more involved in the general healthcare and well being of the public stressing the overall benefits of good health, rather than using these new technologies as a solve-all for preventable health issues.

We recommend that the development of NBIC technologies promote diversity and preserve free choice. We want to avoid homogeneity and over-reliance on these technologies and the creation of a master race. We do not want to live in a world in which everyone is the same or where people have become machine-like, devoid of emotions.

### Inequality

Inequality has always existed. There has never been a modern society where everyone is equal. As nanotechnologies develop, society should try to keep in mind that we do not want to increase the gap between the haves and have-nots. We want to decrease the gap.

If cost alone determines who gets or does not get reparative or enhancement technologies, then it seems obvious that there will be greater inequality in society. Thus, as society tries to compensate for the inequality that may be produced by NBIC technologies, a new healthcare system will be needed. To focus NBIC technologies on lowering inequalities, reparative applications should lead in funding rather than research on enhancements.

### Public Information

We recommend that the regulatory agency described earlier maintain an NBIC website collaboratively with international partners and in multiple languages.

Transparency and open access will help to maintain the accuracy of information on the site. The global website at a minimum will provide the following information:

- All companies involved in manufacture and maintenance of NBIC technologies including the processes they use.
- All NBIC products and components in development and commercially available
- Updated information on risks, side effects, and benefits, with the percentage of people who have experienced each of these risks, side effects, and benefits.
- An Ask the Experts feature where experts can respond to NBIC questions within a timely manner
- All pertinent information for a particular product concerning providers and procedures for installation, maintenance, and reversibility.
- Insurance coverage – percentage exceptions, inclusions, and other financial assistance information
- Percentage of successful installations per doctor per product and public disclosure of non-complying individuals and corporations

Private companies can all link to the web page, and the web site could be expanded to cover all nanotechnologies. The web site should be widely promoted. The information should be available in printed form if requested and available in public facilities such as libraries, hospitals, and doctors' offices.

We are particularly concerned about NBIC advertising focusing more on benefits and not enough on risks. Therefore, advertising of NBIC technologies should be subject to full disclosure of known risks.

#### 17.4.1.3 Conclusion

We understand that investment in NBIC technologies is critical to encourage innovation. We support public funding for research in this area, including opportunities for individuals to donate.

We strongly encourage that the responsible authorities, national and international, consider and implement these recommendations.

## 17.5 Chapter 17e 2008 National Citizens' Technology Forum on Human Enhancement, Identity, and Biology

### *17.5.1 New Hampshire Panelists' Final Report*

New Hampshire Panelists: Cook, Frank; DeGrandpre, Angel; Hahn, Emily; Jones, Catherine; Kavanagh, John; Lapriore, Jane; Leigh, Katherine; Lemieux, Tammy; Murphy, Emily; Sutherland, Marc; Turni, Jennifer; Ward, Daniel

Editor's Note: These reports have been republished in approximately their original form, with only modest reformatting to preserve the visual continuity of this volume. The reports are the product of the participants in the National Citizen's Technology Forum, and we wanted to preserve their authentic voice. No one specific view or conclusion can be attributed to any particular author.

**FINDING 1: The distinction between remediation and enhancement is too subjective to be used as the basis for public policy decisions.**
*RECOMMENDATIONS:*

1. Alternative bases for public policy decisions should be based on a set of ground rules or prime directive including:
    - Most good for the most people
    - Least potential harm (The Prime Directive, again)
    - Favor environmental good over personal good
    - Favor benefits to groups of people over individuals
    - Favor patient autonomy
    - Favor health over military and cosmetic applications
    - First do no harm

**FINDNG 2: Proprietary information precludes transparent knowledge to enable in depth discussion for independent evaluation before the technology reaches the market.**
*RECOMMENDATIONS:*

1. There should be incentives for scientists to share information or collaborate on projects. Researchers and inventors should be urged to patent their discoveries promptly so that not only are they able to profit from their work but colleagues and competitors may openly evaluate those discoveries and build upon them.
2. Perhaps an intermediate waiting period in addition to existing requirements should be instated before a given technology is released to the public, to allow an independent evaluation regarding possible positive and negative implications to society.

**FINDING 3: Existing regulatory agencies and statutes are ill equipped to review and regulate NBIC technologies emerging into the marketplace.**
*RECOMMENDATIONS:*

1. A registry should be created under an appropriate agency in which corporations, universities and government entities (including the military) outline in broadest terms their ongoing NBIC research projects.
2. FDA should restructure to encompass not only its current responsibilities but will also oversee NBIC development and marketing. This will obviously require reorganization and an expanded workforce that will need to become educated in this new field.

3. Mandatory periodic professional development for law-makers to stay abreast of current NBIC-related technologies and how this relates to current statutes and any future statutes that might need to be proposed.

**FINDING 4: Market place incentives will drive development of NBIC to occur outside of the existing US regulatory framework.**
*RECOMMENDATIONS:*

1. Private enterprise will lead the development of new NBIC technologies, though international support for a defined set of guidelines must be developed to ensure social responsibility, thus decreasing the risk of unintended consequences. Appropriate structures must be in place to represent the public interest; these might include strong laws and treaties, but in any case, they should require accountability for companies producing products and people using them.

**FINDING 5: Potential negative impacts of NBIC on society including biological evolution and use of public investments need to be considered in light of ethical considerations rather than solely commercial (cost-benefit).**
*RECOMMENDATIONS:*

1. An emphasis should be placed on teaching ethics and personal responsibility at all levels of education; particularly including business and scientific education.
2. People with ethical training and backgrounds must be well represented on all boards and committees that have grant funding and regulatory powers.
3. There should be a diverse panel including scientists, ethicists, spiritual leaders, philosophers, and not including commercial interests, that will review and evaluate what NBIC technologies are being developed. The purpose of this panel would be to educate the public and recommend policy.
4. That there be firm legal ground for individuals and groups to seek redress for ethical breaches (rather than, or in addition to material damages) as a means to encourage accountability among the proponents of NBIC and provide broad control on the development and deployment of these technologies.
5. In addition to groups being held legally accountable, we propose recognition for ethically responsible practices such as minimizing environmental waste, safe work conditions, etc. Types of recognition may be access to grants, tax breaks, public acknowledgment (e.g. "Energy Star" version for ethics.)
6. The role of the medical doctor in implementation of NBIC technologies with human subjects/patients is of such import, that restatement of the relevant nondiscrimination provision of the Hippocratic Medical Oath produced by the World Medical Association in 1948, following the Nuremberg Nazi Drs. Trials, is imperative: "I will not permit consideration of race, religion, nationality, party politics or social standing to intervene between my duty and my patient."

**FINDING 6: There appears to be no mechanism in place to ensure equitable access to potential beneficial NBIC products and therapies.**
*RECOMMENDATIONS:*

1. Access to emerging technologies must be guaranteed to those who need them most, therefore those with the power to regulate new NBIC technologies coming to market, must develop a framework to define the basis of need. This basis of need should be defined concerning the potential positive impact the therapy will have on the quality of life for that person, and not on desire for enhancement or financial resources available.
2. We believe that the focus should be on fixing the current health care system and making it affordable to the public.
3. Funding and regulatory decisions should favor applications of NBIC that encourage products that are widely distributable and easily affordable. NBIC, as a potential "building block" of our health system, like other medical products, vaccines, and technologies, is essential to a well functioning health system that ensures equitable access to essential medical products, vaccines and technologies of assured quality, safety, efficacy and cost-effectiveness, and that there uses be scientifically sound and cost-effective. (Recommendation substance derived "D. The Building Blocks of a Health System") from United Nations Human Rights Council (UNHRC) Special Rapporteur Paul Hunt's report to the UNHRC on "the right of everyone to the enjoyment of the highest attainable standard of physical and mental health" January 31, 2008
4. To facilitate lower costs and greater accessibility of these NBIC technologies naturally through corporate competition, the patent structure and lifespan should be modified. Patent laws should not allow rights to fundamental NBIC technologies to be maintained for excessive lengths of time.

**FINDING 7: Investment of limited public funds in NBIC diverts those resources from more pressing social (e.g., health related) needs.**
*RECOMMENDATIONS:*

1. Formation of NBIC Council consisting of taxpayers from each state as well as government officials whose responsibility it is to review the national investment in specific NBIC technologies in relation to government spending on other programs such as public health needs, etc, and recommend budget allocations. This council will also generate a publication available to citizens, which will be followed by public hearings where ordinary citizens can give their input to the NBIC Council.

**FINDING 8: Given claims of potential power of NBIC technologies there is insufficient discussion of worst-case scenarios including unintended consequences and abuse and misuse**
*RECOMMENDATIONS:*

1. The nation's scientific community and policy think tanks, both public and private, should develop scenarios, possible responses, and theoretical outcomes. These scenarios and findings must be made available to the public. Potential risks must be evaluated independently from profit motives.

## 17.6 Chapter 17f 2008 National Citizens' Technology Forum on Human Enhancement, Identity, and Biology

### *17.6.1 Wisconsin Panelists' Final Report*

Wisconsin Panelists: Daña Alder, Theresa Behnen, John Bushong, Nathan Comp, Andrea Connell, Madhavi Dodda, John Endres, Abbey Johnson, Leslie Kurabelis, Virginia Pickerell, Joseph Powell, Don Schantz, Marissa Steen, Magda Valdes

Editor's Note: These reports have been republished in approximately their original form, with only modest reformatting to preserve the visual continuity of this volume. The reports are the product of the participants in the National Citizen's Technology Forum, and we wanted to preserve their authentic voice. No one specific view or conclusion can be attributed to any particular author.

#### 17.6.1.1 Introduction

In March 2008, 14 residents of the Madison metropolitan area participated in the "National Citizens' Technology Forum" (NCTF), a project funded by the National Science Foundation through the Center on Nanotechnology and Society at Arizona State University. Teams of researchers from six universities, including the University of Wisconsin-Madison, organized parallel panels of approximately 15 individuals. Each group was charged with the task of developing policy recommendations on the topic of *technologies of human enhancement*, addressing scientific and technical developments in nanotechnology, biotechnology, information technology, and cognitive science (NBIC technologies). Participants undertook a guided process of learning and deliberating in order to create a set of recommendations arrived upon by consensus. Participants were chosen to reflect the diversity of each region and applicants were screened for conflicts of interest and prior affiliation with organizations that had taken a political position on nanotechnology.

#### 17.6.1.2 The Process

Participants began by reading a 60-page packet of background materials compiled by the NCTF coordinating researchers at North Carolina State University. The packet, "Human Enhancement, Identity, and Biology," represented an attempt to assemble the most accurate, current, and non-partisan information available and was reviewed by a number of specialists to ensure its accuracy, balance, and accessibility to non-experts. On March 1 and 2, the Madison participants met together with a team of researchers from the University of Wisconsin-Madison to discuss the background materials and begin to explore the various social, political, technical, economic, environmental, and ethical aspects of human enhancement NBIC technologies. The second phase of deliberation occurred online: Madison participants joined participants from the five other sites across the USA in a series of nine virtual, web-based meetings to develop and pose questions to scientific experts. In the third phase, the Madison group met in person on March 29 and 30 to evaluate what they had learned and deliberate over possible policy recommendations.

#### 17.6.1.3 The Purpose

Challenges such as genetically modified food, climate change, and stem cell research suggest the value of engaging citizens in technological governance. Often, however, citizens are invited to learn and deliberate only *after* a technology has been introduced. In contrast, the NCTF aims to engage laypersons before significant technologies of human enhancement mature and reach the stage of deployment and commercialization. Researchers and participants alike see the value in incorporating the concerns of laypersons in the governance, research, and development of technologies with great potential to affect human society and the environment. The consensus format of the NCTF, in particular, represents one strategy to take advantage of the intersection of lay and expert knowledge – engaging "ordinary" citizens who have invested time and energy to learn from experts and deliberate over possible guidelines for technology.

#### 17.6.1.4 Policy Recommendations

The following recommendations represent the consensus of the 14 members of the Madison panel of the NCTF. For the purposes of this document, consensus indicates not unanimous support, but the wisdom of the group without major objection.

1. FUNDING ACCOUNTABILITY
   Require state and federal agencies that fund or provide partial funding for human enhancement technologies research at either private or public institutions to make the following information available at a centralized, online location:

   - Availability of funding/criteria
   - Agency/researchers being funded
   - Goals of the research
   - Regular status reports
   - Final reports
   - Community notification/outreach

2. PRIVACY CONCERNS

   - Recent erosions of privacy might combine with unprecedented possibilities of nanotech to further endanger privacy rights. We propose that appropriate precautions be taken to safeguard privacy, favoring individual rights.
   - We propose that diagnostic tests or procedures, especially those that could result in denial of health coverage, be kept confidential and private.

- Legal privacy concerns will be more complex as new NBIC technologies emerge. For example, body modifications and modifications of one's children challenge current distinctions between individual rights and the public interest.

3. ETHICAL CONSIDERATIONS

We believe that ethical considerations are integral to the scientific process. To that end, scientists in both the public and private sectors of nanotechnology development should strive to benefit the greatest common good and address basic societal issues. In order to achieve this goal, we recommend the following:

- Ethical concerns should be incorporated in all science curricula.
- All regulatory bodies involved with nanotechnology should include at least one ethicist.
- Policies should be developed to address the possibility that an increase in expensive technology will result in an increase in economic and social division, both nationally and internationally.

4. SAFETY/TESTING CONCERNS

"Prevention is better than cure" holds true for NBIC human enhancement developments. We suggest that it will be more cost efficient to enforce comprehensive and rigorous testing and safety standards before the products are on the market, rather than addressing the various health and legal problems that might arise without extensive and thorough testing of the products.

In order to ensure that the risk/benefit ratio is properly assessed prior to using such products, we recommend that FDA provide effective communication of possible/expected side effects and long term effects of using any nanotech products, not just medicine and food-related, in easy language that is understood by the common people.

5. HEALTH INSURANCE

- Health insurance policies in general should cover any nanotech enhancements and remediation technologies that could be deemed as medically necessary by current medical and insurance industry standards.
- Health insurance providers should be diligent in keeping up with current NBIC procedures and technologies and updating policies with specific inclusions and exclusions to coverage.
- Health insurance providers should include clear statements of their coverage of NBIC technologies in their policies.

6. FDA

- Empower the FDA to accomplish its mission with adequate funding in order to review its mission and to operate successfully to meet the mission.
- The FDA's mission needs to include nanotechnology.
- Ensure enforcement of FDA guidelines for all individuals involved in the development and manufacture of NBIC technologies.

- Ensure that guidelines keep pace with new developments and their social and ethical impacts.
- Make nano-toxicity research a higher funding priority.
- Establish a series of regulatory goals and deadlines for the nanotech industry.

7. EDUCATION

- Include nanotechnology in basic high school science curricula.
- Increase funding for science programs for improving teacher training and recruitment as part of a broad effort to improve K-12 education.

# Part IV
# Nanoparticle Toxicity and the Brain

# Chapter 18
# A Review of Nanoparticle Functionality and Toxicity on the Central Nervous System*

**Z. Yang, Z.W. Liu, R.P. Allaker, P. Reip, J. Oxford, Z. Ahmad, and G. Reng**

## 18.1 Introduction

The advent of nanoparticle systems has had a major impact in a host of scientific areas, opening up new capabilities and functionalities across a wide range of applications. The properties of nanomaterials can differ from those demonstrated by their bulk forms and, in some cases, give completely unexpected physical and chemical properties. For this reason many industries and manufacturers are now introducing nanomaterials and nanotechnologies in their mainstream products so as to exploit these new capabilities.

Many types of nanomaterials are also flourishing in medical science and technological areas while related research and applications are exploring potentials in

---

*Springer Science+Business Media Dordrecht/Journal of the Royal Society Interface published online 2 June 2010, doi: 10.1098/rsif.2010.0158.focus – p. 1–12, A review of nanoparticle functionality and toxicity on the central nervous system, with kind permission from Springer Science+Business Media Dordrecht 2012.

One contribution to a Theme Supplement 'Scaling the heights— challenges in medical materials: An issue in honour of William Bonfield, Part I. Particles and drug delivery'.

Z. Yang • Z.W. Liu
School of Medical Science, Nankai University, Tianjin, China

R.P. Allaker • J. Oxford
Barts and The London School of Medicine and Dentistry, Institute of Dentistry, Queen Mary University of London, Newark Street, London E1 2AT, UK

P. Reip
Intrinsiq Materials Ltd., Farnborough, Hants GU14 0LX, UK

Z. Ahmad
Department of Mechanical Engineering, University College London, Torrington Place, London WC1E 7JE, UK

G. Reng (✉)
School of Engineering and Technology, University of Hertfordshire, Hatfield AL10 9AB, UK
e-mail: g.ren@herts.ac.uk

S.A. Hays et al. (eds.), *Nanotechnology, the Brain, and the Future*, Yearbook of Nanotechnology in Society 3, DOI 10.1007/978-94-007-1787-9_18,

biosensors, biomaterials, tissue engineering, DNA modification, drug and drug-delivery systems (Chen et al. 2006; Lu et al. 2008; Kim et al. 2009; Sun et al. 2009; Kirkpatrick and Bonfield 2010). Another area which has benefited from these advances is microbiology, where the inhibitory effect of nanoparticles on microbes can be seen as a tool to combat and control outbreaks of disease. However the effect on microbes must also be viewed carefully as it demonstrates the potential effect nanoparticle systems can have on living systems; the relatively modest information on nanoparticle toxicity in various human systems means the issue of safety still remains incomplete. Several studies have highlighted this issue further. For example, it has been shown that manmade nanomaterials possess highly activated surfaces which are capable of inducing carcinogens, mutagens or cytotoxic activity (Seemayer et al. 1990; Seaton et al. 2010). Coupled to these surface properties is the direct size comparison; for example, nanoparticles are 100 times smaller than normal red blood cells, which increases the potential for interaction, and there is evidence that nanoparticles interact with proteins, DNA (Seeman 2006), lung cells and viruses. Hence, understanding nanoparticle interaction with living cells and other biological systems is critical, especially as the potential and exploitation of such technologies is rapidly gathering pace, requiring healthcare professionals and the public in general to be in closer contact with such materials. For example, metallic nanoparticle systems are now readily used for their anti-microbial properties in everyday products ranging from deodorants to personalized computer devices (Allaker and Ren 2008).

As was the case with nuclear technology, the personal computer boom and GM crops/animals, more than 68% of people now believe that nanotechnology 'would make things better in the future' (Royal Academy of Engineering 2004).

One class of material which has had such impact is carbon nanotubes (CNTs). Single-wall (approx. 10 nm in diameter) or multi-wall CNTs with extremely high aspect ratios (as high as 1,000) have a much lower percolation threshold. They are composed of many nanotubes and are chemically bonded together, possessing a highly activated surface and exhibiting superior strength, rigidity and electrical conductivity. Figure 18.1 shows standard industrial grade multi-wall CNTs used in high-performance conventional batteries (large scale).

Biocompatible and biomedical materials such as hydroxyapatite (HA) and carbon-coated titanium alloys for bone and hip replacements release debris of nanometre- and micrometre-scaled particles from metallic implants owing to host environment friction, and several studies have pointed out the potential risks in such bone repair and replacement (Moore et al. 2001). Nanoparticles such as $SiO_x$ and CNTs may also act similarly with neurotoxins which affect the central or peripheral nervous system. Although the most common neurotoxin is alcohol, others include heavy metals, organic solvents and rarer "designer" drugs (Cole and Sumnall 2003). As nanoscaled substances have an active surface, they may cause or produce acute neurological complications, or subacute or chronic illnesses. Neuron synaptic transmission and neuron cell membrane with the ionic channels for $Ca^{2+}$, $Na^{+}$, $K^{+}$ and $Cl^{-}$ may also provide a route of entry for CNTs or smaller nanoparticles.

Although there are numerous biological systems which can be investigated for such nanoparticle interactions, this review focuses on studies related to the functionality

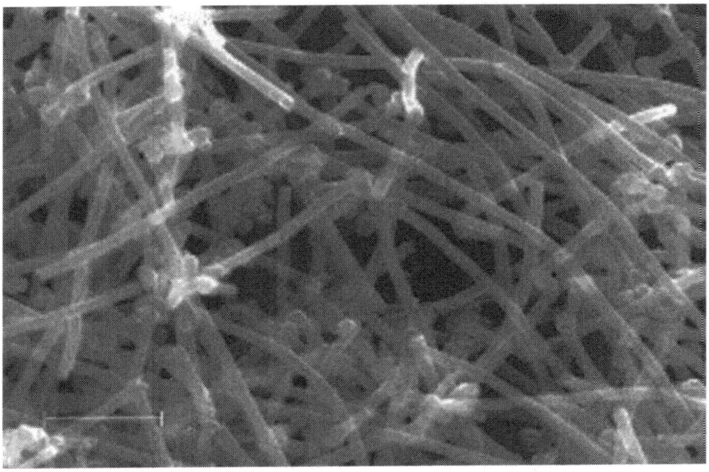

Fig. 18.1 Scanning electron micrograph (*SEM*) showing multi-wall carbon nanotubes

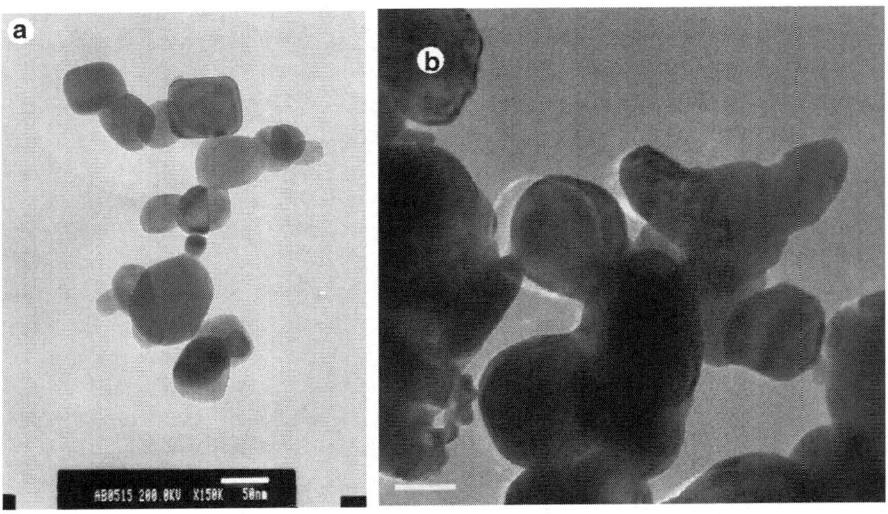


**Fig. 18.2** Transmission electron micrograph (*TEM*) of (**a**) CuO nanoparticles (QinetiQ Nanomaterials Ltd.) and (**b**) agglomerated silver nanoparticles

and toxicity of anti-viral/antimicrobial nanoparticles in the central nervous system (CNS). The current findings show that both negative and positive effects are observed by using selected nanoparticles, typically deployed as anti-bacterial/viral compositions (0.1–1.0%w/w; Xu et al. 2009). Figure 18.2 shows typical copper oxide (CuO) nanoparticles (Fig. 18.2a) and agglomerated silver nanoparticles (Fig. 18.2b) that are used in such studies as model anti-microbial nanomaterials, which provide better anti-bacterial capabilities than their bulk material (Raffi et al. 2008; Fig. 18.3).

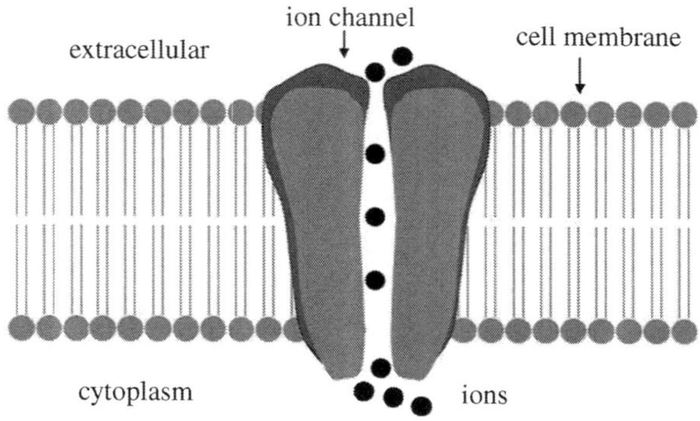


**Fig. 18.3** Neuron synaptic transmission and neuron cell membrane with passes for $Ca^{2+}$, $Na^{+}$, $K^{+}$ and $Cl^{-}$ (interactive physiology), where carbon nanotubes or smaller nanoparticles might be able to pass through easily

### *18.1.1 Important Issues of Nanoparticle Toxicity in General*

A selection of findings on nanoparticle toxicity on a host of living systems can be illustrated to elucidate these points, and also advocate the need to understand these interactions in greater detail. For example, some studies on rats have shown that 15% of the sample population died within 24 h due to blockage of the airways as a result of carbon nanotubes being injected into their lungs (Lecoanet et al. 2004; Warheit et al. 2004). More of a concern is the effect observed from micro- and/or nanoscaled debris of artificial hip replacements as there is a growing demand for such biomaterials (e.g. implantable devices). These loose particulates arise as a result of friction, and travel into the blood stream and eventually lead to the formation of a thrombus (De Jong and Borm 2008). There is also evidence to suggest migration of particles into organs (liver and the spleen) from similar prostheses (Gatti et al. 2004). Moving away from implantable devices, there is a risk posed from inhalation. Research has demonstrated that radio-labelled nanoparticles can reach the blood stream within 60 s via inhalation; and the liver within 60 min (Chunfu et al. 2004).

The current assumption is that smaller sized and highly activated nanoparticles (such as silica featured as hydrophilic, hydrophobic or even amphiphilic) can be taken up by human membranes. However, no response or signal is initiated that leads to the rejection of the particles, and these pass through the membrane passively. Potential health hazards related to such particles are the adsorption and enrichment of various poisonous substances (metals, dioxins, combined with hydrogen chloride (HCl), hydrofluoric acid (HF) on the particle phase which possess a much larger surface area (Robichaud et al. 2005)). Tetra-ethyl lead ($PbC_8H_{20}$), generated by lead petrol (4 Star) from carexhausts, once inhaled could be accumulated in the human brain.

This review considers three key elements of the toxicity screening methodologies or strategies covering physico-chemical characteristics, in vitro assays (cellular and non-cellular), and in vivo assays relevant to CNS cells. In particular, the review intends to concentrate on introducing the current techniques of drug in vitro and in vivo toxicology test methods into nanoparticle toxicology test methods by using CNS cells. These could be considered to determine proposed possibilities that the biological activity of nanoparticles can depend on physico-chemical parameters; however, they are not routinely being considered in toxicity screening studies because of the complexities of the tests of physico-chemical properties of nanoparticles, as recognized by many leading toxicity researchers (Oberdörster et al. 2005; Wang et al. 2009). Although the functionality of nanoparticles is closely linked to their physical status and properties such as the particle morphology, particle interactions of agglomeration and aggregation, this paper is mainly concerned with the identified results of biological interactions between the current industrial nanoparticles and CNS cells.

The general interaction between physical properties and biological functionalities have been investigated and clearly highlighted by Oberdörster et al. (2005) on the basis of physico-chemical interactions with biological cells such as liver, blood, lung, macrophages, spleen and the immune system, CNS and neurons and skin, etc. The importance of physico-chemical properties has been emphasized again in the understanding of the toxic effects on biological cells, which include particle size and size distribution, agglomeration state, shape, crystal structure, chemical composition, surface area, surface chemistry, surface charge, and porosity.

While the size aspect of nanoparticles attracts considerable and rapidly growing attention to several industries, the chemical aspect of the materials should not be overlooked as this has been shown to be an important aspect in several nanomaterial – cell interactions (Thian et al. 2008). This review focuses on nanoparticle functionality with a broad view on materials falling into the nanomaterials range. However, materials chemistry will also have an impact; for example, it has been shown that silver nanoparticles are more anti-bacterial than copper nanoparticles.

## 18.2 Neuron Cells and Central Nervous System

Neurons are nerve cells that, together with neuroglial cells, constitute the nervous tissue making up the nervous system. A neuron consists of a nerve cell body (or soma), axon and dendrites. Neurons receive nerve signals (action potentials), integrate action potentials, and transmit the signals to other neurons. Although the human nervous system is much more specialized and complicated than that of lower animals, the structure and function of neurons is essentially the same in all animals. In vitro systems to study the effects of particles on the nervous system have included neuron and nanoparticle cultures to determine the effects on neuronal functions (Oberdörster et al. 2005).

Ion channels are transmembrane proteins that mediate passive transport of ions, and the channels underlie a broad range of the most basic biological processes, from excitation and signalling to secretion and absorption. Studies of ion channels provide useful and informative clues for understanding the biophysics and pharmacology of these important and ubiquitous membrane proteins. There are many kinds of ion channels, such as sodium channels, calcium channels and potassium channels in rat models e.g. CA1 hippocampal neurons. Voltage-gated potassium ($K^+$) channels can play crucial roles in regulating a variety of cellular processes in both excitable and non-excitable cells, such as setting and re-setting membrane potential, action potential duration, the delay between a stimulus and the first action potential and discharge patterns.

Further research has been carried out using metal nanoparticles such as Ag, Cu and Mn on P12 brain cells to investigate potential neurotoxicity (Wang et al. 2009).

### *18.2.1 Nanoparticles' Interaction with the Central Nervous System*

Nanoparticles have shown biological functions such as killing pathogenic bacteria and viruses (e.g. flu), but research has also shown that nanoparticles may produce adverse effects (dose related) in human cells on contact. Human neural cells, such as hippocampal cells in the CNS, are the most sensitive and delicate cells in bioorganisms, and are responsible for brain functions and emotions. They are vulnerable to ischaemia, oxygen deficiency and external factors. One of the great concerns in science and technological development in the twenty-first century is that nanoparticles may produce potential functional and toxicity effects on human neural cells owing to their ability to pass through biological membranes (Brooking et al. 2001).

The blood – brain barrier (BBB) is a separation of circulating blood and cerebrospinal fluid (CSF) maintained by the choroid plexus in the CNS, which results from the selectivity of the tight junctions between endothelial cells in CNS vessels that restrict the passage of solutes. At the interface between blood and brain, endothelial cells and associated astrocytes are stitched together by tight junctions. Endothelial cells restrict the diffusion of microscopic objects and large or hydrophilic molecules into the CSF, while allowing the diffusion of small hydrophobic molecules (e.g. $O_2$, hormones, $CO_2$). Cells associated with the BBB actively transport metabolic products such as glucose across the barrier with specific proteins (Seidner et al. 1998).

Exposure to nanoparticles (such as Ag) in the body is also becoming increasingly widespread through antibacterial fabrics and coatings. However, effects from the presence (or even accumulation) of metal nanoparticles in the brain and through the BBB have not yet been fully studied. Small-sized particles have better mobility and it is expected that the transportation of nanoparticles across the BBB is possible either by passive diffusion or by carrier-mediated endocytosis (Hoet et al. 2004). In addition, nanoparticles may be taken up directly into the brain by trans-synaptic transport (Oberdörster 2004). For example, Ag nanoparticles can enter via the BBB (Panyala et al. 2008) and accumulate in different regions of the brain (Rungby and Danscher 1983), and this may be beneficial for drug delivery, but may also pose a risk to the patient (Sarin et al. 2008; Muthu and Singh 2009). It has also been reported that nanoparticle exposure can induce impairments to normal neurons

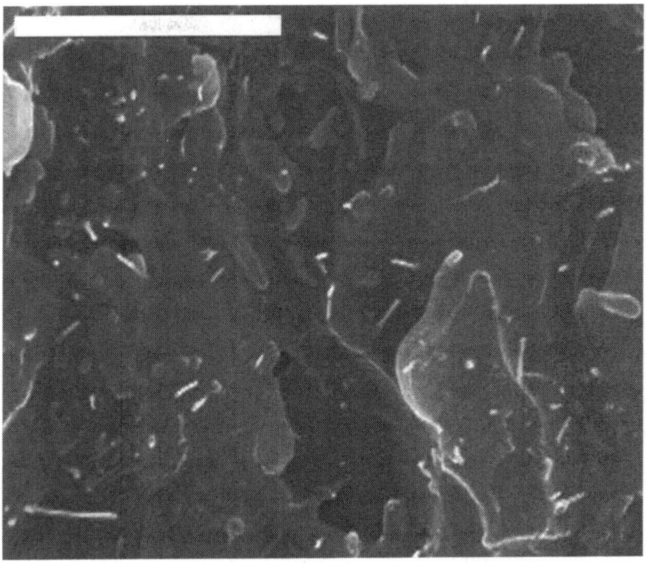

**Fig. 18.4** Scanning electron micrograph (*SEM*) of carbon nanotubes reinforced with epoxy resin (scale bar, 5 μm)

(Tang et al. 2008), microglia (Au et al. 2007) and even aggravate the process of brain pathology (Sharma and Sharma 2007). Ion channels play an important role in cell viability and functionality, especially in the CNS, which serve as a subtle indicator of the condition and viability of the cells.

Voltage-gated sodium currents determine a large number of neuronal properties, such as influencing action potential generation and the propagation of action potentials to synapse terminals. The local depolarization of neurons may also be affected by the existence of nanoparticles. However, what plays a key role remains to be determined in the transportation of amino acid neurotransmitters (e.g. aminobutyric acid (GABA)) and monoamines (i.e. dopamine (DA), norepinephrine, and serotonin). In addition, mutations may also cause changes in voltage-gated $Na^+$ channels, which are associated with a number of neurological diseases, including spontaneous epilepsy and pain conditions, and have been implicated in various psychiatric disorders (Meisler and Kearney 2005; Guo et al. 2008).

The effective nanoparticle content used in applications (0.1–1%w/w) could be well below the toxicity dosage limit by the time nanoparticles reach the CNS from point of contact. This estimation takes into account the fact that released nanoparticles/ions also need to enter the body, then cross the BBB and finally reach the CNS. Most of the nanoparticle release rates from work carried out on solid matrix – particle composites (i.e. epoxy resins, as shown in Fig. 18.4) display the release range of ion/particle concentration to be less than $10^{-5}$ g $ml^{-1}$, which is the minimum non-effective dosage for all the CNS neuron cell tests (drug toxicity tests are around $10^{-6}$ g $ml^{-1}$). These nanoparticle neuron tests also take into account several types of nanoparticles, e.g. Ag, CuO, ZnO, $TiO_2$ (Xu et al. 2009;

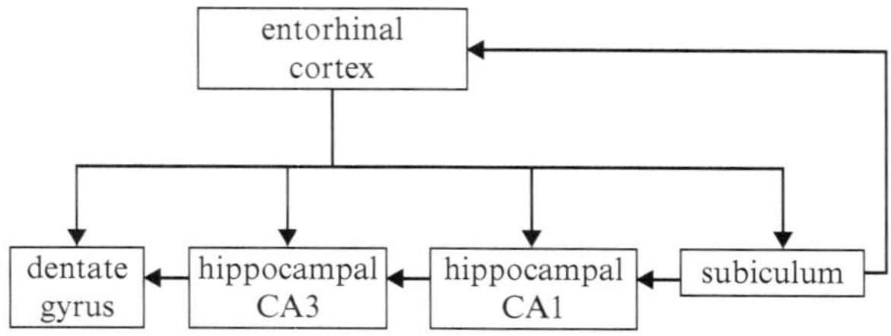


**Fig. 18.5** Schematic of hippocampal pathways

Zhao et al. 2009; Liu et al. 2010). Since metallic particle systems are being used in a host of contact applications such as computers, paints and clothing, research into this area and further parametric variables needs to be considered (e.g. exposure, contact time, strength of binding and weight loading).

Voltage-gated sodium current is responsible for modifying the excitability of neuronal cells and neuronal activity and function in the CNS. Therefore, potential modulation of the current by nanometal particles would be expected, leading to alterations in functionality. Some reports have shown that nanoparticles can impair cell function and even induce certain cell death (Shin et al. 2007; Cha et al. 2008; Tang et al. 2008). In recent studies on the neurotoxicity of metallic nanoparticles, a neuro-endocrine cell line (PC-12 cells) was exposed to nanoparticles such as Ag ($5 \times 10^{-5}$ g $ml^{-1}$), which reduced the level of DA. It was also found that Ag nanoparticles were more toxic than manganese (Mn) nanoparticles to particular cells (Hussain et al. 2006a). These findings suggest that Ag and other nanoparticles might have significant pathological consequences for the brain of mammalians while enhancing or inhibiting some particular functionality (Fig. 18.5).

Nanoparticles have potential functionality and toxic effects on human neuron cells since they can pass through biological membranes (Brooking et al. 2001). It is known that the biological half-life of silver in the CNS is longer than that in other organs, suggesting that there may be some significant physiological functions, consequences and risks to the brain due to prolonged exposure.

However, the effects from the presence (or even accumulation) of such particles, especially Ag, in the CNS are not very well documented.

## 18.3 Current Research Advances in Nanoparticle and Neuron Cell Interaction

### *18.3.1 The Neurotoxicity Research of Nanoscaled Materials In Vivo*

To investigate the potential effects of nanomaterials on the brain, some in vivo tests have been carried out on different animal models. Nanoparticles (50 nm) of silica-coated cobalt ferrite were found in the brain after being administered via an intravenous injection in mice (Kim et al. 2006). In another study, F344 female rats received single or multiple exposures to 20, 100 and 1,000 nm latex fluorospheres by intravenous injection or oral pharyngeal aspiration into the airways. In this instance, the 20 nm spheres were not detected in the brain; however, the 100 nm spheres were detected in the CNS 24 h after administration. The 1,000 nm spheres were detected for up to 28 days and were no longer found in the brain after this time point (Sarlo et al. 2009). Although this study utilized the same material (latex) for the various particle sizes, modest consideration was granted for material physicochemical properties and it would not be representative if other materials, e.g. other polymers, metals, metal oxides, ceramic composites, CNTs, etc., were to be used in the same tests.

In addition, maternal exposure of mice to $TiO_2$ nanoparticles may affect the expression of genes related to the development and function of the CNS. Analysis of gene expression using gene ontology indicated that gene expression levels associated with apoptosis were altered in the brain of newborn pups, and those associated with brain development were altered in early age. The genes associated with response to oxidative stress were changed in the brains of 2- and 3-week- old mice. Changes to gene expression associated with neurotransmitters and psychiatric diseases were found (Shimizu et al. 2009). The results suggest the potential toxicity of nanoparticles on the development of newborns. Nano-$TiO_2$ has also been shown to induce an increase in glial fibrillary acidic protein (GFAP), producing positive astrocytes in the CA4 region, which was in good agreement with higher Ti contents in the hippocampus region. This resulted in various types of oxidative stress in the brain of exposed mice such as lipid peroxidation, protein oxidation and increased activities of catalase, as well as the excessive release of glutamic acid and nitric oxide (Wang et al. 2008).

Nanotoxicology studies on the brain have also focused on fish. For example, in the brain of juvenile largemouth bass, a significant increase in lipid peroxidation was observed due to exposure to fullerenes ($C_{60}$; $0.5 \times 10^{-6}$ g ml$^{-1}$; Oberdörster 2004). In addition, it is conceivable that colloidal fullerenes need to be transported to lipid-rich regions (e.g. brain) before the colloid dissociates and frees individual redox-active fullerenes. It is also possible that there may be an inflammatory response creating reactive oxygen species (ROS) or that a reactive fullerene metabolite is produced. The actual mechanism still needs to be determined, and future research will focus on this question. The depletion of glutathione (GSH) is used as an indication of oxyradical scavenging ability, showing that the antioxidant defence system is overwhelmed by ROS (Oberdörster 2004).

Recent research has focused on the effects of nano-particles on the BBB. According to a study by Sharma et al. (2010), administration of Ag, Cu or Al/$Al_2O_3$ nanoparticles showed disrupted BBB function and induced brain oedema formation. Moreover, silver nanoparticles induced BBB destruction and astrocyte swelling, and caused neuronal degeneration (Tang et al. 2009).

**Fig. 18.6** Confocal image of PC-12 cells used in toxicity studies. The cell lengths are approximately 25–30 μm under the confocal microscope before cell-line proliferation. However, after the cell division their lengths could grow up to several hundred micrometres (scale bar, 10 μm)

### *18.3.2 Neurotoxicity Research of Nanoscaled Materials In Vitro*

Several studies have focused on PC-12 cells, a neuroendocrine cell line with the capability to produce the neurotransmitter DA and contain functional DA metabolism pathways (Fig. 18.6). Normal PC-12 cells are around 25–30 μm; after cell division this could increase to several hundred micrometres. However, with exposure of PC-12 cells to Mn nanoparticles (40 nm), or $Mn^{2+}$ (acetate), or Ag nanoparticles (15 nm) for 24 h, the cells showed contrasting results. Phase-contrast microscopy studies show that exposure to Mn particles or $Mn^{2+}$ does not greatly change the morphology of PC-12 cells. But exposure to Ag particles caused cell shrinkage plus irregular membrane borders compared with the control cells. Further microscopic studies at higher resolution microscopy revealed that Mn nanoparticles and agglomerates were effectively internalized by PC-12 cells (Hussain et al. 2006a, b). Mitochondrial reduction activity, a sensitive measure of particle and metal cytotoxicity, showed only moderate toxicity for Mn compared with similar Ag and $Mn^{2+}$ doses. Mn particles and $Mn^{2+}$ ions depleted DA (dose dependent) and its metabolites, dihydroxyphenylacetic acid (DOPAC) and homovanillic acid (HVA). Ag particles significantly reduced DA and DOPAC only at concentrations of 50 mg $ml^{-1}$. Therefore, DA depletion due to Mn particles was most similar to $Mn^{2+}$ ions, which is known to induce concentration-dependent DA depletion. The significant increase in ROS with Mn particle exposure also suggests that the increased ROS levels may participate in DA depletion (Hussain et al. 2006a).

In another study, the expression of 11 genes associated with the dopaminergic system was examined using real-time reverse transcription polymerase chain reaction (RT-PCR). The results indicated that the expression of Txnrd1 was upregulated after the Cu-90 treatment and the expression of Gpx1 was downregulated after Ag-15 or Cu-90 treatment. These alterations are consistent with the oxidative stress induced by metal nanoparticles. Mn-40 induced a downregulation of the expression of Th; Cu-90 induced an upregulation of the expression of MAOA. Mn-40 also induced a downregulation of the expression of Park2; while the expression of SNCA was upregulated after Mn-40 or Cu-90 treatment (Wang et al. 2009).

PC-12 cells have also been treated with different concentrations of $TiO_2$ nanoparticles (1, 10, 50 and $100\times10^{-6}$ g ml$^{-1}$) and the viability of these cells was significantly reduced, showing a significant dose- and time-dependent effect (Liu et al. 2010). In agreement with earlier findings (Hussain et al. 2006a, b), the flow cytometric assay gave an indication that the $TiO_2$ nanoparticles induced intracellular accumulation of ROS (as shown in Fig. 18.7) and apoptosis of the PC-12 cells with increasing concentration of $TiO_2$. Interestingly, pre-treatment with a ROS scavenger could inhibit PC-12 apoptosis induced by the particles (Liu et al. 2010). Similar findings have been reported by Long et al. (2006), where $TiO_2$ stimulated immediate ROS production.

Zinc (Zn) and Iron (Fe) nanoparticles have also been assessed for their cell interactions using a glioma cell line, A-172. Zn (300 nm), Fe (100 nm), Si (10–20, 40–50, 90–110 nm; 0.24–2,400$\times10^{-9}$ g ml$^{-1}$) and a micro-sized (45 μm) Si (control) were analysed and used in cell cytotoxicity (Cha and Myung 2007). Fluorescence was absent inside the glioma cell line A-172, suggesting that nanoparticles did not alter the membrane permeability and the cytotoxicity of nanoparticles in vitro was low, and it was not dependent on the types and sizes of nanoparticles, showing a low level of toxicity in vivo (Zhao et al. 2009). Here, the toxicity was due to material chemistry rather than size (Cha and Myung 2007). Results obtained using reduced nanoparticle concentrations (0.24–2,400$\times10^{-9}$ g ml$^{-1}$) compared with other studies (Hussain et al. 2006a; Liu et al. 2009) suggest that concentration is an important parameter when assessing exposure to cells.

In a separate study, up to 30 μg ml$^{-1}$ single-walled CNTs (SWCNTs) significantly decreased the overall DNA content in chicken embryonic spinal cord or dorsal root ganglia. This effect was more pronounced when cells were exposed to highly agglomerated SWCNTs than when they were exposed to better dispersed SWCNT bundles (Belyanskaya et al. 2009).

### *18.3.3 The Patch Clamp Technique*

The patch clamp technique is a laboratory technique in electrophysiology that allows the study of single or multiple ion channels in cells. The technique can be applied to the study of excitable cells such as neurons, cardiomyocytes, muscle fibres and pancreatic beta cells.

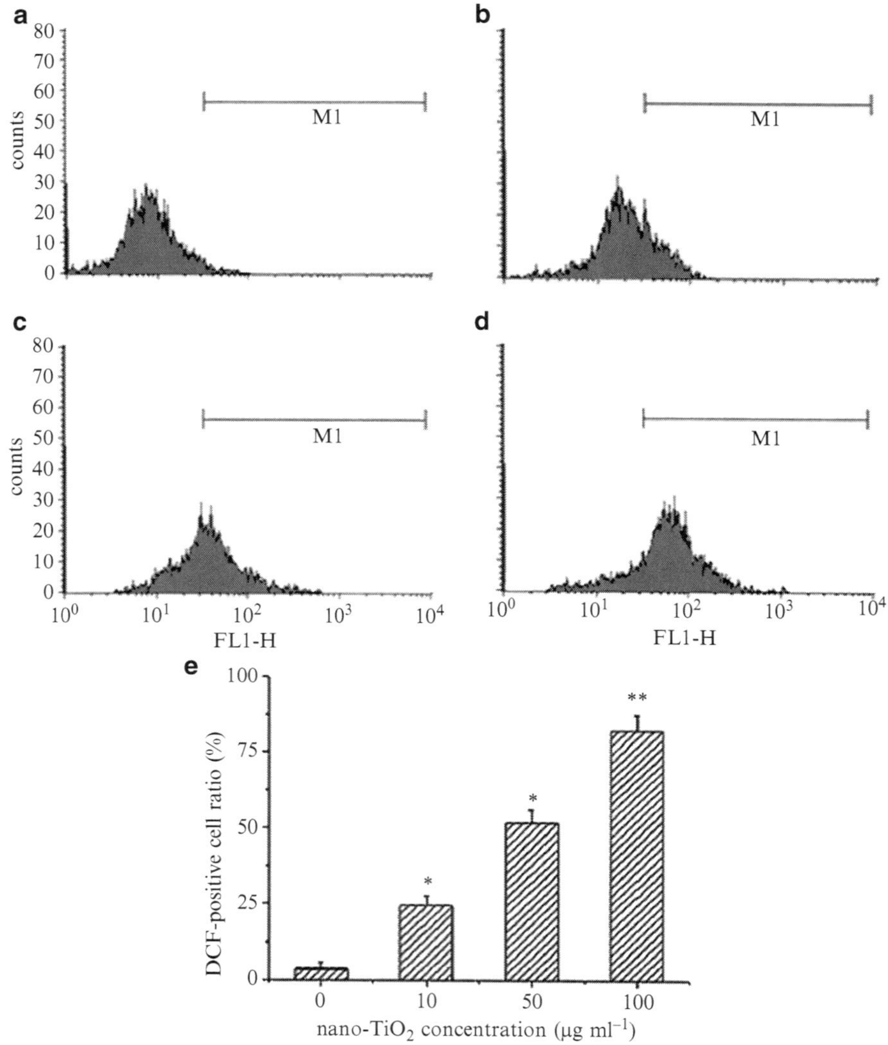


**Fig. 18.7** Measurement of ROS generation in PC-12 cells by flow cytometry. The cells were cultured with nano-$TiO_2$ at concentrations of (**a**) 0 µg ml$^{-1}$, (**b**) 10 µg ml$^{-1}$, (**c**) 50 µg ml$^{-1}$ and (**d**) 100 µg ml$^{-1}$ for 24 h. ROS levels are dose dependent. The corresponding linear diagram of flow cytometry is shown (**e**). n=3; mean+SEM; *statistically significant difference compared with controls ($p<0.05$); **$p<0.01$ (Liu et al. 2010)

This technique has been used in studying the effects of nanoparticles on ion channels (Zhao et al. 2009). This can be demonstrated using single rat hippocampal pyramidal neurons, isolated by enzymatic digestion and mechanical dispersion (according to the method of Zou et al. 2000). Sample preparation includes slicing

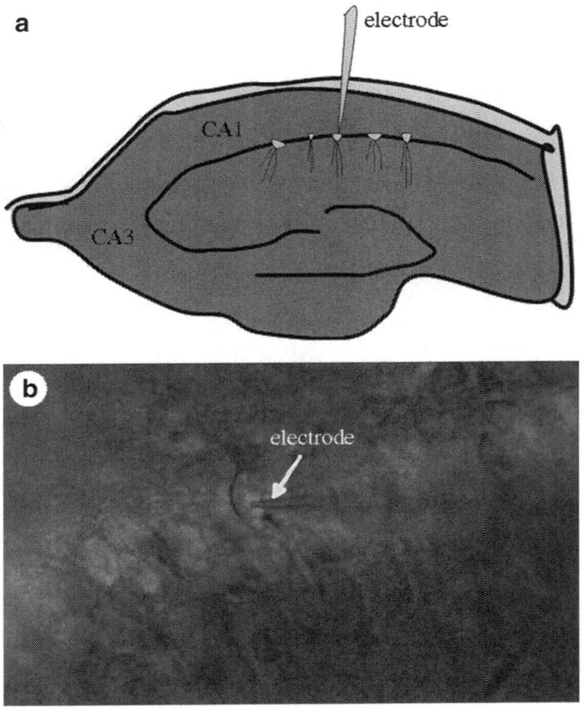


**Fig. 18.8** (**a**) Schematic diagram of hippocampal CA1 pyramidal neurons in the brain. (**b**) Whole cell patch clamp recording in CA1 pyramidal neuron from 14 to 18 Wistar rats

the entire hippocampus and subiculum horizontally (400 μm in thickness) using a vibratome (VT1000M/E, Leica, Germany) and incubating with artificial CSF (ACSF). Hippocampal CA1 neurons were then visualized on a monitor connected to a low light-sensitive charge-coupled device camera (Fig. 18.8; Liu et al. 2009).

Whole-cell currents of pyramidal neurons were recorded using an EPC10 patch clamp amplifier (HEKA, Germany; Fig. 18.9). After the rupture of the membrane and the establishment of a whole cell voltage-clamp configuration, compensation (80%) for series resistance was routinely used. These data were low-pass filtered at 2.9 kHz, digitized at 10 kHz (fourpole Bessel filter) and used PULSE 8.74 software (HEKA, Germany) at the ambient temperature (21–23°C). The effect of metal nanoparticles on voltage-gated channels (hippocampal neurons) can be shown using ZnO nanoparticles (concentration of $10^{-4}$ g ml$^{-1}$; Zhao et al. 2009). Here, the transient outward potassium current ($I_A$) and delayed rectifier potassium current ($I_K$) increased considerably (Fig. 18.10). However, it is apparent that the ZnO solution/suspension did not shift the steady-state activation curve of $I_A$ and $I_K$, and nor did it have a significant effect on the inactivation and the recovery from the inactivation of $I_A$. Peak amplitude and overshoot of the evoked single action potential was increased and the half-width was diminished in the presence of the $10^{-4}$ g ml$^{-1}$ ZnO solution (Zhao et al. 2009). Using different nanoparticles, such as CuO, other studies

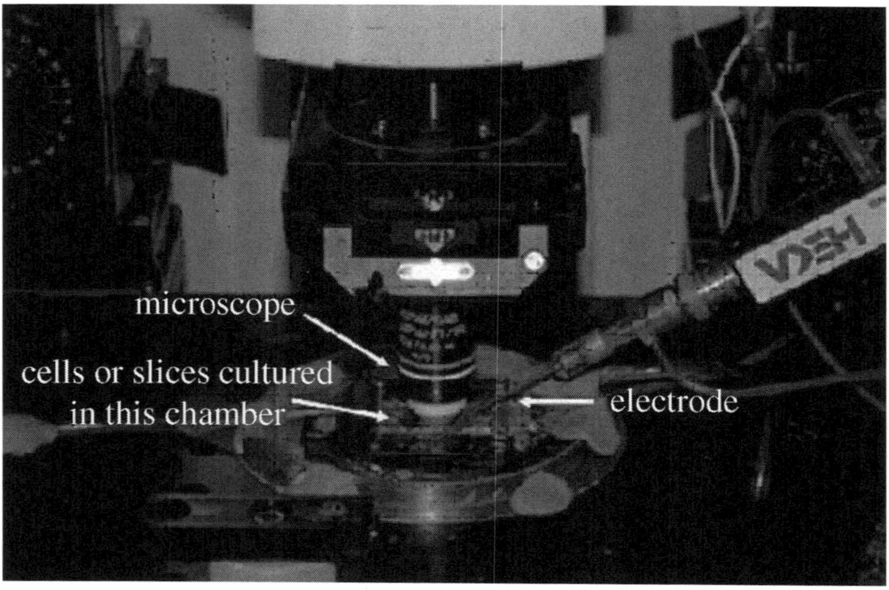


**Fig. 18.9** Experimental set-up for recording the response of neuron cells in ion channel currents

(Xu et al. 2009) have shown that CuO nanoparticles ($5\times10^{-5}$ g ml$^{-1}$) have no effects on $I_A$, but inhibited $I_K$ (Fig. 18.10). Furthermore, CuO nanoparticles did not shift the steady-state activation curve of $I_K$ and $I_A$, but the inactivation curve of $I_K$ was shifted negatively. The effects on the inactivation curve of $I_A$ have no statistical significance (Xu et al. 2009).

More recent work (Xu et al. 2009) demonstrated that ZnO nanoparticles increase the peak amplitudes of the voltage-gated sodium current ($I_{Na}$; Fig. 18.11), while the inactivation and the recovery from inactivation of the $I_{Na}$ are promoted by ZnO. The data also show that the steady-state activation curve of the $I_{Na}$ has not been shifted by ZnO nanoparticles. When the effects of Ag nanoparticles on the $I_{Na}$ were examined with increasing concentrations ($10^{-6}$, $5\times10^{-6}$, $10^{-5}$ g ml$^{-1}$), the results revealed that only concentrations of $10^{-5}$ g ml$^{-1}$ reduced the amplitude of the $I_{Na}$ (Fig. 18.11). Similar to ZnO, Ag particles produced a hyperpolarizing shift in the activation – voltage curve of $I_{Na}$. Ag nanoparticles delay the recovery of the $I_{Na}$ from inactivation (Liu et al. 2009), but the ZnO accelerates the process (Zhao et al. 2009).

ZnO also increases the evoked single action potential and repetitive firing rate. Action potentials are a fundamental property of excitable cells in the mammalian CNS. ZnO enhances peak amplitude and overshoot and demonstrates decreased half-width of the evoked single action potential. Conversely, peak amplitude and overshoot of the evoked single action potential are decreased and half-width is increased in the presence of a $10^{-5}$ g ml$^{-1}$ Ag nanoparticle solution (Liu et al. 2009).

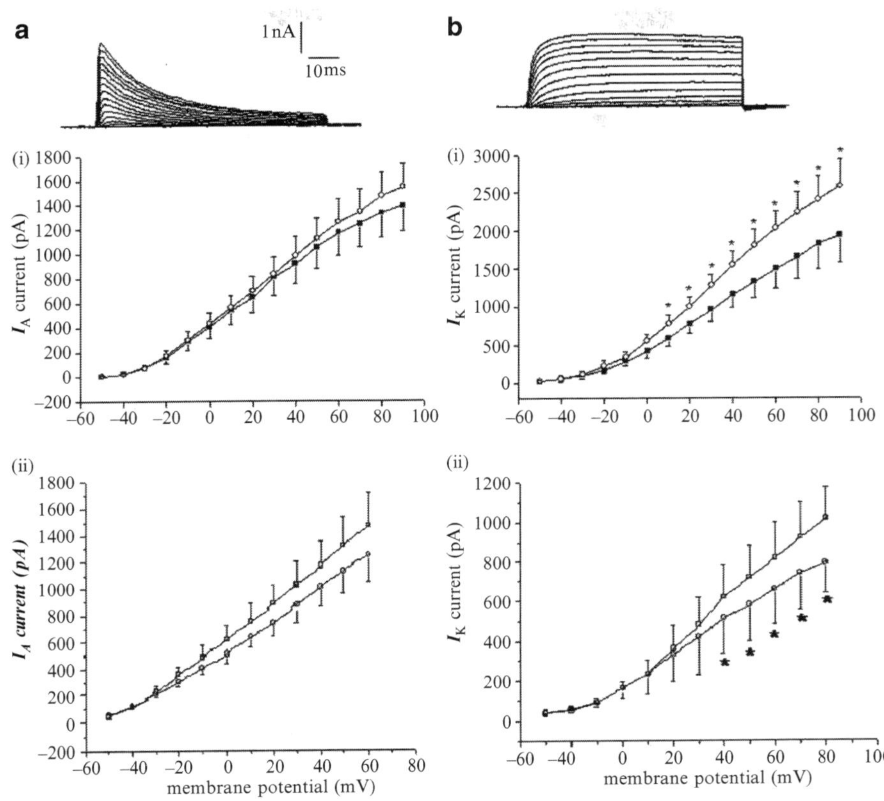


**Fig. 18.10** The original traces of (**a**) (*i*) (*ii*) $I_A$ and (**b**) (*i*) (*ii*) $I_K$. The contrasting effects of ZnO and CuO nanoparticles on *I*–*V* currents and peak amplitudes of $I_A$ and $I_K$ on hippocampal pyramidal neurons from Wistar rats (Xu et al. 2009; Zhao et al. 2009). (**a**, **b**) (*i*) *Filled square*, control; *open circle*, nano-ZnO; (**a**, **b**) (ii) *open square*, control; *open circle*, nano-CuO

## 18.4 Conclusion

In conclusion, most studies on the interaction between CNS neuronal cells and nanoparticles have used metal or metal oxides (including Cu, CuO, Zn and Ag) with selected neuronal cell lines (PC-12, CA1 and CA3). Neurologists have an interest in both functionality and toxicity with regard to the effects of nanoparticles, with the more recent studies focussing on the interaction with hippocampal cell membranes, as carried out for CNS drug toxicity. The effects of nanomaterials on ion channels within neurons may specifically relate to $Na^+$ [$I_{Na}$ (A)] and $K^+$ [$I_K$(A)] channels, as shown by a number of studies. It is possible that such ion channel effects may not apply to some other nanoparticles such as gold, which has been reported to possess unique biological properties and positive functionalities.

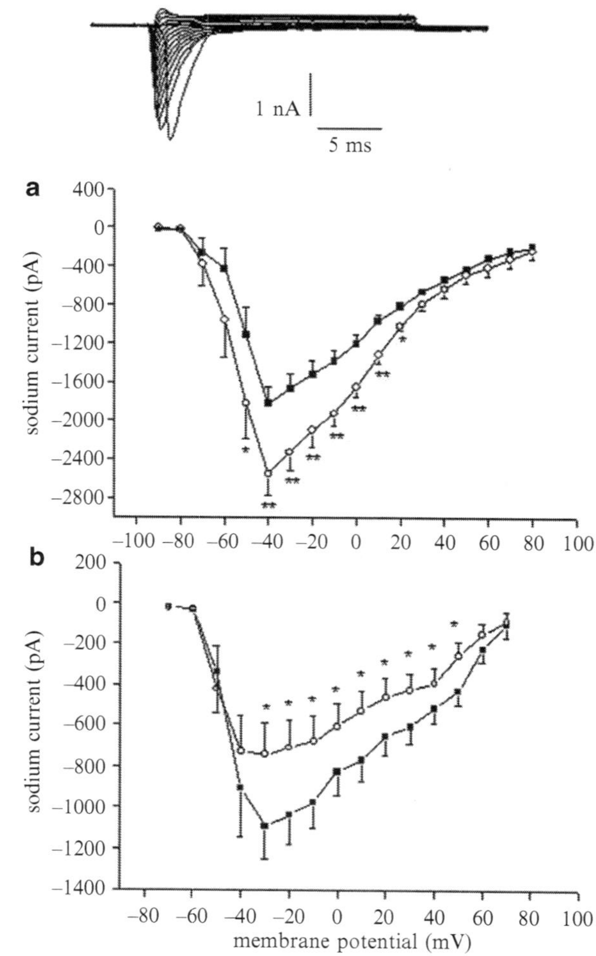


**Fig. 18.11** The original traces of the $I_{Na}$. (**a**) The increased effects of the nano-ZnO (*filled square*, control; *open circle*, nano-ZnO) and (**b**) decreased effects of nano-Ag on $I-V$ currents and peak amplitudes of $I_{Na}$ on hippocampal pyramidal neurons from Wistar rats ((Liu et al. 2009; Zhao et al. 2009) *filled square*, control; *open circle*, nano-Ag)

Neurologists have an equal interest in both areas of positive functionality and negative toxicity of nanoparticles on the human neuron cells as well as the interactions when passing through the BBB. Studies now have focused on biological membranes on the hippocampal cells, as carried out previously for drug toxicities within the CNS.

The range of applications continues to grow for nanomaterials at a rapid rate. The potential of individual nanoparticles and carbon nanotubes as constituents of toothpastes, beauty products, sunscreens, coatings, drug delivery systems, sensors, building materials, and textiles are being explored. Thus, a complete understanding of the mechanisms of interaction between nanoparticles and target cells that may lead to local and systemic effects within the CNS is required.

## 18.5 Future Research Work and Potential Research Directions

Nanotoxicity research can be applied to a number of applications, such as determining composition levels in coatings for medical devices, medical-grade sheet moulding compounds for hospital equipment, aircraft filter fabrics, printing-coat films/inks and compositions for high-performance aviation gas turbine lubricants.

The direction provided from previous work on functionality/nanotoxicity of nanoparticles on CNS cells should be focused towards further understanding the mechanism of action, and their neurological and circulatory effects using animal and in vitro models. Rat models, such as ischaemia, vascular dementia, epilepsy and diffuse axonal injury, are the first step for further assessment of functionality. Methods to address current problems in such tests also need to be developed. The existing problems in biological tests include nanoparticle agglomeration and aggregation within both liquid and airborne forms. In particular, nanoparticle dispersion in air with different sizes, materials and morphologies with controlled agglomeration involving aerosol delivery for in vivo and in vitro studies is the most challenging work in the field of nanoparticle toxicology due to difficulties in nanoparticle measurements, generation and observation (Kim et al. 2010), although some technological advances have been made on the proof of concept stage.

Also, the current knowledge on engineered nanoparticles and their interactions with the CNS cells is extremely limited and traditional drug toxicology studies may not be ideal models to draw comparison with due to the special nanofunctions and features. Further research on nanotoxicity as well as functionality will allow the expansion of the much needed understanding in this area, with a build-up of the physical and chemical properties of nanostructures influencing in vivo and in vitro behaviour towards CNS neural cells. In the CNS, microglial cells are a type of macrophage found in the brain, and they may be involved in handling any nanoparticles that reach the brain, and these cellular responses to nanoparticles should be investigated. Biological (CNS cells) interactions linked to particle size, surface energy, composition and aggregation will form a focal point of some future studies. Many biological properties of nanoparticles (i.e. Ag, Cu, $Fe_2O_3$, $Al_2O_3$, ZnO, $SiO_2$, $TiO_2$, CuO, $Cu_2O$, and WC, etc.) have been investigated in terms of the aetiology, pathology, physiology and epidemiology; however, no report has been obtained on CNS neurons owing to the complexity and high costs associated with assessments.

This future work will be supported by a grant from the UK Royal Academy of Engineering (ref. 5502) on a Major Research Exchanges Award.

## References

Allaker, R.P., and G. Ren. 2008. Potential impact of nanotechnology on the control of infectious diseases. *Transactions of the Royal Society of Tropical Medicine and Hygiene* 102: 1–2. doi:10.1016/j.trstmh.2007. DOI:dx.doi.org 07.003 DOI:dx.doi.org.

Au, C., L. Mutkus, A. Dobson, J. Riffle, J. Lalli, and M. Aschner. 2007. Effects of nanoparticles on the adhesion and cell viability on astrocytes. *Biological Trace Element Research* 120: 248–256. doi:10.1007/s12011-007-0067-z DOI:dx.doi.org.

Belyanskaya, L., S. Weigel, C. Hirsch, U. Tobler, H.F. Krug, and P. Wick. 2009. Effects of carbon nanotubes on primary neurons and glial cells. *Neurotoxicology* 30: 702–711. doi:10.1016/j.neuro.2009.05.005 DOI:dx.doi.org.

Brooking, J., S.S. Davis, and L. Illum. 2001a. Transport of nanoparticles across the rat nasal mucosa. *Journal of Drug Targeting* 9: 267–279. doi:10.3109/10611860108997935 DOI:dx.doi.org.

Cha, K.E., and H. Myung. 2007. Cytotoxic effects of nanoparticles assessed in vitro and in vivo. *Microbial Biotechnology* 17: 1573–1578.

Cha, K., H.W. Hong, Y.G. Choi, M.J. Lee, J.H. Park, H.K. Chae, G. Ryu, and H. Myung. 2008. Comparison of acute responses of mice livers to short-term exposure to nano-sized or micro-sized silver particles. *Biotechnology Letters* 30: 1893–1899. doi:10.1007/s10529-008-9786-2 DOI:dx.doi.org.

Chen, J., C.M. Han, X.W. Lin, Z.J. Tang, and S.J. Su. 2006. Effect of silver nanoparticle dressing on second degree burn wound. *Zhonghua Wai Ke Za Zhi* 44: 50–52.

Chunfu, Z., C. Jinquan, Y. Duanzhi, W. Yongxian, F. Yanlin, and T. Jiaju. 2004. Preparation and radiolabeling of human serum albumin (HSA)-coated magnetite nanoparticles for magnetically targeted therapy. *Applied Radiation and Isotopes* 61: 1255–1259. doi:10.1016/j.apradiso.2004.03.114 DOI:dx.doi.org.

Cole, J.C., and H.R. Sumnall. 2003. Altered states: The clinical effects of ecstasy. *Pharmacology & Therapeutics* 98: 35–58. doi:10. DOI:dx.doi.org 1016/S0163-7258(03)00003-2 DOI:dx.doi.org.

De Jong, W.H., and P.J. Borm. 2008. Drug delivery and nanoparticles: Applications and hazards. *International Journal of Nanomedicine* 3: 133–149.

Gatti, A.M., S. Montanari, E. Monari, A. Gambarelli, F. Capitani, and B. Parisini. 2004. Detection of micro- and nano-sized biocompatible particles in the blood. *Journal of Materials Science. Materials in Medicine* 15: 469–472. doi:10.1023/B:JMSM. DOI:dx.doi.org 0000021122.49966.6d DOI:dx.doi.org.

Guo, F., N. Yu, J.Q. Cai, T. Quinn, Z.H. Zong, Y.J. Zeng, and L.Y. Hao. 2008. Voltage-gated sodium channels Nav1.1, Nav1.3 and b1 subunit were up-regulated in the hippocampus of spontaneously epileptic rat. *Brain Research Bulletin* 75: 179–187. doi:doi:10.1016/j.brainresbull.2007.10.005 DOI:dx.doi.org.

Hoet, P.H., I. Bruske-Hohlfeld, and O.V. Salata. 2004. Nanoparticles—known and unknown health risks. *Journal of Nanobiotechnology* 2: 12. doi:10.1186/1477-3155-2-12 DOI:dx.doi.org.

Hussain, S.M., A.K. Javorina, A.M. Schrand, H.M. Duhart, S.F. Ali, and J.J. Schlager. 2006a. The interaction of manganese nanoparticles with PC-12 cells induces dopamine depletion. *Toxicological Sciences* 92: 456–463. doi:10.1093/ DOI:dx.doi.org toxsci/kfl020 DOI:dx.doi.org.

Hussain, S.M., A.K. Javorina, A.M. Schrand, H.M. Duhart, S.F. Ali, and J.J. Schlager. 2006b. The interaction of manganese nanoparticles with PC-12 cells induces dopamine depletion. *Toxicological Sciences* 92: 456–463. doi:10. DOI:dx.doi.org 1093/toxsci/kfl020 DOI:dx.doi.org.

Kim, J.S., et al. 2006. Toxicity and tissue distribution of magnetic nanoparticles in mice. *Toxicological Sciences* 89: 338–347. doi:10.1093/toxsci/kfj027 DOI:dx.doi.org.

Kim, K.J., W.S. Sung, B.K. Suh, S.K. Moon, J.S. Choi, J.G. Kim, and D.G. Lee. 2009. Antifungal activity and mode of action of silver nano-particles on Candida albicans. *Biometals* 22: 235–242. doi:10.1007/s10534-008-9159-2 DOI:dx.doi.org.

Kim, S.C., D.-R. Chen, C. Qi, R.M. Gelein, J.N. Finkelstein, A. Elder, K. Bentley, G. Oberdörster, and D.Y.H. Pui. 2010. A nanoparticle dispersion method for in vitro and in vivo nanotoxicity study. *Nanotoxicology* 4: 42–51. doi:10.3109/17435390903374019 DOI:dx.doi.org.

Kirkpatrick, C.J., and W. Bonfield. 2010. NanoBioInterface: A multidisciplinary challenge. *Journal of the Royal Society, Interface* 7: S1–S4. doi:10.1098/rsif.2009.0489.focus DOI:dx.doi.org.

Lecoanet, H.F., J.Y. Bottero, and M.R. Wiesner. 2004. Laboratory assessment of the mobility of nanomaterials in porous media. *Environmental Science & Technology* 38: 5164–5169. doi:10. DOI:dx.doi.org 1021/es0352303 DOI:dx.doi.org.

Liu, Z., G. Ren, T. Zhang, and Z. Yang. 2009. Action potential changes associated with the inhibitory effects on voltage- gated sodium current of hippocampal CA1 neurons by silver nanoparticles. *Toxicology* 264: 179–184. doi:10. DOI:dx.doi.org 1016/j.tox.2009.08.005 DOI:dx.doi.org.

Liu, S., L. Xu, T. Zhang, G. Ren, and Z. Yang. 2010. Oxidative stress and apoptosis induced by nanosized titanium dioxide in PC12 cells. *Toxicology* 267: 172–177. doi:10. DOI:dx.doi.org 1016/j.tox.2009.11.012 DOI:dx.doi.org.

Long, T.C., N. Saleh, R.D. Tilton, G.V. Lowry, and B. Veronesi. 2006. Titanium dioxide (P25) produces reactive oxygen species in immortalized brain microglia (BV2): Implications for nanoparticle neurotoxicity. *Environmental Science & Technology* 40: 4346–4352. doi:10.1021/es060589n DOI:dx.doi.org.

Lu, S., W. Gao, and H.Y. Gu. 2008. Construction, application and biosafety of silver nanocrystalline chitosan wound dressing. *Burns* 34: 623–628. doi:10.1016/j.burns.2007. DOI:dx.doi.org 08.020 DOI:dx.doi.org.

Meisler, M.H., and J.A. Kearney. 2005. Sodium channel mutations in epilepsy and other neurological disorders. *The Journal of Clinical Investigation* 115: 2010–2017. doi:10.1172/JCI25466 DOI:dx.doi.org.

Moore, W.R., S.E. Graves, and G.I. Bain. 2001. Synthetic bone graft substitutes. *ANZ Journal of Surgery* 71: 354–361. doi:10.1046/ DOI:dx.doi.org j.1440-1622.2001.02128.x DOI:dx.doi.org.

Muthu, M.S., and S. Singh. 2009. Targeted nanomedicines: Effective treatment modalities for cancer, AIDS and brain disorders. *Nanomedicine (London, England)* 4: 105–118. doi:10. DOI:dx.doi.org 2217/17435889.4.1.105 DOI:dx.doi.org.

Oberdörster, E. 2004. Manufactured nanomaterials (fullerenes, C60) induce oxidative stress in the brain of juvenile largemouth bass. *Environmental Health Perspectives* 112: 1058–1062.

Oberdörster, G., ILSI Research Foundation/Risk Science Institute Nanomaterial Toxicity Screening Working Group, et al. 2005. Review: Principles for characterizing the potential human health effects from exposure to nanomaterials elements of a screening strategy. *Particle and Fibre Toxicology* 2: 8. doi:10.1186/1743-8977-2-8 DOI:dx.doi.org.

Panyala, N.R., E.M. Pena-Mendez, and J. Havel. 2008. Silver or silver nanoparticles: A hazardous threat to the environment and human health? *Journal of Applied Biomedicine* 6: 117–129.

Raffi, M., F. Hussain, T.M. Bhatti, J.I. Akhter, A. Hameed, and M.M. Hasan. 2008. Antibacterial characterization of silver nanoparticles against. *Journal of Materials Science and Technology* 24: 192–196.

Robichaud, C.O., D. Tanzil, U. Weilenmann, and M.R. Wiesner. 2005. Relative risk analysis of several manufactured nanomaterials: An insurance industry context. *Environmental Science & Technology* 39: 8985–8994. doi:10.1021/es0506509 DOI:dx.doi.org.

Royal Academy of Engineering. 2004. Nanotechnology: Views of the general public. Quantitative and qualitative research carried out as part of the Nanotechnology Study. BMRB Social Research Report for the Royal Society and Royal Academy of Engineering Nanotechnology Working Group, BMRB/45/101-666. London: BMRB Social Research.

Rungby, J., and G. Danscher. 1983. Localization of exogenous silver in brain and spinal cord of silver exposed rats. *Acta Neuropathologica* 60: 92–98. doi:10.1007/BF00685352 DOI:dx.doi.org.

Sarin, H., et al. 2008. Effective transvascular delivery of nanoparticles across the blood-brain tumor barrier into malignant glioma cells. *Journal of Translational Medicine* 6: 80. doi:10. DOI:dx.doi.org 1186/1479-5876-6-80 DOI:dx.doi.org.

Sarlo, K., et al. 2009. Tissue distribution of 20 nm, 100 nm and 1000 nm fluorescent polystyrene latex nanospheres following acute systemic or acute and repeat airway exposure in the rat. *Toxicology* 263: 117–126. doi:10.1016/j.tox. DOI:dx.doi.org 2009.07.002 DOI:dx.doi.org.

Seaton, A., L. Tran, R. Aitken, and K. Donaldson. 2010. Nanoparticles, human health hazard and regulation. *Journal of the Royal Society, Interface* 7: S119–S129. doi:10.1098/rsif.2009.0252. focus DOI:dx.doi.org.

Seeman, N.C. 2006. DNA enables nanoscale control of the structure of matter. *Quarterly Reviews of Biophysics* 6: 1–9.

Seemayer, N.H., W. Hadnagy, and T. Tomingas. 1990. Evaluation of health risks by airborne particulates from in vitro cyto- and genotoxicity testing on human and rodent tissue culture cells: A longitudinal study from 1975 until now. *Journal of Aerosol Science* 21(Suppl. 1): 501–504. doi:10.1016/0021-8502(90)90290-E DOI:dx.doi.org.

Seidner, G., et al. 1998. GLUT-1 deficiency syndrome caused by haploinsufficiency of the blood-brain barrier hexose carrier. *Nature Genetics* 18: 188–191. doi:10.1038/ng0298-188.

Sharma, H.S., and A. Sharma. 2007. Nanoparticles aggravate heat stress induced cognitive deficits, blood-brain barrier disruption, edema formation and brain pathology. *Progress in Brain Research* 162: 245–273. doi:10.1016/S0079-6123(06)62013-X.

Sharma, H.S., S. Hussain, J. Schlager, S.F. Ali, and A. Sharma. 2010. Influence of nanoparticles on blood-brain barrier permeability and brain edema formation in rats. *Acta Neurochirurgica. Supplement* 106: 359–364. doi:10.1007/978-3-211- DOI:dx.doi.org 98811-4_65 DOI:dx.doi.org.

Shimizu, M., H. Tainaka, T. Oba, K. Mizuo, M. Umezawa, and K. Takeda. 2009. Maternal exposure to nanoparticulate titanium dioxide during the prenatal period alters gene expression related to brain development in the mouse. *Particle and Fibre Toxicology* 6: 20. doi:10.1186/1743-8977-6-20 DOI:dx.doi.org.

Shin, S.H., M.K. Ye, H.S. Kim, and H.S. Kang. 2007. The effects of nano-silver on the proliferation and cytokine expression by peripheral blood mononuclear cells. *International Immunopharmacology* 7: 1813–1818. doi:10.1016/j.intimp. DOI:dx.doi.org 2007.08.025 DOI:dx.doi.org.

Sun, H., T.S. Choy, D.R. Zhu, W.C. Yam, and Y.S. Fung. 2009. Nano-silver-modified PQC/DNA biosensor for detecting E. coli in environmental water. *Biosensors and Bioelectronics* 24: 1405–1410. doi:10.1016/j.bios.2008.08.008 DOI:dx.doi.org.

Tang, M., et al. 2008. Unmodified CdSe quantum dots induce elevation of cytoplasmic calcium levels and impairment of functional properties of sodium channels in rat primary cultured hippocampal neurons. *Environmental Health Perspectives* 116: 915–922. doi:10.1289/ehp.11225 DOI:dx.doi.org.

Tang, J., L. Xiong, S. Wang, J. Wang, L. Liu, J. Li, F. Yuan, and T. Xi. 2009. Distribution, translocation and accumulation of silver nanoparticles in rats. *Journal of Nanoscience and Nanotechnology* 9: 4924–4932. doi:10.1166/ DOI:dx.doi.org jnn.2009.1269 DOI:dx.doi.org.

Thian, E.S., et al. 2008. The role of electrosprayed nanoapatites in guiding osteoblast behaviour. *Biomaterials* 29: 1833–1843. doi:10.1016/j.biomaterials.2008.01.007 DOI:dx.doi.org.

Wang, J., et al. 2008. Potential neurological lesion after nasal instillation of TiO(2) nanoparticles in the anatase and rutile crystal phases. *Toxicology Letters* 183: 72–80. doi:10. DOI:dx.doi.org 1016/j.toxlet.2008.10.001 DOI:dx.doi.org.

Wang, J., M.F. Rahman, H.M. Duhart, G.D. Newport, T.A. Patterson, R.C. Murdock, S.M. Hussain, J.J. Schlager, and S.F. Ali. 2009. Expression changes of dopaminergic system-related genes in PC12 cells induced by manganese, silver, or copper nanoparticles. *Neurotoxicology* 30: 926–933. doi:10.1016/j.neuro.2009.09.005 DOI:dx.doi.org.

Warheit, D.B., B.R. Laurence, K.L. Reed, D.H. Roach, G.A. Reynolds, and T.R. Webb. 2004. Comparative pulmonary toxicity assessment of single-wall carbon nanotubes in rats. *Toxicological Sciences* 77: 117–125. doi:10.1093/toxsci/ DOI:dx.doi.org kfg228 DOI:dx.doi.org.

Xu, L.J., J.X. Zhao, T. Zhang, G.G. Ren, and Z. Yang. 2009. In vitro study on influence of nano particles of CuO on CA1 pyramidal neurons of rat hippocampus potassium currents. *Environmental Toxicology* 24: 211–217. doi:10.1002/ DOI:dx.doi.org tox.20418 DOI:dx.doi.org.

Zhao, J., L. Xu, T. Zhang, G. Ren, and Z. Yang. 2009. Influences of nanoparticle zinc oxide on acutely isolated rat hippocampal CA3 pyramidal neurons. *Neurotoxicology* 30: 220–230. doi:10.1016/j.neuro.2008.12.005 DOI:dx.doi.org.

Zou, B., Y. Chen, C. Wu, and P. Zhou. 2000. Blockade of U50488H on sodium currents in acutely isolated mice hippocampal CA3 pyramidal neurons. *Brain Research* 855: 132–136. doi:10.1016/S0006-8993(99)02360-4 DOI:dx.doi.org.

# Recommendations for a Municipal Health & Safety Policy for Nanomaterials

## A Report to the Cambridge City Manager

July 2008

Submitted by:
Cambridge Nanomaterials Advisory Committee
Cambridge Public Health Department

**CAMBRIDGE PUBLIC HEALTH DEPARTMENT**

Cambridge Health Alliance

Springer Science+Business Media Dordrecht/A Report to the Cambridge City Manager, Cambridge Nanomaterials Advisory Committee and Cambridge Public Health Department, 2008, p. 1–14, Recommendations for a Municipal Health & Safety Policy for Nanomaterials, Mark Griffin, with kind permission from Springer Science+Business Media Dordrecht 2012.

S.A. Hays et al. (eds.), *Nanotechnology, the Brain, and the Future*, Yearbook of Nanotechnology in Society 3, DOI 10.1007/978-94-007-1787-9_19,

# Cambridge Nanomaterials Advisory Committee

**Carol Lynn Alpert**, Director of Strategic Projects Museum of Science
*Cambridge resident*

**David Bright**, Attorney
*Cambridge resident*

**Tom Brown**, Process Engineering
Hyperion Catalysis

**Daniel Gilden**
Health Statistics Consultant
*Cambridge resident*

**Pamela Greenley**, Deputy Director
MIT Industrial Hygiene Program

**Joe Griffin**, Director
Harvard Environmental Health & Safety Program

**Liz Gross**, Certified Industrial Hygienist
Safety Partners, Inc.

**Matt Henshon**, Attorney

**Bob Hoch**, Technology Director
Hyperion Catalysis

**Michael Huguenin**, Former Executive Director, Harvard Center for Risk Analysis *Cambridge resident*

**Igor Linkov**, PhD, Research Scientist, U.S. Army Engineer Research and Development Center and Adjunct Professor, Carnegie-Mellon University

**Chris Long**
Senior Environmental Health Scientist
Gradient Corporation

**Eric Martin**, Laboratory for Integrated Science & Engineering (LISE)
Harvard University

**John C. Monica**, Attorney
Porter Wright Morris & Arthur

**Dave Rejeski**, Director
Foresight and Governance Project
Woodrow Wilson International Center for Scholars

**Martin Schmidt**
Professor of Electrical Engineering
Massachusetts Institute of Technology

**Anant Singh**, PhD, Associate Principal
TIAX, LLC

**Terrence Smith**
Director of Government Affairs
Cambridge Chamber of Commerce

***Sam Lipson, Director of Environmental Health, Cambridge Public Health Department*** *facilitated the meetings of the Cambridge Nanomaterials Advisory Committee.*

# Acknowledgments

*Recommendations for a Municipal Health & Safety Policy for Nanomaterials* was written by:

**Sam Lipson**
Director of Environmental Health
Cambridge Public Health Department

The report was edited by:

**Susan Feinberg, MPH**
Communications Specialist
Cambridge Public Health Department

A special thanks to the members of the **Nanomaterials Advisory Committee** for their comments, guidance, perseverance and patience; to **Captain Gerard Mahoney** (Cambridge Fire Department) for his readiness to assist; and to **Claude-Alix Jacob** (Cambridge Public Health Department) for his support of this extended health policy review process.

# Table of Contents

Cambridge Nanomaterials Advisory Committee Members .... 335

Acknowledgements .... 337

Table of Contents .... 339

Preface .... 341

Executive Summary .... 343

Introduction .... 345

Findings .... 347

Recommendations .... 353

References .... 355

# Preface

In January 2007, the Cambridge City Council adopted the following policy order:

> *That the City Manager be and hereby is requested to examine the nanotechnology ordinance for Berkeley, California, and recommend an appropriate ordinance for Cambridge.*

At the request of the City Manager, the Cambridge Public Health Department (CPHD) reviewed the Berkeley ordinance and related issues. In its written response to the City Manager, the public health department described the limited scientific consensus available to characterize the health risks posed by engineered nanoscale materials. Prior to making any regulatory or policy recommendations, the department proposed that an advisory committee be established so that city decision makers could learn more about the potential impact of the nanotechnology sector on public health and the impact of regulations on research and manufacturing. The proposed advisory committee would include experts in the field, as well as representatives from the universities, the community, and the nanotechnology manufacturing, research, and consulting sectors.

In summer 2007, the City Manager convened the Cambridge Nanomaterials Advisory Committee, which was charged with developing recommendations for oversight of local nanotechnology activities to protect human health. On behalf of the City Manager, the public health department facilitated six monthly meetings of the committee through January 2008. The committee developed a series of recommendations, which are described in this report.

The Cambridge Public Health Department endorses these recommendations and is prepared to implement them in collaboration with other city departments and with institutions and companies that conduct nanoparticle research and manufacturing.

# Executive Summary

The Cambridge Public Health Department, in collaboration with the Cambridge Nanomaterials Advisory Committee, recommends that the City of Cambridge take several positive steps to gain a better understanding of the nature and extent of nanotechnology-related activities now underway within the city, to encourage research institutions and firms within the growing nanotechnology sector to share and improve practices leading to safe management of engineered nanomaterials, and to improve community access to the best available health and safety information as it relates to consumer products containing engineered nanomaterials.

In recognition of the limited health effects data and the absence of a clear consensus on best practices and standards for engineered nanomaterials, the Cambridge Public Health Department, in collaboration with the Cambridge Nanomaterials Advisory Committee, **does** not **recommend** that the City Council enact a new ordinance regulating nanotechnology at this time.

The Cambridge Public Health Department, in collaboration with the Cambridge Nanomaterials Advisory Committee, does **recommend** that the City of Cambridge take the following steps:

- Establish an inventory of facilities that manufacture, handle, process, or store engineered nanoscale materials in the city, in cooperation with the Cambridge Fire Department and the Local Emergency Planning Committee.
- Offer technical assistance, in collaboration with academic and nanotechnology sector partners, to help firms and institutions evaluate their existing health and safety plans for limiting risk to workers involved in nanomaterials research and manufacturing.
- Offer up-to-date health information to residents on products containing nanomaterials and sponsor public outreach events.
- Track rapidly changing developments in research concerning possible health risks from various engineered nanoscale materials.

- Track the evolving status of regulations and best practices concerning engineered nanoscale materials among state and federal agencies, and international health and industry groups.
- Report back to City Council every other year on the changing regulatory and safety landscape is it relates to the manufacture, use, and investigation of nanomaterials.

# Introduction

Nanotechnology is the art and science of manipulating matter at the molecular level to create new and unique materials and products.

Materials engineered at the nanoscale measure between 1 to 100 nanometers in at least one dimension (width or length). A nanometer is a billionth of a meter, which is larger than most atoms but smaller than most molecules. At this small size, materials can have different electrical, mechanical, and light-reflecting properties that can be harnessed to produce useful devices in areas as diverse as medicine, alternative energy, agriculture, and consumer goods. Nanoscale research and manufacturing is a small, but rapidly expanding sector in North America, Europe, Asia, and Australia. The City of Cambridge is home to dozens of scientific and medical research laboratories, as well as several industrial producers working with engineered nanoscale materials.

Researchers worldwide are currently investigating the use of nanotechnology to perform atom-by-atom assembly of specific molecules and to mimic self-assembly found in biological systems. Most applications of this research are either years or decades away from practical benefit. However, engineered nanoscale materials are already being incorporated into an array of industrial and consumer products, including cosmetics and personal care products, sunscreens, paints, coatings, sporting goods, stain-resistant clothing, and light emitting diodes used in computers and cell phones.[1] Today, more than 600 "nanoproducts" are on the market globally.[2]

Despite that these products are already commercially available, the effects of engineered nanoscale materials on human health and the environment are largely unknown. Some of the same properties that make nanoscale materials useful may also pose risks to people and the environment, under specific conditions. Two significant

[1] Consumer products inventory, Project on Emerging Nanotechnologies, Woodrow Wilson Institute. Available at: www.nanotechproject.org/inventories/consumer.

[2] "New Nanotech Products Hitting the Market at the Rate of 3-4 Per Week," Project on Emerging Nanotechnologies, Woodrow Wilson Institute, April 24, 2008. Available at: http://www.nanotech-project.org.

areas of concern are (1) risk to people who manufacture, process, or conduct research on engineered nanomaterials, or reside close to facilities where these activities take place and (2) risk to the general population and the environment. The first concern is the central focus of this report, although consumer education is addressed in the recommendations chapter.

Several government-funded laboratories are currently researching the toxicity and safety of nanoscale materials, and some cautionary procedures have been developed for the safe storage and handling of these materials. There is general consensus among toxicologists that further research is needed regarding the characterization, safety, and handling of various engineered nanoscale materials.

## *Information Gathering and Committee Deliberation*

The Cambridge Nanomaterials Advisory Committee (NAC) was convened in summer 2007 to assist the public health department in reviewing options for local oversight of facilities that handle or process engineered nanoscale materials.

The 19-member committee included individuals with professional expertise in the legal, scientific, and public policy disciplines related to environmental, occupational, and public health. A number of committee members are either employed as materials scientists and have detailed knowledge of engineered nanoscale materials or work in the nanotechnology manufacturing sector. Four committee members are also Cambridge residents, and represented the interests of their fellow citizens in this effort.

Sam Lipson, director of environmental health for the Cambridge Public Health Department, facilitated all six meetings of the Cambridge Nanomaterials Advisory Committee, which were held between August 2007 and January 2008.

At these meetings, NAC members discussed the present state of scientific knowledge about occupational and environmental health risks from engineered nanoscale materials, various oversight approaches that Cambridge might consider, and existing and likely future actions by regulatory authorities in other cities and at the state, federal, and international levels. These discussions helped clarify complex safety issues, risk management frameworks, and current efforts to understand engineered nanoscale materials, their potential impacts on human health, and their fate in the environment.

The following presentations from NAC members and the discussions that ensued provided the knowledge base and conceptual framework for the report recommendations:

- *Nanotoxicology overview*
  Dr. Chris Long, Gradient Corporation
- *Oversight of nanomaterials safety in an academic setting*
  Marilyn Hallock, MIT (presented in brief; slides made available to committee)

- *Oversight of existing EPA, OSHA, and FDA regulations with implications for nanomaterials*
  John Monica, Porter-Wright (a Washington, D.C. law firm)
- *Overview of risk management frameworks that could be used to address nanomaterials*
  Dr. Igor Linkov, US Army Engineer Research and Development Center and Carnegie-Mellon University

In addition, Captain Gerard Mahoney of the Cambridge Fire Department provided detail to NAC members about ongoing efforts by the Fire Department and the Local Emergency Planning Committee to gather and maintain a citywide inventory of hazardous chemicals found in high volume or presenting specials hazards.

## Findings

In developing its recommendations, the Nanomaterials Advisory Committee limited the scope of its discussion to the potential health effects of engineered nanoparticles on people who manufacture, process, or conduct research on engineered nanomaterials, or those who reside close to facilities where these activities take place. This constraint was placed on the committee by the public health department, and reflects the practical and historic role that local public health agencies have played in protection of individuals in their places of work and residence.

Larger regulatory questions pertaining to the impact of these materials on the environment and on consumers of nanomaterial-containing products need to be addressed at the state or federal level where such oversight responsibilities traditionally and appropriately sit. This does not preclude the City of Cambridge from helping to improve consumer access to updated information about the safety of nanotech products.

### *Overview of Nanotoxicology*

Nanotoxicology is an emerging subdiscipline of toxicology that explores whether and to what extent nanomaterials may adversely impact human health and the environment. Until quite recently, the toxicological assessment of materials on this scale primarily focused on "ultrafine particles," which include ***naturally occurring nanoparticles*** (e.g., volcanic ash) as well as ***incidental nanoparticles*** (e.g., diesel exhaust, welding fumes) generated as by-products of industrial and commercial processes.

In the past few years, however, some toxicologists have begun to focus on ***engineered nanoparticles***, which are purposely manufactured or created for their desirable physical and chemical properties. Engineered nanoscale materials assume a variety of structural forms, from self-assembling nucleic acids and semiconducting alloys to ornate forms of pure-carbon structures that are produced in several distinct shapes.

It is essential to note that various engineered nanoparticles can differ significantly from one another with regard to their physical and chemical properties, and thus are likely to differ significantly from one another in their toxic potential. Like any broad class of substances, it is expected that some engineered nanoparticles will be found to be relatively non-toxic, while others (including those having the same chemical composition as their less pernicious cousins) will be found to be of much greater toxicity.

A proper assessment of the potential risk to humans requires evaluating the likelihood that a significant exposure will occur when these materials or compounds are produced, processed, or used in an expected manner. A responsible risk management process must be tailored to each worksite to understand the potential health risks that may be present at that location. This "exposure assessment," which is critical to the total estimation of risk, is quite specific to each situation and cannot be evaluated generically.

## *What Is Known About the Potential Health Effects of Engineered Nanoparticles*

The examination of potential health risks from exposure to newly developed engineered nanoscale materials requires researchers to stretch beyond existing toxicology models and published data. Despite concerns raised by individuals (both scientists and non– scientists), the few studies that have been conducted specifically with engineered nanoscale materials do not yet suggest a clear pattern of harm. Some evidence of biological response or elevated reactivity has been presented, but it is not appropriate to use these narrow experimental observations alone to support a conclusion that such materials and products pose a threat under real-world conditions.

One related area of investigation concerns the toxicological effects of "ultrafine" particles. These are the nanoscale soot particles commonly found in exhaust from combustion processes, such as motor vehicle emissions, chimney smoke, and cooking fumes. Over the past three decades, research has shown there are several distinct types of damage to the human body that can occur when very small particles are inhaled at elevated concentrations. Studies indicate that some very small particles may gain entry to areas of the lung that are physically impossible for larger particles to reach and may then fail to be taken away (or cleared) during exhalation.

While engineered nanoparticles, such as carbon nanotubes, may share the nanoscale size with combustion-generated ultrafines, studies that associate chronic respiratory and cardiac problems to combustion-related particle exposures are assessing exposures to heterogeneous particles (comprised of organic compounds, metals, and other impurities) and hazardous gases that are simply not present in engineered nanoscale materials. Despite these differences, many valuable evaluation methods and principles describing transport (movement into and through the body) derived from this body of work have contributed to the study of nanotoxicology.

Although research is ongoing, there are a few important observations about engineered nanomaterials that can be found in the existing toxicological literature:

1. There is emerging evidence that biological effects observed in some studies are tied to various properties, including size, surface area, shape, surface chemistry, and electric charge of nanoscale particles. Some nanoscale particles that do not have special surface charges or reactive sites have been found to elicit inflammation at lower concentrations than would be expected with similar materials produced in larger dimensions (e.g. bulk graphite vs. carbon nanotubes). This has led to the observation that *total surface area* can sometimes be a better predictor of toxicity with certain classes of nanomaterials than mass concentration.
2. Some nanoscale materials appear to be associated with cellular oxidative stress (free radical mechanisms) once inside a cell. This process results in the release of unstable forms of charged molecules and is tied to genetic damage and cellular dysfunction. While this insight may become important in understanding the precise molecular mechanism of harm, it will not help scientists predict the likelihood of these materials finding their way from the place of initial contact into the bloodstream and then inside certain cells.
3. Important questions remain about the ability of inhaled nanoscale materials to be transported into the bloodstream and then to specific organs, or for nanoscale materials to penetrate the skin directly. Specific concerns have been raised about the possibility that engineered nanoparticles, like other nano-sized particles such as viruses, welding fumes, and diesel exhaust particulates, may be able to *translocate* directly into the bloodstream from the surface of the skin or along the olfactory nerve into the brain. There is only limited evidence that these uptake routes may be of potential significance for humans, and some evidence that direct translocation through the skin is not taking place.
4. It is essential <u>not</u> to presume that effects exhibited by "parent" materials (large-scale) can be extrapolated to the effects of the derived nanoscale equivalent (e.g. engineered nano-gold vs. simple gold dust). Efforts to predict nanoscale effects from existing toxicology data have been found to be less than useful in many cases.

## *Gaps in Knowledge About Health Effects of Engineered Nanoparticles*

While ultrafine particle studies, along with earlier toxicological and clinical investigations, have laid the foundation for the nascent field of nanotoxicology, there are some fundamental questions about the impact of engineered nanoparticles on human health that earlier research did not resolve, such as:

- Can human exposures to engineered nanomaterials be prevented? If not, what is a safe threshold for exposure and what are the likely exposure levels that might be encountered by workers, researchers, and consumers?

- How are engineered nanoparticles taken up by the human body and how are they metabolized? Do they reach organs and tissues that larger, less reactive particles are not able to reach? Do they interfere with cellular signaling in consequential ways? Does the immune system treat materials on that scale differently?
- Are there chronic (long-term) health effects associated with exposure to engineered nanoparticles that cannot be evaluated with acute (short-term) and subacute (medium-term) studies? How can long-term effects be assessed without allowing people to be exposed to uncertain risks?
- Are there meaningful differences between the health effects of *engineered* nanoscale materials and those of *naturally occurring* or *incidental* materials with similar chemical composition? If there are different risks posed by engineered nanomaterials, then these materials must be evaluated separately and researchers cannot rely on toxicity values and mechanisms identified in previous studies.
- How do physical, chemical, and electrostatic forces alter the transport dynamics, physical separation, and surface charge of these materials after they are released during processing or manufacturing? How can experimental conditions be defined and controlled so that health effects studies are measuring exposure to discrete nanoscale materials rather than larger clumps of material likely to form over time?
- What is the probability (or risk) that a given individual will experience adverse health effects after being exposed to engineered nanoparticles for a limited period of time? What latent (or delayed) health risks might emerge years after the exposure has ended?

Studies addressing these questions may take many years to complete because it is difficult to rely on experimental controls and conditions traditionally employed in toxicology. The nanomaterials and nanoparticles that are the focus of these studies are derived from previously studied materials and compounds. But much of the complexity underlying the questions posed in this section stems from the fact that researchers are trying to identify hazards that are tied to the special properties exhibited by these derivatives. This requires a deeper understanding of how these materials and compounds differ from their "parent" materials and whether new tools need to be developed to observe these effects.

## *Risk Management*

At present, regulatory agencies, industry groups, and health organizations at the state, national, and international levels are considering how to regulate or monitor possible health risks from engineered nanoscale materials using a full life-cycle (cradle-to-grave) assessment approach. It becomes a great practical challenge to establish an evidence- based risk management framework for the safe production, manipulation, and disposal of engineered nanoscale materials given the large number of questions that remain unresolved.

It is worth noting that regulatory agencies, manufacturers, and academic institutions in the U.S. and Europe have already taken the lead in incorporating a precautionary approach (avoidance of exposure) into a risk management framework to manage persistent uncertainties associated with many nanomaterials. Several creative frameworks have been developed that balance existing technical knowledge, expert judgment, and the use of precautionary policies in the face of larger gaps in knowledge. A responsible framework for managing risk in the face of basic uncertainties must balance the value of the enterprise or research against the cost of taking protective precautionary measures that meet the standards of the company or institution and the community.

Here in Cambridge, the facilities and institutions most actively engaged in nanotechnology research and manufacturing (all of whom participated on the Nanomaterials Advisory Committee) already have highly protective procedures in place. What remains unclear is whether similar practices are being observed at all sites in the city where nanoscale materials are being handled in significant quantities.

## *Oversight*

For any public agency considering regulations, evidence-based safety standards derived from a well-constructed risk framework are essential for meeting the public's expectation that government will provide reasonable assurance to workers and residents. Specific risk-based standards will also help establish credibility in the nanotechnology sector and delineate clear and reasonable expectations for the regulated community. As this technology becomes commonplace it will be important to identify, quantify, and avoid (when feasible) potential health risks. At this time, however, widely accepted standards are not available.

The Cambridge Public Health Department believes that inaction is not good public health policy in the face of persistent gaps in health effects data and uncertainty about how engineered nanoscale materials are being managed by firms not participating in this advisory committee process. In the absence of exposure standards and minimum safety practices, the Cambridge Public Health Department and the Nanomaterials Advisory Committee have concluded that active and constructive collaboration with firms and institutions in Cambridge that currently manufacture, process, or conduct research on engineered nanoscale materials is the most reasonable and effective strategy **at this time.**

This cooperation should be offered in the form of an elective review of risk management practices. Collection of data on the manufacture and use of nanoscale materials should be **strongly** encouraged, with appropriate assurances offered that safety and proprietary information gathered by the city will be protected under state law. Through this collaborative effort, employers would be encouraged to develop and implement precautionary procedures aimed at minimizing exposures to workers and releases to the environment. Such an effort to evaluate and share best health and safety practices would serve to improve the safety culture at Cambridge facilities and laboratories.

Once policies, standards and safety regulations have been developed for nanomaterials by federal, state, academic, and industry partners, Cambridge will then be able to recognize existing gaps that could be addressed with enhanced local oversight.

# Recommendations

The Cambridge Public Health Department, in collaboration with the Cambridge Nanomaterials Advisory Committee, recommends that the City of Cambridge take several positive steps to gain a better understanding of the nature and extent of nanotechnology-related activities now underway within the city, to encourage research institutions and firms within the growing nanotechnology sector to share and improve practices leading to safe management of engineered nanomaterials, and to improve community access to the best available health and safety information as it relates to consumer products containing engineered nanomaterials.

At this time, in recognition of the limited health effects data and the absence of a clear consensus on best practices and standards for these engineered nanomaterials, the Cambridge Public Health Department, in collaboration with the Cambridge Nanomaterials Advisory Committee, **does not recommend** that the City Council enact a new ordinance regulating nanotechnology at this time.

The Cambridge Public Health Department, in collaboration with the Cambridge Nanomaterials Advisory Committee, **does recommend** that the City of Cambridge take the following steps:

- Establish an inventory of facilities that manufacture, handle, process, or store engineered nanoscale materials in the city, in cooperation with the Cambridge Fire Department and the Local Emergency Planning Committee.
- Offer technical assistance, in collaboration with academic and nanotechnology partners, to help firms and institutions evaluate their existing health and safety plans for limiting risk to workers involved in nanomaterials research and manufacturing.
- Offer up-to-date health information to residents on products containing nanomaterials and sponsor public outreach events.
- Track rapidly changing developments in research concerning possible health risks from various engineered nanoscale materials.
- Track the evolving status of regulations and best practices concerning engineered nanoscale materials among state and federal agencies, and international health and industry groups.

- Report back to City Council every other year on the changing regulatory and safety landscape is it relates to the manufacture, use, and investigation of nanomaterials.

The following section describes these recommendations in greater detail.

1. **The City of Cambridge should develop an inventory of commercial, industrial, and research facilities in Cambridge that manufacture, process, handle, or store engineered nanoscale materials (excluding nanomaterial-containing consumer products).**

   Current knowledge about possible human health effects from engineered nanoscale materials in Cambridge is incomplete. Basic information should be collected from each facility, and should include sufficient detail to identify potential risks, exposures, and exposure mitigation strategies.

   To minimize the reporting burden on facilities engaged in nanotechnology research processing or manufacturing, a survey should be developed in cooperation with the Cambridge Fire Department and the Cambridge Local Emergency Planning Committee (LEPC) as part of their ongoing emergency planning and data collection efforts. The survey should be sent to the fire department's list of laboratories (approximately 75 facilities), selected SARA Tier II facilities (approximately 35 facilities), and facilities with flammables permits where it is thought that engineered nanoscale materials may be present (a currently unknown subset of approximately 700 facilities with flammables permits).

   The advisory committee and the public health department envision the survey as the starting point of an effort to reach out to and learn more about Cambridge organizations and firms working with or manufacturing engineered nanoscale materials. Lessons learned from the information gathered through this survey will be incorporated into further efforts to provide technical assistance to encourage best practices for health and safety. Information collected though the survey and other technical assistance activities are strictly protected under state public records laws as confidential business information (CBI).

2. **The City of Cambridge should implement a *voluntary* engineered nanoscale materials technical assistance program.**

   The public health and fire departments should establish a voluntary working relationship with nanomaterials researchers and manufacturers to (1) share information about scientific and regulatory developments and (2) develop best management practices intended to minimize occupational and environmental health and safety concerns.

   The City should offer positive acknowledgement for facilities willing to participate in this effort if such recognition is desired. Development of this technical assistance resource should take advantage of existing efforts at identifying best management practices for health and safety in this sector. Organizations that have supported this policy review process and who have indicated an interest in supporting a "best practices" initiative in Cambridge include the Massachusetts

Institute of Technology, Harvard University, National Science Foundation-funded programs at Northeastern University and the University of Massachusetts at Lowell, the Project on Emerging Nanotechnologies at the Woodrow Wilson Center (Washington, D.C.), and the Toxic Use Reduction Institute.

3. **The City of Cambridge should increase efforts to educate the public about engineered nanoscale materials.**

   The committee recommends two approaches for enhancing public knowledge about engineered nanoscale materials:

   a. Post basic information about engineered nanoscale materials on the Cambridge Public Health Department website. The web pages should include links to other well-vetted governmental and well-regarded Internet sites with information about nanomaterials in the workplace, the environment, and consumer products.
   b. Sponsor or co-sponsor a public forum to discuss the best strategies for informing residents about commercially available products that contain engineered nanoscale materials. The first such public forum in Cambridge on public perceptions and information needs will be co-hosted by the Museum of Science on May 22, 2008 at MIT. Feedback from this public event will help guide the public health department in its future efforts to provide public information about products containing nanoscale materials.

4. **The City should instruct the public health department to provide a report to the City Council summarizing progress with the three recommendations stated above. This report should also provide an update on major changes in the scientific consensus on health risks and state or federal regulatory oversight regarding engineered nanoscale materials. This report should be presented to the City Council every two years.**

   This report should include a brief review of both scientific and regulatory developments relevant to the safe manufacture, handling, and use of engineered nanoscale materials in Cambridge; and an update on regulatory and consensus-based standards developed to promote safety of engineered nanoscale materials. In the event that new state or federal regulations are deemed insufficient to address the understood risks in this community, a review of local oversight options would be recommended. In the event that new, previously unrecognized risks are identified, with or without state or federal action, a review of local oversight options would also be recommended.

## References

Carole Bass (2008). "As Nanotech's Promise Grows, Will Puny Particles Present Big Health Problems?" Scientific American (February, 2008).

Center for Nanotechnology in Society (2007). Proceedings of the 2007 Nanotechnology Occupational Health and Safety Conference. University of California at Santa Barbara.

Rhitu Chatterjee (2007). The challenge of regulating nanomaterials. Environmental Science & Technology Online. November, 2007.

Consumer Reports (2007). Nanotechnology: Untold Promise, Unknown Risk. Consumer Reports. July, 2007.

Antonio Franco, Stephen Foss Hansen, Stig Irving Olsen, and Luciano Butti (2007). "Limits and prospects of the "incremental approach" and the European legislation on the management of risks related to nanomaterials." Regulatory Toxicology and Pharmacology 48(2007): 179–183.

Ron Hardman (2006). "A Toxicologic Review of Quantum Dots: Toxicity Depends on Physicochemical and Environmental Factors." Environmental Health Perspectives 114(2): 165–172.

Suellen Keiner (2008). ROOM AT THE BOTTOM? Potential State and Local Strategies for Managing the Risks and Benefits of Nanotechnology. Washington, DC, Woodrow Wilson International Center for Scholars - *Project on Emerging Nanotechnologies*.

John E. Lindberg & Margaret M. Quinn (2007). A Survey of Environmental, Health and Safety Risk Management Information Needs and Practices among Nanotechnology Firms in the Massachusetts Region. Washington, DC, Woodrow Wilson International Center for Scholars - *Project on Emerging Nanotechnologies.*

Andre Nel, T. X., Lutz Mädler, Ning Li (2006). "REVIEW: Toxic Potential of Materials at the Nanolevel" Science 311: 622–627.

NIOSH, Nanotechnology Research Center (2007). Progress Towards Safe Nanotechnology in the Workplace. Washington, DC, US Government.

Günter Oberdörster, Eva Oberdörster, Jan Oberdörster (2005). "Nanotechnology: An Emerging Discipline Evolving from Studies of Ultrafine Particles." Environmental Health Perspectives 113(7): 823–839.

Paul A Schulte and Fabio Salamanca-Buentello (2007). "Ethical and Scientific Issues of Nanotechnology in the Workplace." Environmental Health Perspectives 115(1): 5-12.

Stephan T Stern and McNeil, Scott E (2008). "REVIEW: Nanotechnology Safety Concerns Revisited." Toxicological Sciences 101(1): 4–21.

Rolf Tolle, Paul Nunn, Trevor Maynard, and David Baxter (2007). Lloyd's of London Report on Nanotechnology: Recent Development, Risks and Opportunities. London, England, United Kingdom, Lloyd's of London Emerging Risk Team.

UK Department for Environment, F. R. A. D. (2007). The UK Voluntary Reporting Scheme for Engineered Nanoscale Materials: Third Quarterly Report. Glasgow, Scotland, United Kingdom, Advisory Committee for Hazardous Substances (ACHS).

Zheng Li, T. H., Rebecca Salmen, Rebecca Chapman, Stephen S. Leonard, Shih-Houng Young, Anna Shvedova, Michael I. Luster, and Petia P. Simeonova (2007). "Cardiovascular Effects of Pulmonary Exposure to Single-Wall Carbon Nanotubes." Environmental Health Perspectives 115(3): 377–382.

*Copies of these materials can be obtained by contacting the Cambridge Public Health Department, 617-665-3800*

# Nanotechnology in Cambridge: What Do You Think?

## May 22, 2008 7-9pm

## AGENDA

7:00 Welcome/Intro (David Sittenfeld, moderator)

7:05 Nanotechnology Overview,
*Tim Miller,*
*Museum of Science*

7:20 Nanotechnology and Consumer Products,
*Todd Kuiken,*
*Project on Emerging Nanotechnologies*

7:35 Cambridge and Emerging Technologies,
*Sam Lipson,*
*Cambridge Public Health Department*

7:50 Questions and Answers/Panel Discussion

8:00 Introduction of Discussion Activity

8:10 Group discussion

8:45 Report out from tables

8:55 Wrap up

## Background Information on Nanotechnology

***We will be discussing the emerging field of nanotechnology throughout this forum. The information below provides a brief introduction to the potential benefits and drawbacks of nanotechnology, and their place in the world of consumer products.***

---

*Springer Science+Business Media Dordrecht/Conference: Nanotechnology in Cambridge: What Do You Think? (May 22, 2008), Hosted by the City of Cambridge and the Museum of Science and initiated by The Project on Emerging Nanotechnologies, with kind permission from Springer Science+Business Media Dordrecht 2012.

S. Hays et al. (eds.), *Nanotechnology, the Brain, and the Future*,
Yearbook of Nanotechnology in Society 3, DOI 10.1007/978-94-007-1787-9_20,

**The field of nanotechnology presents special great potential in energy, computing, medicine, manufacturing, and defense.** By engineering tiny structures and devices on the scale of atoms and molecules, nano researchers are exploring new technologies that may help slow down global warming, provide clean water to millions, build next-generation computers, carry cancer therapies directly into sick cells, repair nerve damage, and sense the presence of even a few molecules of deadly disease agents or toxins.

**However, there is still uncertainty about the safety of some materials produced with nanotechnology**. Tiny nano-particles may present health or environmental risks not present in larger particles of the same materials, either because of their super small size or because of new properties that emerge at such small sizes.

Buckyballs, which are one type of nanoparticle, have caused brain damage in fish, according to research done at Southern Methodist University[1], and carbon nanotubes, which are another, have caused respiratory disease in mice and rats in several studies in recent years. A study published this week in the online journal *Nature Nanotechnology* found that an elongated form of carbon nanotubes injected into the abdominal cavity of mice can mimic the behavior of asbestos, a known carcinogen[2]. However, these findings are very much in the preliminary stages of scientific and public understanding about the potential hazards of nanotechnology upon human and environmental health, because of variables in experimental conditions and unknowns about how to measure the harms that these materials may present. More consistent conditions between these environmental and health studies are needed, but governments at the national, state, and local levels are beginning to consider policies to regulate these materials because of these potential harms.

**Sunscreens, cosmetics, textiles, washing machines, car wax, adhesive bandages, and other nano-particle containing consumer items are already in stores**. However, unlike nano-materials used in carefully controlled laboratory or medical research, consumer products are largely unregulated by the government and could be available in unlimited quantities. The number of known consumer products made using nanotechnology in the world market has tripled in the last two years[3], and consumers may not be aware of their presence, or of the potential harmful effects of the physical and chemical properties of these substances.

**The City of Cambridge is one of the first municipalities in the United States considering measures to make the public aware of the use of nanomaterials in consumer products.** Tonight, you will be given the opportunity to share your thoughts and recommendations: what actions, if any, should the city take with regard to municipal oversight of consumer products made through nanotechnology?

- *Should citizens/consumers be made more aware of the lack of research on the safety of some nanoparticles in consumer goods?*
- *Do nanoparticles differ from other unregulated ingredients in over-the-counter consumer items?*
- *Should there be warning signs or labels?*
- *Should residents be required to consult with a store employee before buying nano- particle-containing products, even if there's no evidence of risk?*

- *Should a public awareness campaign highlight the lack of health and safety studies about these products?*
- *What messages should we send the state and the federal government?*

**Thanks for joining us for this exciting discussion!** We look forward to hearing your ideas and responses.

***The Museum's Forum team would like to thank the following people for making this event possible:***

- *Carol Lynn Alpert, Director of Strategic Projects at the Museum of Science and a member of the Cambridge Nanomaterials Advisory Committee.*
- *Todd Kuiken and David Rejeski of the Project on Emerging Technologies at the*
- *Woodrow Wilson Center for International Scholars, and*
- *Sam Lipson, Director of Environmental Health, Cambridge Public Health Department.*

## References

1. Oberdörster E. (2004) "Manufactured nanomaterials (Fullerenes, C60) induce oxidative stress in brain of juvenile largemouth bass." *Environmental Health Perspectives.*
2. *Nature Nanotechnology*, 5/20/2008, "Carbon nanotubes introduced into the abdominal cavity of mice show asbestos-like pathogenicity in a pilot study,"http://www.nature.com/nnano/journal/vaop/ncurrent/abs/nnano.2008.111.html.
3. Project on Emerging Nanotechologies at the Woodrow Wilson Center for International Scholars, "New Nanotech Products Hitting the Market at the Rate of 3-4 Per Week", http://www.nanotechproject.org/news/archive/6697.

**Product: Nanowhitening Toothpaste**
**Company: Swissdent**
**Country of origin: Switzerland**

**From the product website:**

*"In the development of the active substance Nanoxyd® Dr. Velkoborsky used state-of-the-art technologies.*

The *tinier the active substance used, the easier it gets into places where it is intended to develop its effect.*

*The* ***calcium peroxide*** *used in miniature form even penetrates into the tiniest ci gaps and the interdental spaces, ensurlrg effective bleaching. Thanks to nano technology,* SWISS*DENT is able to generate an ideal result tor your teeth with a small amourt of IJ/each (0.1%), also permitting an extremely gertle and soft use. All Swissdent products have been clinically tested and can be used daily like conventional toothpaste."*

**Source**: http://1\Nww.swissdent.com

# Product: Acticoat® Wound Dressings
# Company: Smith & Nephew
# Country of origin: United Kingdom

**From the product website:**

"Description: A dressing utilizing **advanced silver technology** to help create an optimal wound environment.

- Rayon/polyester core helps manage moisture level and control silver release.
- Silver-coated high-density polyethylene mesh facilitates the passage of silver through the dressing.
- The nanocrystalline coating of pure silver delivers antimicrobial barrier activity within 30 minutes - faster than other forms of silver.
- NUCRYST Pharmaceutical's antimicrobial technology is able to produce silver-coated polyethylene films that can release an effective concentration of silver over several days. Thus, as silver ions are consumed, additional silver is released from the dressing to provide an effective antimicrobial barrier. This patented silver-based antimicrobial technology can be applied to a wide range of medical devices including wound dressings, certain types of catheters and various implants to prevent infection. This technology was first applied to burn wound dressings because burns present a very severe risk of infection."

**Source**: http://global.smith-nephew.com/us/9650.htm

**Product: Solar Rx SPF 30+ Nano-Zinc Oxide Sunblock**
**Company: Keys Soap**
**Country of origin: USA**

**From the product website:**

**"Zinc Oxide that is 10 times smaller** than micronized zinc oxide provides more even coverage to reflect uva and UVb radiation. The nanotechnology 25 nanometer particle is the perfect foundation for makeup.Add moisturizer and makeup to make you look great and be protected from damaging sun…it is chemical free! No Titanium Dioxide or Iron Oxide!

Our broad spectrum uvA and UVB blocking 30+ sPF formulation combines cosmetically clear transparent nano-zinc oxide with therapeutic oils formulated in a lightweight lotion.

To provide more complete broad-spectrum skin protection from antioxidant free radicals, we employ proven zinc oxide in a nano- scale particle size that is 10 times smaller than micronized zinc oxide products. The narrow particle size distribution of the zinc oxide is more effective in providing broad spectrum coverage from damaging UVA and UVB radiation."

**Source**: http://www.keys-soap.com/solarrx.html

## Product: Moisture-protecting underwear
## Company: MyLacys, Inc.
## Country of origin: USA

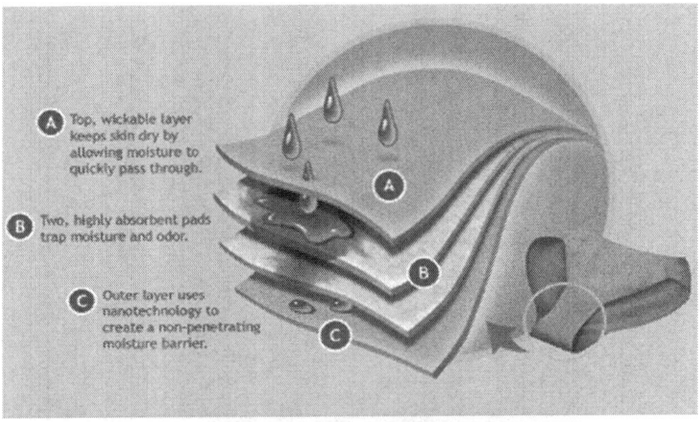


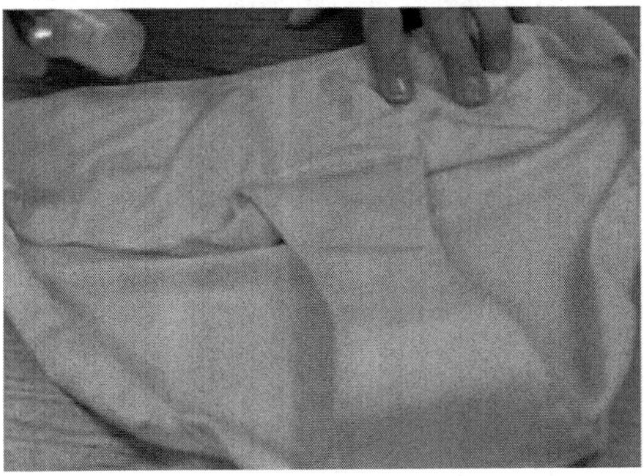

**From the product website:**

*"A moisture barrier technology is infused into the fabric via nanotechnology which means that the molecules are small enough to adhere to the fibers of the fabric, and can't be washed out or worn off…*

*The outer crotch piece (that which touches your outer clothing) is treated with a moisture barrier application utilizing nanotechnology. This nanotechnology bonds the application to the fibers of the fabric, thereby preventing leaks. The moisture barrier application is undetectable and cannot be laundered out of the fabric."*

**Source**: http://www.mylacys.com/technology.php

## Product: Nano Silver Socks and Shoe Pads
## Company: SongSing NanoTechnology Co., Ltd.
## Country of origin: Taiwan

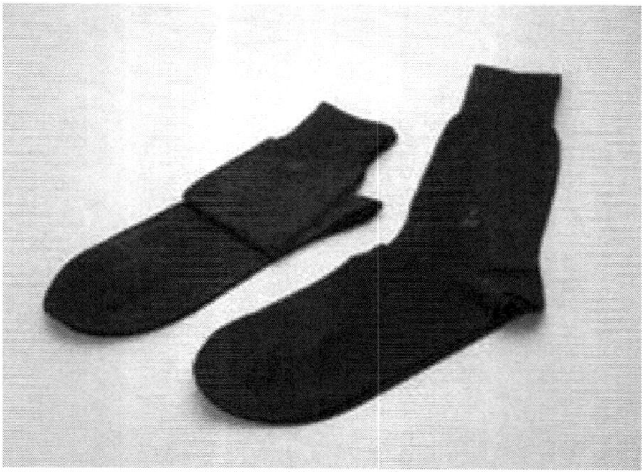

**From the product website:**

"Sterilization and Deodorization; enables feet to be dry and comfortable:
The socks contain the functions of nano silver…can effectively restrain foot mould, carry-over effect of anti-bacteria, enables the foot cool and clean, release the uncomfortable condition."

## Product: Elements Nano-Tex® Jacket
## Company: Jack Wolfskin®
## Country of origin: Germany

**From the product website:**

"For the first time ever, we have applied leading-edge nano- technology to a TEXAPORE base fabric to create an apparel solution that benefits from effective and lasting protection from dirt, dampness and unpleasant odours. The keynote performance advantages of the ELEMENTS jacket are longer wearability and shorter drying times and the treatment also dispenses with the need for reimpregnation. The TEXAPORE membrane also renders the jacket breathable and provides a total waterproof and windproof spec, while a SYSTEM ZIP allows rapid combination with a compatible inner garment."

**From the nano-tex® website:**

Nano-Tex Resists Spills provides breakthrough spill resistance. Each fiber has been fundamentally transformed through nanotechnology, and the result is a fabric that:

- Repels liquids
- Outperforms conventional fabric treatments
- Provides long lasting protection
- Extends the life of the fabric
- Retains fabric's natural softness
- Allows fabric to breathe naturally"

**Sources**: http://nanotechproject.org, http://nano-tex.com

## Product: Nansulate® - in beer bottles
## Company: Industrial Nanotech
## Country of Origin: USA

"Voridian, the company that made Imperm nano-composite barrier technology in collaboration with Nanocor (and since gave its patents over to the nano-center at the University of South Carolina). Imperm technology is currently used by Miller Brewing (specifically Miller Lite, Miller Genuine Draft and Ice House brands) in plastic beer bottles. Imperm is a plastic imbued with clay nano-particles that are as hard as glass but far stronger, so the bottles are less likely to shatter. The layout of the nano-particles is designed to provide a stricter barrier between the carbon dioxide molecules that are trying to escape the beverage and the oxygen molecules that are trying to sneak in, keeping the beer fresher and giving it up to a six-month shelf life."

**– Forbes.com, "Safer And Guilt-Free Nano Foods", 8/10/05, http://www.forbes.com/investmentnewsletters/2005/08/09/nanotechnolo gy-kraft-hershey-cz_jw_0810soapbox_inl.html**

"Industrial Nanotech said in an announcement that the maker of Corona, the fourth most popular beer in the world, is using Nansulate High Heat for thermal insulation and corrosion protection on an interchanger, a common piece of industrial equipment found in the industry.

The interchanger showed a 20 degrees Centigrade (36 degrees Fahrenheit) difference after a three coat application of Nansulate, at a thickness of approximately 7 mils (seven one thousands of an inch)."

**- Nanotech Buzz, 7/1/06, "Better Beer Through Nanotechnology" http://www.nanotechbuzz.com/50226711/better_beer_through_nanotechn ology.php**

## Product: Nanoceuticals™ Slim Shake Chocolate
## Company: RBC Life Sciences®, Inc.
## Country of Origin: USA

**From the product website:**

With slim shake, you can indulge in a rich, delicious shake that has been formulated to help you Iose weight while satisfying your cravings for the sweet taste of chocolate…

The natural health benefits of cocoa have been combined with RBCs Nano-Cluster™ "delivery system to give you CocoaClusters a technologically advanced form of cocoa that offers enhanced flavor without the need for excess sugar".

The natural benefits of cocoa have now been combined with modern technology to create CocoaClusters. RBC's NanoClusters are tiny particles, 100,000th the size of a single grain of sand, and they are designed to carry nutrition into your cells. During the process of creating NanoClusters, pure Cocoa is added to the "Cluster" formation to enhance the taste and the benefits of this treasured food.

**Source:** http://813312.royalbodycare.com/Products.aspx?ltemiD=38

**Nano Silver Baby Mug Cup**
**Baby Dream® Co., Ltd.**
**Country of Origin: Korea**

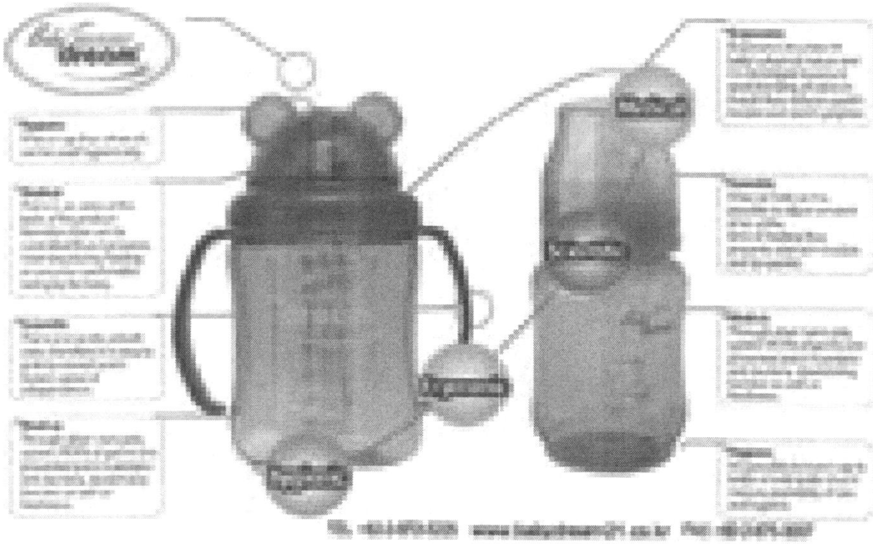

"Through silver nano poly system, 99.9% of germs are prevented and it maintains anti-bacteria, deodorizing function as well as freshness."

**Source**: http://babydream.en.ec21.com/

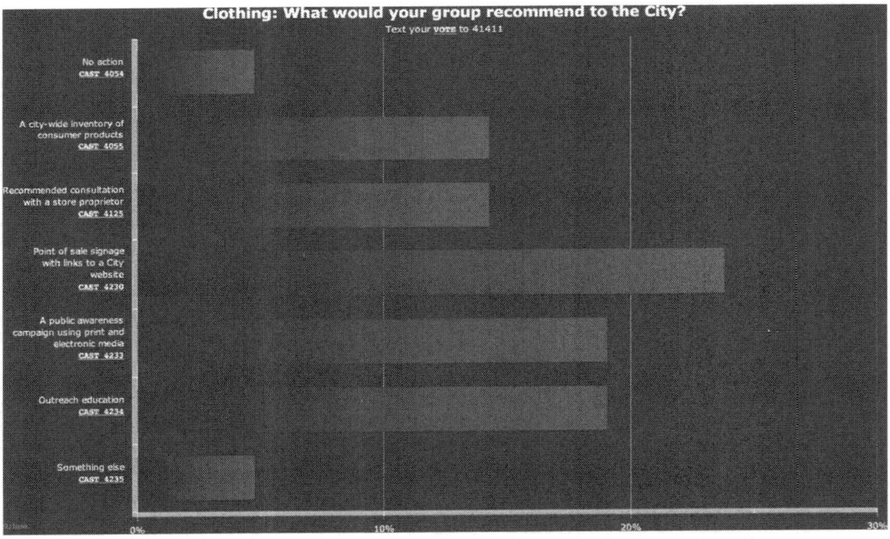
Clothing: What would your group recommend to the City?
Text your VOTE to 41411
No action
CAST 4054
A city-wide inventory of
consumer products
CAST 4055
Recommended consultation
with a store proprietor
CAST 4125
Point of sale signage
with links to a City
website
CAST 4230
A public awareness
campaign using print and
electronic media
CAST 4233
Outreach education
CAST 4234
Something else
CAST 4235
0%
10%
20%
30%

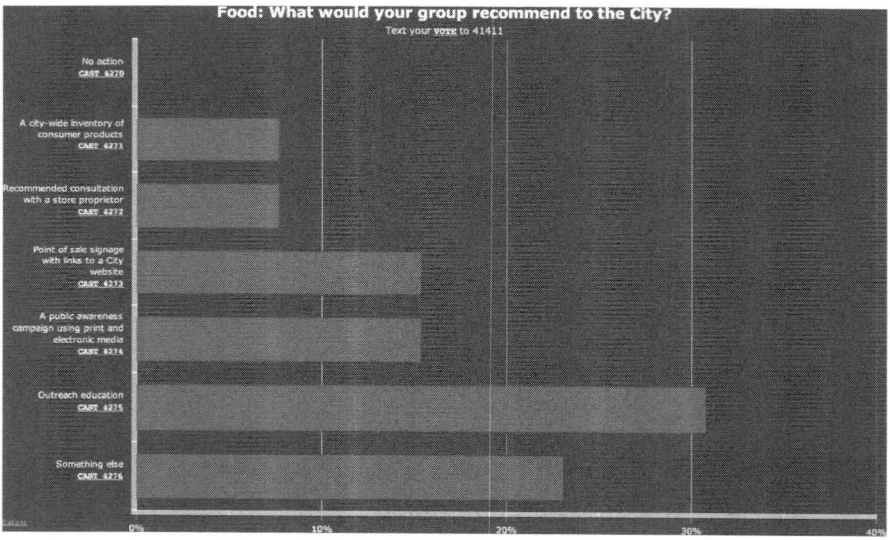
Food: What would your group recommend to the City?
Text your VOTE to 41411
No action
CAST 4270
A city-wide inventory of consumer products
CAST 4271
Recommended consultation with a store proprietor
CAST 4272
Point of sale signage with links to a City website
CAST 4273
A public awareness campaign using print and electronic media
CAST 4274
Outreach education
CAST 4275
Something else
CAST 4276
0%
10%
20%
30%
40%

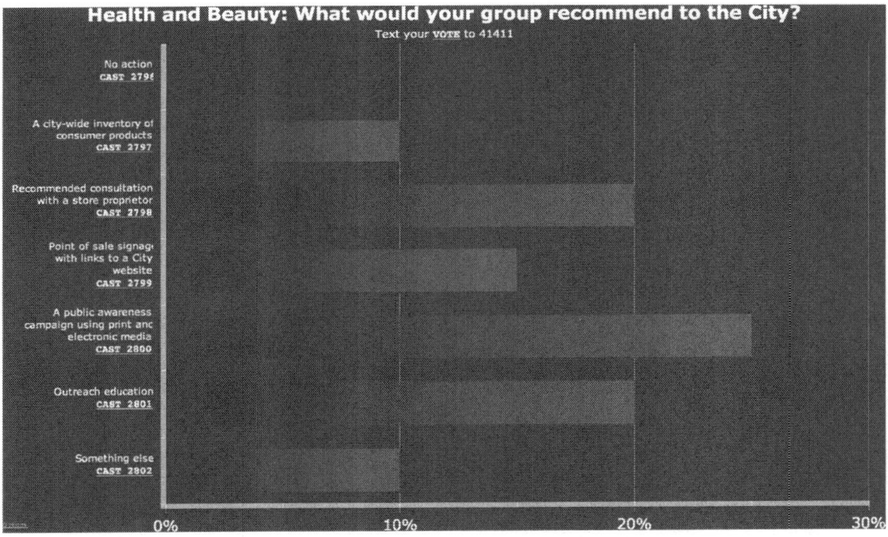
Health and Beauty: What would your group recommend to the City?
Text your VOTE to 41411
No action
CAST 2796
A city-wide inventory of consumer products
CAST 2797
Recommended consultation with a store proprietor
CAST 2798
Point of sale signage with links to a City website
CAST 2799
A public awareness campaign using print and electronic media
CAST 2800
Outreach education
CAST 2801
Something else
CAST 2802
0%
10%
20%
30%

# Chapter 21
# Anticipatory Governance in Practice? Nanotechnology Policy in Cambridge, Massachusetts

**Shannon N. Conley**

## 21.1 Introduction

Anticipatory governance has emerged as an important new concept for understanding the governance of science and technology (Guston 2008; Barben et al. 2008). Anticipation, as discussed in this literature, moves away from the idea of "prediction," and "denotes building the capacity to respond to unpredicted and unpredictable risks" (Guston 2008, 940). The dictionary defines *predict* (2009) as an ability to *know* the future, to "foretell on the basis of observation, experience, or scientific reason." Given the uncertain nature of future technological trajectories and lack of certainty regarding risks related to emerging technologies (such as nanotechnology), however, it is generally impossible to "predict" the future of technologies on any but the shortest of time scales. Instead, anticipatory governance seeks to build society-wide capacities for governing science and technology: to inquire into, assess, and deliberate (1) what new and emerging technologies might mean for society; (2) how they might contribute to enhancing societal outcomes or to creating novel risks; and (3) what kinds of future technological societies people might desire to inhabit. At the same time, anticipatory governance seeks mechanisms to feed insights from these assessments and deliberations back into the process of scientific and technological innovation to help inform the construction of technological futures (Guston and Sarewitz 2002).

This paper has developed out of a concern for two important questions. *First, what does anticipatory governance look like in practice, if we look in existing social and institutional activities rather than in theoretical models? Second, how are*

S.N. Conley (✉)
The Center for Nanotechnology in Society, Arizona State University,
P.O. Box 875603, AZ 85287-5603, Tempe, USA
e-mail: shannon.n.conley@asu.edu

S.A. Hays et al. (eds.), *Nanotechnology, the Brain, and the Future*,
Yearbook of Nanotechnology in Society 3, DOI 10.1007/978-94-007-1787-9_21,

*anticipatory elements of governance situated within larger frameworks for regulating science and technology?* In asking these questions, we acknowledge and seek to respond to an important critique of theories of anticipatory governance, namely that all governance activities are anticipatory. This may be true, but if it is, it nonetheless remains to assess how anticipation is practiced and what role it plays in regulatory governance. To date, the existing literature and research on anticipatory governance have focused largely on theoretical and experimental work (on the latter, see, e.g., Fisher 2007; Fisher et al. 2006). Little work yet examines to what extent various science and technology innovation and regulatory processes already incorporate elements of anticipatory governance, or how existing processes of anticipatory governance could be strengthened.

To begin to explore these questions, this paper examines a case study of local regulatory processes in the city of Cambridge, Massachusetts, as regulators sought to understand and respond appropriately to the potential risks posed by exposure to engineered nanoparticles, including risks to the brain (Lipson 2008b). The case study focuses on the work of the Cambridge Public Health Department in developing a suite of policy recommendations for nanotechnology. The Cambridge policy experiences with nanotechnology lend insight into how local agencies are making efforts to pursue their commitment to innovation as an important source of economic growth while simultaneously trying to avoid unintended health, environmental, and social impacts on local communities. The period covered by this case study, 2007–2009, includes the creation of the initial mandate from the Cambridge City Council, exploratory work by public health officials, formation of a stakeholder advisory board, and the hosting of two public engagement exercises by the Museum of Science, Boston, in collaboration with the Cambridge Public Health Department.

In analyzing this case study, the chapter offers three insights. First, it argues that local regulatory activity can be seen as, if not an explicit implementation of anticipatory governance, at least a cognate activity that shares many of its same objectives, techniques, and sensibilities. Second, the chapter argues, as a corollary, that scholars of anticipatory governance could look to regulatory agencies as examples of what anticipatory governance might look like in practice – not only to illustrate anticipatory governance but also to explore the challenges it may face and techniques and strategies for advancing its goals. Third, the paper suggests that anticipatory governance can be used to suggest ways of improving the practice of science and technology regulation.

The paper is organized into three sections. This paper begins with an overview of theoretical perspectives on anticipatory governance, including a synthesis of three key concepts in the conceptual framework – foresight, integration, and engagement – and their relation to other ideas, such as reflexivity and deliberation. I then apply the three key concepts of anticipatory governance to the Cambridge policy experiences with nanotechnology as an illustration of anticipatory governance in practice, showing how this conceptual model fits well the activities of the Cambridge Public Health Department and thus demonstrating a version of anticipatory governance in practice. Finally, the paper explores how the case can further inform the theory of anticipatory governance and, additionally, provide suggestions from anticipatory governance theory to enhance the work of local regulatory governance.

## 21.2 Theoretical Perspectives on Anticipatory Governance

The conceptual literature on anticipatory governance calls for the development of an approach to governing emerging technologies that prospectively grapples with the societal meanings and implications of new technologies *before* they make their way into the market, in opposition to retroactive approaches that address unintended impacts *after* they have already occurred (Barben et al. 2008). In addition, in concert with work on reflexive governance, anticipatory governance imagines governance as an ongoing process permeating society, as a whole, rather than something that is limited exclusively to either the innovation or government sectors (Guston 2008; Voss and Kemp 2006). The goal, then, of anticipatory governance is to create distributed capacities for anticipating, assessing, and deliberating what new technologies mean for society and how to apply them to achieve desired societal goals and objectives (Guston and Sarewitz 2002). Guston and Sarewitz imagine these capacities as adaptive, evolving alongside developing technologies.

Currently, the literature emphasizes three conceptual lenses for understanding and deploying anticipatory governance at the level of research and development (R&D) activities: foresight, integration, and engagement (Barben et al. 2008).

- *Foresight* focuses on orienting governance towards the future through the creation of capacities for anticipating what future societal and technological trajectories might bring. Tools can include the analysis of the plans, expectations, and values of diverse stakeholders regarding future technological trajectories; developing scenarios to help stakeholders consider the upsides and downsides of multiple possible futures; and assessing how changes in society may reconfigure societal attitudes towards technology and its outcomes. Foresight is deliberately not viewed as prediction, but rather as the exploration of how societies and technologies might evolve together in the future, to inform current technical and policy choices (Guston 2008). Current foresight activities primarily focus on imagining possible futures via scenario development and public deliberation, examining how stakeholders understand and engage in thinking about the future (Rip and Te Kulve 2008; Selin 2006).
- *Integration* highlights the need to create mechanisms through which insights that arise in processes of foresight and engagement can be integrated back into the process of scientific and technological innovation. The strategies of such work include both expanding the kinds of factors taken into account in designing, developing, and implementing new technologies as well as creating deliberative opportunities for those involved in innovation processes to dialogue with one another, citizens, policy officials, and others about the governance of science and technology. Currently, integration activities center around the laboratory. Fisher's (2007) laboratory engagement work serves as a prominent example of present-day integration work. In Fisher's laboratory engagements, social scientists and natural scientists engage in collaborative, reflexive exercises that seek to "broaden...the scope" of scientific decision-making processes.

- *Engagement* hones in on ways of engaging the public in decision-making processes related to R&D activities. Engagement activities, such as deliberative forums and consensus conferences, seek to make the public active participants, rather than passive observers, in technological governance processes. Hamlett and Cobb's (2006) work highlights engagement efforts such as the National Citizen's technology forms, funded by the National Nanotechnology Initiative and the National Science Foundation. Barben et al. (2007) discuss other engagement activities, such as deliberative, participatory public forums currently being held at the University of South Carolina and the University of California, Santa Barbara. Engagement embraces the notion that members of the public can contribute meaningful feedback to decision-makers, even if they are not experts in the particular technological field that is being governed. More broadly, it embraces the idea that understanding and evaluating what new technologies might mean – *and should mean* – in and for society can only come through broad engagement with many diverse publics.

In building on these ideas, it is useful to compare anticipatory governance to two other, potentially similar ideas. The first is *reflexive governance* (Voss and Kemp 2006). The theory of reflexive governance portrays governance not as a series of isolated, independent exercises in problem solving but rather as a set of ongoing processes that grapple with uncertain policy challenges in the face of an unknowable future. Like anticipatory governance, reflexive governance starts from an assumption that governance must "take account of the complexity of interlinked social [and] technological [...] systems," "fundamental uncertainty" regarding the future, as well as deep ambiguity and a lack of good criteria and assessment mechanisms for evaluating the risks and impacts of new and emerging technologies (Voss and Kemp 2006, 7). Likewise, reflexive governance is "geared towards continued learning in the course of modulating ongoing developments, rather than towards complete knowledge and maximization of control" (Voss and Kemp 2006, 7). Thus, while reflexive governance is closely related to ideas such as mid-stream modulation that are associated with anticipatory governance (see Fisher et al. 2006), it tends to stress a more systematic confrontation with uncertainty and therefore be more skeptical of techniques such as foresight methodologies that, while acknowledging some uncertainty, nonetheless strive to hold it within limits.

The second related concept is *deliberative democracy*. Theories of deliberative democracy share with anticipatory governance a strong emphasis on processes of discourse and engagement with members of society. While anticipatory governance has seen public engagement as a tool for dialogue, exchange, and learning among citizens about new technologies and their potential implications for society, deliberative theories of democracy have tended to place greater emphasis on questions of power and accountability.

Deliberative democracy is not a new idea. In fact, its roots can be traced as far back as fifth century BC Athens (D'Entrèves 2002). Pericles, in the eulogy of Athens, characterized deliberation and discussion as an "indispensable" component of "wise action" in public decision-making (Elster 1998, 1). The notion that deliberation is a feature of "wise action" continues to pervade the political theory literature of the new millennium. Gutmann and Thompson (2004, 3–4), for example, highlight deliberative democracy as an essential facet of a just decision-making process. Deliberative democracy provides a venue for agents, both the governors the governed, to offer and respond to the reasons behind a decision. This "reason giving requirement" of deliberative democracy then recognizes persons as "autonomous agents who take part in the governance of their own society."

King (2003) also offers an account of deliberative democracy based on the idea of reason giving. From this theoretical perspective, accountability occurs when decision makers are obliged to justify their choices, making the reasons for taking particular actions open and transparent. As a result, citizens are then positioned to debate and challenge the adequacy of the proffered reasoning and to use the resulting insights to hold policy officials accountable not only for their choices but how they make and defend them.

This view of deliberation could potentially alter the ideas of public engagement embedded in anticipatory governance in several subtle and not so subtle ways:

- It calls for transparency in innovation and governance decision processes, going beyond some of the relatively simple, contained techniques of public engagement currently used in anticipatory governance.
- It would focus deliberation on decisions, including allowing the public to question and challenge a wide range of technical and policy choices involved in technology innovation and governance, as well as to express their concerns upstream in the innovation process.
- It would highlight the need for mechanisms by which the public – and especially those impacted by new technologies – could hold accountable those who make innovation and regulatory decisions for the consequences of their choices.
- At the same time, it also offers a means by which citizen engagement could enhance the legitimacy of innovation processes.

## 21.3 Cambridge, MA: Boots on the Ground

The following sections explore the dynamics of anticipatory governance at work, as they manifest in the efforts of both local government officials and citizens. What does anticipatory governance look like to those who have their "boots on the ground" of local governance? How do individuals and institutions grapple with the big picture of emerging technologies while simultaneously attempting to deal with the day-to-day details of local government? What considerations do officials take into account when making decisions? How can local governments build capacity for dealing with new and emerging technologies? Additionally, this chapter seeks to understand the potential value of deliberative processes and stakeholder committees in the context of the local governance of emerging technologies. I explore these questions in a case study of environmental health officials in the city of Cambridge, Massachusetts, grappling with the regulation of nanotechnology.

### 21.3.1 *Anticipatory Governance: The City of Cambridge and the Governance of Emerging Technologies*

The City of Cambridge has a population of approximately 105,000 people. It stands as a culturally and ethnically diverse neighbor to its larger sister, the City of Boston, across the Charles River. Established as a city in 1846, Cambridge has a reputation for excellence in higher education, research, and scientific innovation, as well as political progressivism. Harvard, Radcliffe, the Massachusetts Institute of Technology (MIT), and Lesley College are nestled within the city boundaries. Industry is primarily based on technology, such as biotechnology, electronics, and software research (Cambridge Historical Commission 2008), although the city also has a long history based in textiles, rubber, tanning, ice, clay pits, and automotive industries. There is a burgeoning nanotechnology sector. The city is also known for pioneering political innovation alongside scientific and intellectual innovation. In one of its best known cases, the Cambridge City Council passed a recombinant DNA ordinance in the 1970s and still prides itself on taking a proactive approach to health and safety issues related to new and emerging technologies. Nanotechnology is the most recent technology that the city considered regulating.

To some observers, it is not a surprise that Cambridge, after Berkeley, California, would be one of the first American cities to create a nanotechnology policy mandate (Rabinovici et al. 2007, 3). Rather than following Berkeley's lead by choosing to move quickly to regulate nanotechnology and establish a detailed health and safety documentation plan, however, Cambridge adopted a more deliberate approach. City council members and officials probed a wide range of questions. Should Cambridge follow Berkeley's approach? Alternatively, should the city avoid regulating nanotechnology altogether? How much knowledge existed about the health and safety risks of nanotechnology, and what was the appropriate relationship between that knowledge and the degree and pacing of regulation?

Officials in the Environmental Health Unit (EHU) played a leading role in developing Cambridge's policy efforts. EHU officials had begun, over 2 years before Berkeley adopted its regulations, to follow the emerging debate about nanotechnology health risks. The EHU scoured the literature, examining studies on this emerging and seemingly ubiquitous technology and its risks to human and environmental health, and its Director concluded the issue was important. Without a policy mandate, however, the unit's capacity and resources to systematically evaluate those risks or to advocate for regulatory action was limited. In the process, EHU officials looked to the model provided by the city's successful efforts to provide regulatory oversight for biotechnology in the 1970s. At the same time, they concluded that significant differences exist between the potential risk posed by recombinant research on highly infectious organisms and the uncertain public health challenges associated with nanotechnology and nanomaterials. On the one hand, they were concerned that knowledge about the health risks of nanotechnology was still too uncertain. On the other hand, in the case of nanotechnology, neither the federal nor the state government yet provided a comprehensive regulatory framework that the city could use as a model.

The context shifted in December 2006, when the City of Berkeley's regulations for nanotechnology took affect, in response to an increasing body of scientific literature detailing the risks related to nanoparticle exposure. Prompted by Berkeley's actions, the Cambridge City Council put nanotechnology on its agenda for January 2007 and voted that the City Manager ask the city's Director of Environmental Health to examine and respond to Berkeley's regulatory efforts (Bray 2007).

WHEREAS: The use of subatomic [*sic*] materials as microscopic building blocks for thousands of consumer products has turned into a big business so quickly that few are monitoring nanotechnology's effects on health and the environment; and

WHEREAS: The city of Berkeley, California, has amended the hazardous materials section of its Municipal Code in order to monitor those impacts; now therefore be it

ORDERED: That the City Manager be and hereby requested to examine the nanotechnology ordinance for Berkeley, California, and recommend an appropriate ordinance for Cambridge (Cambridge City Council Meeting 2008, 6).

The result gave EHU officials the mandate they needed to dedicate new time and resources to exploring the issue and determining the steps necessary for the formation of a robust policy (Lipson 2008a).

Local movement towards nanotechnology policy comes at a time when public trust in the federal government's ability to handle domestic problems is at an all-time low (Jones 2008). The local regulation of nanotechnology in Berkeley was framed as a symbolic differentiation of the local government from the federal government. It served as an effort to convey a message of trustworthiness to the local community. The Mayor of Berkeley, Tom Bates, went on record, stating that the city's actions were "groundbreaking," and that it was the job of local government to fill in where the federal government had failed: "If the federal government isn't going to do anything, it's up to us to step up" (Snapp 2008).

Berkeley's regulations amended the city's municipal code to add a health and safety disclosure section for nanoparticles. The Community Environmental Advisory Commission (CEAC) and the Hazardous Materials Manager determined that the disclosure of nanoparticle use and manufacturing is necessary. Title 15 of the code requires disclosure information for hazardous materials once a certain level of the material has been exceeded. The amended code also requires that any business that uses nanoparticles must submit a written report of the nanomaterial's toxicology, as well as describe their business practices for "safe handling, monitoring, containing, disposing, and tracking the inventory," in order to help prevent unintended releases of nanoparticles (Al-Hadithy 2006, 1). Berkeley's ordinances were born within a local political rhetoric that simultaneously encouraged local action and criticized the federal government non-action.

### 21.3.2 *Foresight in Action*

Here we see several aspects of foresight in action. One important aspect is the cultivation of a political culture that expects to be proactive on emerging technologies.

An official from the EHU noted that one of the reasons Cambridge decided to address the issue was due to its acknowledged reputation and a local culture that prides itself on both innovation and good public policy:

> "I think that there is some feeling of a special role in this city of having good public policy around emerging technologies, because we have played that role in the past, not just that we are an academic city, we are a city with a lot of start-up companies, a lot of emerging technology companies. However, we are also a city with a political history that demands a high bar for public accountability and transparency." (Lipson 2008a)

At the same time, we also see concrete strategies of foresight being adopted by city officials in several capacities. City Council members, for example, were clearly engaged in at least some degree of horizon scanning. However they became aware of the Berkeley effort, they clearly took it to be a signal that they needed to look more closely at an issue that was emerging as a potentially significant policy issue. According to the same EHU official, the Berkeley regulations served the primary impetus for the Cambridge City Council's interest in the issue. Likewise, the Cambridge EHU was also engaging in proto-horizon scanning in its early efforts, before 2007, to be aware of and develop at least a rudimentary understanding of the emerging issue of health risks and nanotechnology.

While such proto-horizon scanning efforts might not have been explicitly oriented towards a "foresight" approach, as it is currently understood within the anticipatory governance framework, the activities of the EHU before 2007 had an eye turned towards the future of nanotechnology in Cambridge. While these efforts differ from some of the foresight work done by anticipatory governance scholars, such as Selin, in that they were the result of an organic governance process, rather than an explicitly intended "foresight" activity, the EHU's experiences serve as evidence of real-world anticipatory governance activity – anticipatory efforts that arise naturally, apart from efforts within academia.

Finally, we see evidence in the City's ultimate legislative action of another foresight capacity – namely, the ability to understand and interpret new developments in light of broader political sensibilities and culture, recognizing that political culture plays a significant role in shaping the trajectories of new technologies. In responding to the City Council, the Director of Environmental Health proposed that Cambridge should not attempt to mimic the approach that Berkeley took. The official final report emphasizes that the situation is not a matter of "one size fits all," and relates a sophisticated understanding of local policy implementation. The director recognized that a "cookie-cutter" policy mold was undesirable, and that a robust policy must account for local conditions and circumstances. This perspective is similar to Jasanoff's (2006, 255) discussion on local modes of understanding, or "tacit knowledge-ways through which [the public] assess[es] the rationality and robustness of claims that seek to order their lives." EHU officials emphasize that Cambridge's approach was *not* prompted by material or economic differences between the cities, but was instead based on a history and culture of the oversight of emerging technologies and a local political culture that emphasizes "'good government' versus radical politics" (Lipson 2008a). Jasanoff, for example, argues that it is these "collective knowledge-ways" that comprise a society's civic epistemology; "they are distinctive, systematic, often institutionalized, and articulated through practice rather than formal rules" (Jasanoff 2006, 255). Cambridge officials' sensitivity to local culture and expectations provides illumination to this concept. EHU officials recommended that Cambridge take a "longer view" on the issues, rather than quickly pass a reporting or documentation requirement like Berkeley did (Lipson 2008a, b). The "longer view" advocated by EHU leadership, in light and in knowledge of emerging health and safety issues and the distinct socio-political Cambridge culture, represents foresight as a capacity for developing a long-term governance approach that takes into account the role that local political culture can play in how emerging technologies play out on the local level.

### 21.3.3 Integration: Connecting Science, Business, and Regulation

Anticipatory governance calls for reflections about the potential societal outcomes of emerging technologies to be integrated back into the innovation process, either in the lab or in industrial production settings. In the case of nanotechnology policy development in Cambridge, we see both the importance of integration and one approach to it in practice. This emerges, especially, in the comparison between Cambridge's efforts to regulate nanotechnology and its efforts, three decades earlier, to regulate biotechnology.

In deciding not to regulate, EHU officials reasoned that a nanotechnology *regulation* might not necessarily be appropriate now, even given the city's history of regulating another emerging technology, recombinant DNA. The city's recombinant DNA ordinance was passed in 1977. Despite potential similarities in the ethical, legal, and social implications (ELSI) of nanotechnology and biotechnology (Moore 2002, 13), Environmental Health Unit officials argued that the differences between them are much greater than their similarities (Lipson 2008a). This conclusion emerged from the EHU's prior experiences with biotechnology and warned officials that the comparison was not enough, by itself, to justify adopting a local regulatory framework for nanotechnology.

The reasoning behind this conclusion stemmed, especially, from very different circumstances vis-à-vis what anticipatory governance terms integration. When Cambridge chose to regulate recombinant DNA, the city based its ordinance off a set of guidelines from the National Institutes of Health (NIH). An EHU official noted that there is not yet an adaptable comprehensive framework for nanotechnology, therefore making Cambridge's task more complex, as there is no wholesale guide for nanotechnology governance at either the federal or state levels:

There needs to be a single framework so you can understand that there are different types of risks and different levels of risk. Some very low, some potentially higher, which would in turn result in converged best practices around personal protection and engineering controls, practices that are appropriate for different kinds of laboratory processes. This sort of insight was already widely shared when Cambridge passed its biotech ordinance, because the year before that ordinance was enacted (1977), the NIH guidelines were approved, which did a lot of this work – establishing the rules of the road, so to speak (Lipson 2008a).

The official compared the necessary creation of a framework for emerging technologies such as biotechnology and nanotechnology to the necessity of having a common set of rules to abide by while driving a car:

> "Imagine when cars were first constructed, and available to individuals in the early part of the twentieth century. People were never trained to drive the cars, so eventually some system of rules, a framework, had to be created and people had to be trained in that framework. The NIH guidelines went a very long way in creating a single set of rules that were, in principle, adopted by people in both the private and public sector. We do not have that in nanotech." (Lipson 2008a)

What is important here, however, is more than just the political and policy foundations provided by a pre-existing policy framework; it is also that this pre-existing framework reflected a deep integration of ELSI ideas into the scientific community. From Lipson's (2008a) perspective, in the 1970s, the main stakeholders and proponents of a recombinant DNA ordinance were researchers at Harvard and MIT, who had also been active in pushing for the development of the NIH guidelines.

> "[W]hen that first version of the [DNA] ordinance passed in Cambridge, there really was no private sector. It was all being done through MIT and Harvard. The people who raised the concerns in the city initially were in fact researchers; they were not people in the public." (Lipson 2008a)

In their nanotechnology work, by contrast, officials assert that they faced a significantly different context. Given this perspective, nanotechnology research scientists at Harvard and MIT remain important but, for the most part, nanotechnology researchers have not been widely involved in the effort to confront nanotechnology risks; nor have they pushed for federal or state regulatory frameworks. Largely, local Cambridge EHU officials grappling with these issues feel that nanotechnology researchers have resisted the need for new regulations.

At the same time, university researchers are only one group involved, today, in nanotechnology innovation processes; a burgeoning private sector also exists that would be impacted by a nanotechnology policy. Officials found it a difficult task to engage meaningfully with this already existent nanotechnology private sector and "convince different firms within that sector to essentially join a public process."

The process [for nanotechnology] took a different route, and that may be one of the reasons we found it a little more difficult to engage with those doing work in nanotechnology. There already was a private sector, and not only that, there are already firms that have been engaged in work that we would now classify as the processing or manufacturing of engineered nanomaterials, some for decades (Lipson 2008a).

It was evident, therefore, to EHU officials that new trust relationships would have to be forged with both academic researchers and an intrinsically wary private sector in the case of nanotechnology. Brian Wynne (1996, 20) reflects that trust and credibility are not simply "intrinsic or inevitable characteristics of knowledge or institutions" but are instead entrenched within "changing social relationships."

EHU officials acknowledged the important role of fostering trust relationships with academic and private sector stakeholders and recognized that if the policy were to be widely accepted by the university and industry communities, they would have to understand their perspectives and include their voices in the conversation from the very beginning.

> "A large part of the problem was that many businesses had long histories of work with what would today be considered nanotechnology, including businesses in the chemical manufacturing, pharmaceutical, and semiconductor industries. Hence, many of them do not think of themselves as part of a new sector with a new label attached to it. They feel that they know their business and that they do not necessarily want to have these rules imposed upon them. Perhaps they have not been involved in a process like the one that is taking place in Cambridge in the past, so perhaps there is some anxiety or fear of the unknown. They might feel that they do not have anything to gain by joining the process. It has been remarkably different in that way." (Lipson 2008a)

To give voice to a multiplicity of academic and industrial perspectives before making its policy recommendations, the EHU assembled a committee of stakeholders to "spend a little more time addressing some of the [consumer and manufacturing health and safety] areas that came up" (Lipson 2008a). The committee of 19 stakeholders, the Cambridge Nanomaterials Advisory Committee (CNAC), included scientific experts, university representatives, community members, and representatives from the manufacturing and research sectors (Lipson 2008b, 1). CNAC members were primarily on the committee to share their academic or professional insight, rather than to represent the average person. However, CNAC did have four members that were Cambridge residents, although their main role was to share their technical expertise (Lipson 2008a). The EHU sought to comprise the CNAC with individuals from an array of professional backgrounds. For example, in addition to academic and industry stakeholders, committee members had specialties in community outreach and education, the law, and one member worked in epidemiology and statistics.[1]

The CNAC convened for the first time in summer 2007. The Director of Environmental Health facilitated six meetings, one per month, through January 2008. In the meetings, CNAC members dialogued about the current state of knowledge regarding the health and safety effects of manufactured nanoparticles, and CNAC members with specific professional expertise gave presentations to the rest of the group on topics such as such as nanotoxicology, nanomaterial oversight, regulatory policy, and risk management. Additionally, the fire department and the emergency planning committee updated the CNAC on efforts to build a citywide inventory of manufactured nanoparticles.

CNAC examined a broad range of topics related to nanotechnology in Cambridge. The issues can be broken down into two broad topic areas. The first topic was related

[1] For a list of the names CNAC members and details on their professional backgrounds, please see Recommendations for a Municipal Health & Safety Policy for Nanomaterials: A Report to the Cambridge City Manager, also included in this volume.

to consumer awareness and risk. CNAC, in collaboration with EHU officials, considered a consumer education program as part of a proposed policy. The second initiative was to determine the feasibility of establishing an inventory of nanomaterials being worked with in Cambridge. On this effort, CNAC worked with the fire department. Collaboration on this front was relatively easy, as the fire department was already gathering information in response to potential safety hazards as part of its existing responsibilities. The EHU and CNAC also approached the local emergency planning committee for essentially the same purpose of achieving preparedness, "so the city, as first responder to public safety, is aware of what is going on around the city should they have to respond" (Lipson 2008a). Note the foresight element here, as well.

The Director of Environmental Health authored a report, *Recommendations for a Municipal Health & Safety Policy for Nanomaterials: A Report to the Cambridge City Manager*, which detailed CNAC and the Health Departments' policy recommendations. The report offers six initial policy recommendations, based on CNAC's research and the deliberations amongst its members and with various city officials.

CNAC's first recommendation states that Cambridge should develop an inventory of all facilities that work with nanomaterials, whether they engage in research, store, or manufacture the materials. In order to alleviate some of the reporting burden for those facilities, the report encourages the development of a survey in collaboration with the local fire department and emergency planning committee as part of a pre-existing emergency planning and data collection initiative. This survey would be a "starting point" for the city to reach out to and learn more about Cambridge institutions that engage with nanomaterials. The report envisions this knowledge ultimately helping foster best practices for health and safety (Lipson 2008b, 11).

The second recommendation is the creation of a voluntary "engineered nanoscale materials technical assistance program." This program would encourage policymakers, researchers, and manufacturers to engage in dialogue and share knowledge about the latest scientific and regulatory developments in nanotechnology, as well as collaborate on the development of occupational and environmental health and safety best practices (Lipson 2008b, 11).

CNAC's third recommendation focuses on public outreach and awareness activities. The report suggests public engagement in both active and passive contexts. The report suggests that deliberative citizens' forums are an important part of active engagement. Cambridge hosted two such forums, in May 2008 and January 2009. The deliberative forums as a study in the engagement aspect of anticipatory governance theory are examined in the next section.

CNAC's final three recommendations consist of a suite of horizon-scanning mechanisms. The policy requires the health department to provide a progress report to the city council every 2 years. The progress report would examine work on the other recommended policy steps and would explore new policy and scientific research on nanomaterial health and safety (Lipson 2008a, b). Given the rapid speed at which new research and policy developments can occur, CNAC sees the reporting process as providing an opportunity to adjust or augment the local policy framework in light of changing contexts.

### 21.3.4 Public Engagement and Deliberation

The third element of anticipatory governance theory is public engagement and deliberation. Traditionally, regulation involves periods of public comment, however the EHU wanted to engage with and educate Cambridge citizens in a venue that would go beyond the bounds of a traditional public comment period and allow citizens and policymakers to engage on a deeper, more meaningful level. Building on past technology education and engagement efforts by the Museum of Science, Boston (see Bell 2008), the vision was to create a format in which citizens could both learn about the nanotechnology policy and provide feedback on it. A venue such as a deliberative forum would allow the EHU to directly engage with citizens from a wide variety of backgrounds, including those particularly passionate or knowledgeable citizens, the ones most likely to attend a traditional public comment period, as well as those citizens who might be curious about the topic, but have little knowledge about nanotechnology. Officials would have an opportunity to "tease out the most meaningful and important public policy issues" affecting citizens from a multiplicity of intellectual, social, and economic backgrounds (Lipson 2008a).

In order to engage with members of the public, the City of Cambridge collaborated with the Museum of Science, Boston and the Project on Emerging Nanotechnologies at the Woodrow Wilson International Center for Scholars to structure two educative, deliberative, and participatory forums for Cambridge citizens. The Museum of Science Boston has been proactive in engaging citizens with technology through a variety of forums and activities, with the philosophy that such efforts should be "two-way" – bringing together members of the public with policymakers and scientists for dialogue and mutual learning (Bell 2008). Both the EHU and the forum organizers at the Museum of Science envision the forums serving as a new type of "public comment period," an alternative space in which citizens can voice their concerns, opinions, and hopes about the city's nanotechnology policy. The May 2008 forum had 31 attendees, and the January 2009 forum had 44 attendees (Kunz Kollmann 2008, 2009). The majority of participants at the May 2008 forum heard about the event directly through museum recruitment materials, however most of attendees at the January 2009 forum became aware of it via word-of-mouth sources, either through a friend, at work, or on a social networking website (Kunz Kollmann 2009, 3). Most of the participants were residents of Cambridge, Boston, or lived in nearby areas. A number of attendees made the decision to attend at the last minute and dropped by after work, curious to hear about the policy and what their fellow citizens thought about it.

The forums were 2 h long and integrated three expert speakers, a question and answer session between the experts and the citizen participants, small group deliberations, and time at the end for sharing results of group deliberation. The forum agenda and final reports characterize the time breakdown as such:

- 5 min for a welcome and introduction to forum,
- 45 min for the live speaker presentations,

- 10 min for questions and answers,
- 10 min for an introduction to the small group discussion,
- 35 min for the small group discussion,
- 10 min for the report-out, and
- 5 min for wrapping up (Kunz Kollmann 2008, 1).

At each forum, three live speakers gave presentations. The first speaker was a nanotechnology "101" educator from the Museum of Science. In both the May and January forums, this speaker gave an overview of nanotechnology and the potential risks and benefits of having nanotechnology in consumer products. The second presentation was by a representative from the Project on Emerging Nanotechnologies who discussed the multitude of consumer products containing engineered nanoparticles that are already on the market, as well as the low levels of public awareness of nanotechnology in consumer products. The last speaker was a representative from the Cambridge Public Health Department (2008). This representative discussed how Cambridge has handled emerging technologies in the past, reflecting on Cambridge's experiences regulating recombinant DNA. After the presentations, forum participants asked questions to the speakers to clarify any confusion they had (Kunz Kollmann 2008).

For the deliberative portion of the night, attendees were split into groups of five or six, and were given a series of three scenarios. The scenarios can be considered a foresight activity, as they were future-oriented in the sense that each dealt with a future policy action, such as product labeling or signage requirements. Each scenario was meant to "provoke" a policy recommendation regarding consumer products containing nanotechnology (Kunz Kollmann 2008). In both forums, scenarios dealt with clothing containing nanotechnology, such as nanosilver in underwear and socks, nanotechnology in personal care products, such as sunscreen, lotions, and toothpastes, and nanotechnology in food and food containers, such as in milkshake mixes and baby bottles. For each scenario, participants in each group were asked to come to a consensus and as a group, vote on a policy recommendation. They could choose from a series of possible policy actions, such as:

- Take no action,
- Create a city-wide inventory of nano-consumer products,
- Recommend that store employees provide consultations or handouts about nano-consumer products,
- Have point-of-sale signage with links to a city website about nano-consumer products,
- Create a public awareness campaign using print and electronic media,
- Have outreach at schools, libraries, or community events, or
- Do something else (and explain this something else to the other participants) (Kunz Kollmann 2008, 2).

According to participant-observer data and a Museum of Science report (Kunz Kollmann 2008) detailing the activities of the forums, each group used a text-messaging system to vote on their recommended policy action after each scenario.

After all of the scenarios were vetted, each group nominated a member to describe the group's decision-making processes and why the group decided on particular policy recommendations. Forum organizers recorded participants' responses and recommendations and provided the data to local policymakers.

When the time came for the small groups to provide feedback to the larger group, it was observed that the groups had incorporated information provided by the live speakers into their deliberations. Respondents stated that the live speakers' presentations on nanomaterial risk and regulation were helpful in informing the small group dialogues. Nanosilver, a nanomaterial discussed in the presentations, was used as an example at many of the tables. When discussing consumer product labeling requirements, nanosilver was used as an illustration of why product labeling would be desirable (Kunz Kollmann 2009, 7). Participants discussed the positives and negatives of labeling products containing manufactured nanomaterials. Most participants made efforts to connect scenarios to their everyday lives, asking each other questions such as "Wouldn't you want to know if there was nanosilver in your child's teddy bear?"

When providing feedback to the larger group, the tables explained the thought processes that led to their final recommendations. When discussing labeling of products containing manufactured nanomaterials, for example, the pros and cons of each option were considered. The labeling choices consisted of: "labeling of all products," "labeling of only certain products," and "no labeling." Some participants felt that product labeling might cause "unnecessary panic," others felt that labeling is necessary for consumers to be aware of potential risks. Five of seven (5/7) tables felt that the labeling of "only certain products" would be appropriate, and two of seven (2/7) voted for "labeling of all products" (Kunz Kollmann 2009, 12). However, the predominant focus on nanosilver in the presentations might have affected the groups' overall decisions. One table was confused by the references to nanosilver in the presentations, and wondered whether nanosilver is the only nanomaterial they should be concerned about (Kunz Kollmann 2009, 9).

In addition to educating and engaging the citizen participants, the deliberative forums offered an engaging exercise for officials and visiting scholars. In the second forum, an EHU official worked through the scenarios with a group comprised of representatives from the Woodrow Wilson Center and a visiting academic. The benefit of such an exercise was twofold: First, officials and visiting scholars experienced the scenarios and deliberations in the same manner that citizen participants did; second, it provided useful feedback to museum organizers for future forums. Instead of simply observing the deliberations, this exercise was a firsthand opportunity for the "experts" to work through the scenarios alongside the citizen participants. By participating, the official and visiting scholars contributed to an atmosphere of camaraderie and community at the forum, and it was instrumental in fostering the notion that the "experts" were not separate or apart from the citizen participants. Because they participated in the deliberative exercises, the official and visiting scholars were afforded the opportunity to comment on the structure of the scenarios and questions. The table's feedback to museum organizers focused primarily on issues related to question clarity and the subject matter of the scenarios.

## 21.4 Conclusions: Putting Anticipatory Governance to Work in Local Regulation

The Cambridge case serves as an illustration of local efforts that mirror, in many respects, theoretical ideas of anticipatory governance. Themes of anticipatory governance appear in at least three distinct parts of the policy process, including developing foresight techniques; integrating social and ethical considerations into the awareness of academics, scientists, professionals, and industry officials working in nanotechnology innovation; and engaging citizens in the policy process via deliberative forums. In this sense, Cambridge offers at least one model for how communities might begin to imagine their regulatory efforts in anticipatory terms – helping highlight what may already be in place and identify places for improvement.

A look at the Cambridge process for grappling with nanotechnology also offers potentially important insights into anticipatory governance. First, and most obvious, is that regulation is a possible site for anticipatory governance activities. Interestingly, for example, if one looks at seminal presentations of anticipatory governance (e.g., Barben et al. 2008; Guston 2008), they remain largely focused on laboratories, universities, and, perhaps, businesses and NGOs as sites for anticipatory governance. The Cambridge example demonstrates that traditional governance institutions should offer an important site for learning lessons and for innovation as we imagine what anticipatory governance might look like in the future.

A second contribution to anticipatory governance theory from this case involves the role of ongoing, iterative processes of review and revision. As I discussed at the beginning of this essay, the idea of bringing perspectives from reflexive governance and anticipatory governance together may be useful in this regard. Current conceptions of anticipatory governance are compatible with the notion of ongoing, reflexive processes designed to ensure that changes over time in the assumptions, values, and knowledge available in initial foresight, integration, and engagement activities are adequately addressed in ongoing governance processes. After all, they often call for capacity building throughout society. Nonetheless, they have tended to stress the functional elements of anticipatory governance over any discussion of its long-term procedural dimensions.

Because of their oversight function, Cambridge officials see themselves as operating within a dynamic, changing temporal process in which it is important for policy to remain flexible and able to evolve alongside the technology. In their view, uncertainty is pervasive and needs to be accompanied by periodic revisiting of regulatory practices. When the unit issued its final nanotechnology policy recommendations, *Recommendations for a Municipal Health & Safety Policy for Nanomaterials: A Report to the Cambridge City Manager*, they emphasized that preemptive regulation would be unwise due to a general lack of knowledge regarding nanomaterial risk and widely accepted risk-based standards (Lipson 2008b, 9). The report noted that efforts to extrapolate the effects of nanoparticles from "parent" materials has been found to be "less than useful" in a number of cases. The report also detailed the "gaps" in knowledge about the health risks of engineered nanoparticles

(Lipson 2008b, 7). These gaps included the lack of knowledge on safe thresholds for nanomaterial exposure, nanomaterial uptake and metabolizing by the human body, chronic health effects that cannot be evaluated with short and medium-term studies, meaningful differences in the health risks of engineered versus naturally occurring nanomaterials, and exposure risks to individuals (Lipson 2008b, 8).

EHU officials consider the final recommendations representative of a nanotechnology "blueprint," rather than a precise regulation, allowing policy to evolve with the technology as it emerges over time (Lipson 2008a). The report makes clear that it does not perceive Cambridge as a standard-maker or regulator, instead, the development of standards and safety regulations should fall to the federal and state governments, academic researchers, and industry. After those policies have been created, Cambridge can "recognize existing gaps that could be enhanced with local oversight" (Lipson 2008b, 9).

To set up this iterative dimension, the report's fourth recommendation suggests that the health department provide a progress report to the city council every 2 years. This report would account for any new regulatory actions taken by the state and federal government and detail novel research on previously unknown health and safety risks. In light of new information on either front, "a review of local oversight options would also be recommended" (Lipson 2008b, 12). EHU officials consider this reporting period an ideal time to engage with the public and bring it "up to speed" on any new developments (Lipson 2008a).

Public attitudes may also shift. Hence, in addition to the May 2008 and January 2009 forums, museum representatives, and EHU officials plan to continue engagement efforts by holding forums in concert with the iterative policy re-visitation process every 2 years. This effort is enshrined in the report's third recommendation, which suggests both passive and active public engagement. Passive engagement methods include posting basic information about nanotechnology on the Public Health Department's website. This website would also direct individuals to other resources with information on the health and safety issues associated with nanotechnology. Active engagement consists of sponsoring public forums that solicit feedback from community-members on the public awareness campaign as well as engage citizens in a discourse with scientists and local policymakers (Lipson 2008b, 12). Opportunities such as deliberative forums are envisioned to give the public an opportunity to learn about the latest issues in nanotechnology as well as voice their concerns to public officials.

The third and final benefit to bringing Cambridge's efforts to grapple with nanotechnology policy development into dialogue with ideas of anticipatory governance is to identify opportunities for enhancing local regulatory work. Let me offer two examples. One relates to the challenge of deliberation. One difficult hurdle to overcome for those striving to create a democratic setting for deliberation is that of representation. Museum of Science forum organizers reflect that creating broad, diverse forms of public engagement are not easy, even with the visibility of the museum:

It can be argued that with different [discussion] topics that the danger is that the same group of 12 people will show up to everything and it happens to be the same 12 people that often have a lot of time on there hands, so you want to try to make sure you get people who are doing this because [they believe that it is] their civic action to show up and talk to other people and that it's not just the same people doing it over and over again (Sittenfeld and Bell 2008).

In order to overcome this hurdle, museum organizers are considering bringing deliberation into the community itself – into bars, restaurants, and other community settings. The key, museum organizers reflect, is determining the best methods for engaging those who would not typically come to museum events:

The biggest challenge so far has been getting audiences that wouldn't be at the museum already, [we have to] try to engage people that may have very divergent opinions because they haven't been asked in the past when we have developed exhibits and programs, and because they may not think it is appropriate for them. Therefore, we have been doing a lot of work on that and need to do more work on that but one of things we are really hoping to do is bring these events physically, virtually, and problematically into the community where people are considering the issues themselves (Sittenfeld and Bell 2008).

The challenge, however, is more than just getting diverse participation at deliberative events, but also, as I discussed above, to imagine deliberative mechanisms not just as opportunities for public learning and public input into policy decisions but also as elements for helping create transparency and accountability in nanotechnology decision making. Indeed, we might expand ideas from deliberative theory here not only to public officials in Cambridge, who would be expected to justify and open for public engagement the rationales behind their policy choices, but also to nanotechnology innovators. After all, as wielders of power to shape future societies, they, too, might be considered subject to the need for democratization.

The second example involves Cambridge's 2-year iterative reviews. Such reviews are an important step towards fostering an ongoing reflexive nanotechnology governance process, and at first thought, it would appear that 2 years would be a short enough time between policy reviews. However, 2 years in the world of technological development can be quite a long time. As new information comes out on health and safety risks, it would be worthwhile for the health department to maintain a standing stakeholder committee that can respond to new, salient, data that the health department collects, rather than addressing the developments all at once at the end of the 2 years. The stakeholder committee could be "on call" and convene if there are particularly important or crucial developments in the realms of nanotechnology risk research or nanotechnology policy. Alternatively, another configuration might be to have the stakeholder committee meet more frequently. For example, a meeting every 6 months in which health department officials update the committee would be useful because it would help the health department stay abreast of new developments, and would provide a space for addressing the adequacy of the policy for addressing those new developments. Such configurations would assist in breaking up the 2-year intervals into smaller, perhaps more manageable, segments of time.

Now, where the federal and state governments have not acted to formulate standards, Cambridge exists as an important ongoing experiment for nanotechnology policy in local contexts. The final report of policy recommendations recognizes that policy goals related to nanotechnology, an emerging and ubiquitous technology currently in an embryonic form, are not and cannot be final. Moving away from efforts to predict, and moving towards anticipation, the Cambridge nanotechnology policy possesses a capacity for being flexible and evolving new capacities for systematic, ongoing, reflective foresight, integration, and engagement.

## References

Al-Hadithy, Nabil. 2006. *Manufactured nanoparticle health and safety disclosure*. Community Environmental Advisory Council, Berkeley City Council.

Barben, Daniel, et al. 2008. Anticipatory governance of nanotechnology: Foresight, engagement, and integration. In *The handbook of science and technology studies*, 3rd ed, ed. Edward J. Hackett et al. Cambridge: MIT Press.

Bell, Larry. 2008. Engaging the public in technology policy: A new role for science museums. *Science Communication* 29: 386–398.

Bray, Hiawatha. 2007. Cambridge considers nanotech curbs: City may mimic Berkeley bylaws. *The Boston Globe,* 26 Jan 2007, 25 Nov 2008. http://www.boston.com/business/technology/articles/2007/01/26/cambridge_considers_nanotech_curbs/.

Cambridge City Council Meeting – January 8, 2008. *Cambridge Civic Journal* 1–7, 1 Dec 2008. http://www.rwinters.com.

Cambridge Historical Commission. 2008. A brief history of Cambridge, Massachusetts, USA. Cambridge Historical Commission, 25 Nov 2008. http://www.cambridgema.gov/~historic/cambridgehistory.html.

Cambridge Public Health Department. 2008. Recombinant DNA. *Cambridge Public Health Department*. Cambridge Health Alliance, 19 Nov 2008. http://www.cambridgepublichealth.org/services/regulatory-activities/rdna/overview.php.

D'Entrèves, Maurizio P. 2002. *Democracy as public deliberation: New perspectives*. Manchester: Manchester University Press, Distributed exclusively in the USA by Palgrave.

Elster, Jon. 1998. *Deliberative democracy*. Cambridge: Cambridge University Press.

Fisher, E., R.L. Mahajan, and C. Mitcham. 2006. Midstream modulation of technology: Governance from within. *Bulletin of Science, Technology & Society* 26(6): 485–496.

Fisher, Erik. 2007. Ethnographic intervention: Probing the capacity of laboratory decisions. *NanoEthics* 1(2): 155–165.

Guston, David H. 2008. Innovation policy: Not just a jumbo shrimp. *Nature* 454(7207): 940.

Guston, David H., and Daniel Sarewitz. 2002. Real-time technology assessment. *Technology in Society* 24(1–2): 93–109.

Guttmann, Amy, and Dennis F. Thompson. 2004. *Why deliberative democracy?* Princeton: Princeton University Press.

Hamlett, Patrick W., and Michael D. Cobb. 2006. Potential solutions to public deliberation problems: Structured deliberations and polarization cascades. *Policy Studies Journal* 34(4) (11): 629–648.

Jasanoff, Sheila. 2006. *Designs on nature: Science and democracy in Europe and the United States*. Princeton: Princeton University Press.

Jones, Jeffrey M. 2008. Trust in government remains low. *Gallup*, 18 Sept 2008, 17 Nov 2008. http://www.gallup.com/poll/110458/trust-government-remains-low.aspx.

King, Loren A. 2003. Deliberation, legitimacy, and multilateral democracy. *Governance* 16: 23–50.

Kunz Kollmann, Elizabeth. 2008. *Museum of science forum 3.3 results (draft)*. Boston: Forum 3.3 Museum of Science.

Kunz Kollmann, Elizabeth. 2009. *Museum of science forum 4.1 results (draft)*. Boston: Forum 4.1 Museum of Science.

Lipson, Sam. 2008a. Cambridge nanomaterial policy. *Interview*, 4 Nov 2008.
Lipson, Sam. 2008b. *Recommendations for a municipal health & safety policy for nanomaterials: A report to the Cambridge city manager*. Cambridge: Cambridge Public Health Department, Cambridge Health Alliance.
Moore, Fiona N. 2002. Implications of nanotechnology applications: Using genetics as a lesson. *Health Law Review* 10: 9–15.
Predict. Merriam-Webster Online Dictionary. 2009. Merriam-Webster Online, 3 June 2009. http://www.merriam-webster.com/dictionary/predict.
Rabinovici, Sharyl, Javiera, Barandiaran, and Margaret Taylor. 2007. Local disclosure ordinance as regulatory catalyst: Early insights from the Berkeley, California manufactured nanoscale materials health and safety disclosure ordinance. In *Working Draft – Presented at a Symposium of the Northern California Chapter of the Society for Risk Analysis*, 4 Oct 2007.
Rip, A., and H. Te Kulve. 2008. Constructive technology assessment and socio-technical scenarios. In *The yearbook of nanotechnology in society*, Presenting futures, vol. 1, ed. E. Fisher, C. Selin, and J.M. Wetmore, 49–70. Dordrecht: Springer.
Sittenfeld, David, and Larry, Bell. 2008. Policy, public engagement, and the role of museums. *Telephone interview*, 6 Nov 2008.
Selin, Cynthia. 2006. Trust and the illusive force of scenarios. *Futures* 38(1): 1–14.
Snapp, Martin. 2008. City council gives OK to landmarks preservation ordinance. *East Bay Daily News*, 12 Sept 2006, 1 Dec 2008. http://www.ebdailynews.com/article/2006-12-9-eb-council.
Voss, Jan-Peter, and Rene Kemp. 2006. Sustainability and reflexive governance: Introduction. In *Reflexive governance for sustainable development*, ed. Jan-Peter Voss, Dierk Bauknecht, and Rene Kemp, 3–30. Cheltenham/Northampton: Edward Elgar.
Wynne, Brian. 1996. Misunderstood Misunderstandings: Social Identities and the Public Uptake of Science. In *Misunderstanding science? The public reconstruction of science and technology*, ed. Alan Irwin and Brian Wynne, 19–46. Cambridge: Cambridge University Press.

# Index

**A**
Ablative surgery, 126
Adaptive brain interface, 131
Adaptive machines, 1
Adderall, 61, 235–237
ADHD. *See* Attention-deficit hyperactivity disorder (ADHD)
AFM. *See* Atomic-force microscopy
ALS. *See* Lou Gehrig's disease
Alzheimer's, 9, 23, 27–29, 33–35, 120, 141, 188, 237, 280, 281, 298
American Foundation for the Blind, 159
Andersen, I.E., 272
Aneurysm, 23, 129
Animal welfare, 248, 252, 256–260
Anthrax, 30
Anticipatory governance, 1–16, 43, 97–100, 104, 105, 109, 110, 147–157, 369–393
Anticipatory research, 2, 7–8
Anticipatory technology assessment, 7
Anti-microbial, 314, 315
Architecture models, 212
Argus II, 160
Aricept, 237
Artificial vision, 139, 163, 165
Asbestos, 71, 82, 354
Atomic-force microscopy (AFM), 27
Attention-deficit hyperactivity disorder (ADHD), 223, 237, 240, 242
Automated sewer surveillance, 279–280

**B**
Barless prison, 281–282
Bauby, J.D, 130
BBB. *See* Blood-brain barrier (BBB)
Behavioral control, 7
Berger, H., 118
Berger, T.W., 141
Berne, R.W., 5, 85, 98, 101, 104
Berry, R.M., 271
Binet, A., 219, 228
Biocompatibility, 26, 35–37, 134, 151, 170, 180, 188, 189, 281, 314
Bioengineered scaffold, 278
Biofeedback, 120, 121
Bionic, 115, 142, 144, 150, 160, 161, 164, 177, 274, 282, 296
  eye, 160, 161, 164, 274, 282, 296
  man, 115, 142
Biosensors, 9, 26–29, 36, 314
Bioterrorism, 82
Birbaumer, N., 131, 132
Blind chickens, 14, 248–250, 252, 254, 256, 257, 259–261
Blind identity, 162, 165
Blindness, 12, 52, 63, 68, 128, 159–165
Blood-brain barrier (BBB), 5, 15, 68, 82, 87, 88, 170
Board of technology, 268, 270
Boltzmann machines, 212
Bond, P.J, 4, 5, 82, 85
Boston Museum of Science, 16, 350, 355, 370, 381, 382, 385
Braille, 12, 160, 162
Brain
  cancer, 9, 28, 29, 37
  chip, 106, 109, 237, 278–270
  enhancement, 13, 187
  imaging, 6, 196
  tumor, 28, 37

S.A. Hays et al. (eds.), *Nanotechnology, the Brain, and the Future*, Yearbook of Nanotechnology in Society 3, DOI 10.1007/978-94-007-1787-9,

Brain Fingerprinting Laboratories, 120
BrainGate, 13, 105, 133, 136, 148, 149, 153, 169, 171, 176–177, 222
Brindley, G., 168
Buckyballs, 75, 354
Build Up approach, 251

**C**
Cambridge, MA, 15, 373–383
Cambridge Nanomaterials Advisory Committee (CNAC), 15, 333, 335, 337–339, 342, 348, 355, 379, 380
Cambridge Public Health Department, 15, 16, 333–336, 338, 339, 342, 347, 348, 350, 353, 355, 370, 382
Carbon nanotubes, 29, 35, 36, 68, 180, 279, 314–316, 319, 321, 328, 344, 345, 354
Carson, J., 14, 61, 220, 221, 226–232
Caton, R., 118
Center for Nanotechnology in Society at Arizona State University (CNS-ASU), 8, 9, 15, 22, 23, 97, 98, 102, 265, 269, 270, 276
Chapin, J.K., 133
Children, 43, 55, 56, 119, 121, 143, 150–154, 160, 161, 165, 169, 173, 219, 228, 234, 239, 240, 242, 308
Chorost, M., 150, 177
CIA mind control, 125
Citizen science, 109
Civil rights, 267, 275
Climate change, 51, 268, 307
Clinical trials, 13, 105, 139, 152, 153, 160, 161, 164, 182–185, 188, 189, 237, 295, 296
Clinton, B., 67, 69
Cloning, 250, 257
CNAC. *See* Cambridge Nanomaterials Advisory Committee (CNAC)
CNS-ASU. *See* Center for Nanotechnology in Society at Arizona State University
Cobb, M.D., 10, 14, 43, 68, 265, 269, 372
Cochlear implant, 12, 13, 139, 147–157, 162, 167–169, 175–177
Cognitive enhancement, 6, 10, 11, 13, 14, 43–65, 219–233, 235, 236, 238–244
Cognitive prosthesis, 141
Cognitive psychology, 206, 210
Commercialization, 152, 300, 307
Commodification, 85
Common morality, 181
Competition, 54, 55, 188, 198, 219, 224, 226, 229–233, 240, 260, 291, 305
Complexity, 3, 136, 185, 197–208, 220, 222, 329, 346, 372
Complex systems problem, 196–207, 210–212, 215, 216
Computationally irreducible, 201
Computed tomography (CT), 117
Connectionism, 196, 197, 211–213
Consensus conference, 153, 268, 372
Consumer products inventory, 341
Converging technologies, 4, 249, 265, 267, 286, 290
Cortical visual stimulator (CVS), 160, 164
Crichton, M., 105, 125, 126
Cyberkinetics, 169
Cybernetic organism, 142
Cybernetics, 142
Cyberonics, 128
Cyborg, 115, 142–144, 168, 171, 173

**D**
Danish Parliament, 268
DARPA, 137, 140
Data protection, 185
DBS. *See* Deep brain stimulation (DBS)
Deaf community, 151, 154–157, 162, 175
Declaration of Helsinki, 184
Deep brain stimulation (DBS), 13, 86, 126–128, 140, 169, 170, 173, 176, 185, 223
Delgado, J., 122–125, 127, 129
Deliberation, 5–8, 10, 22, 97, 99, 102–104, 109, 266–268, 270, 273–275, 289, 298, 306, 342–343, 369, 373, 380–386
Deliberative democracy, 372, 373
Democracy, 1, 11, 43, 225, 232, 233, 242, 372, 373
Democratizing nanotechnology, 103
Demos, 98, 101
Denmark, 24, 142, 268
Dennett, D., 130
Depressed, 68, 119, 123, 127, 128, 140, 161, 162, 169, 173, 180, 181, 185–187, 223
Diagnosis, 6, 33, 37, 180
Dignity, 144, 254–256, 258, 260, 261
Direct current polarization, 140
Disease detector, 280–281
Disenhancement, 14, 247–250, 253, 256–259, 261
Diverse social groups, 97

Doc in a box, 280, 281
Donchin, E.E., 120, 131
Donoghue, J.P., 133, 136
Drexler's gray goo, 3
Drug delivery, 6, 26, 30, 33, 68, 85–87, 314, 318, 328
Drug development, 6
Dumb Down approach, 250, 253, 259

**E**
Echo-planar magnetic resonance spectroscopic imaging (EP-MRS), 140
Economics, 26, 45, 142, 205, 232, 275
Electrical stimulation, 122–125, 135
Electroencephalograms (EEG), 118–122, 130–132, 135, 137
Electro-mechanical therapies, 12
Electroshock therapy, 117, 140
Endangerment of organisms, 289
End-to-end (E2E), 8, 9
Engagement, 7, 9, 11, 15, 35, 97–110, 147, 220, 221, 370–372, 380–387
Engineered nanoparticles, 329, 343–346, 370, 382, 384
Engineered tissues, 278
Entrepreneurship & development, 272
Environmental consequences, 88, 266
Epilepsy, 29, 32, 68, 119, 125, 127, 128, 180, 223, 319, 329
EP-MRS. *See* Echo-planar magnetic resonance spectroscopic imaging (EP-MRS)
Equity, 81, 105, 153, 156, 157, 267, 272, 275
Essential tremor, 126, 127
Estranged, 188
Ethical challenges, 2, 11, 12, 82, 98
Ethical consideration, 7, 79–91, 102, 181, 185, 266, 272, 304, 308, 384
Exoskeletons, 137
Expert panels, 100, 228, 229

**F**
Farah, M.J., 14, 86, 87, 172, 235, 236, 238, 240
Farwell, L.A., 120, 131
FES. *See* Functional electrical stimulation (FES)
Firefox, 138
Food and Drug Administration (FDA), 126, 128, 132, 139, 150–153, 155, 164, 272, 292, 297, 303, 308, 343
Forecasting, 22
Foresight, 80, 97, 181, 189, 335, 370–372, 375, 377, 380, 382, 384, 387
Forums, 15, 16, 91, 99, 100, 109, 265–282, 285–289, 291, 295–309, 350, 353, 354, 372, 380–385
Foundation for Retinal Research, 161, 164
Free-will, 106
Fukuyama, 241
Functional electrical stimulation (FES), 135, 136
Functional magnetic resonance imaging (fMRI), 1, 23, 117
Funding accountability, 272, 307

**G**
Gage, P., 117
Gardner, H., 219, 222, 225
Gender, 10, 53, 56, 57, 60, 107, 163, 186, 289
  gap, 55, 60
Generation, 7, 127, 141, 144, 180, 278, 279, 289, 319, 324, 329, 354
Gene therapies, 32, 87, 160, 161, 164, 165
Genetic engineering, 248, 250, 253–257
Genetically modified foods, 61, 268, 307
Georgia Institute of Technology, 8, 22, 23, 269, 271, 276
Giffords, G., 1
Gillett, G., 130, 167, 168, 171, 172
Glannon, W., 86
Global, 25, 81, 82, 84, 86, 89, 91, 153, 154, 198, 203, 212, 250, 287, 289, 290, 297, 302, 341, 354, 357
Gould, S.J., 14, 220–222, 225–232
Greenfield, S., 87
Grey goo, 3, 84

**H**
Hamlett, P.W., 265, 269, 372
Haptic technology, 138
Health insurance companies, 189
Health services, 274
Heath, R., 122–127, 289
Hess, W.R., 122
Horch, W.K., 138
Human
  enhancements, 10, 13–15, 21, 43–65, 81, 83, 89–91, 105, 154, 220, 224–226, 238, 247–261, 265–282, 285–309
  nature, 85, 103, 105–107, 110, 129, 173, 225
  personhood, 13, 167
  suffering, 106, 107, 291

Human-machine interaction, 11, 81, 91
Human-machine interfaces, 5, 83, 85, 89–91
Humanness, 79, 82–85, 88, 91, 177

I
Imagination, 3, 7, 11, 105, 107, 135, 141, 142
Imaging technologies, 27, 34, 117, 149
Incompetent patients, 185, 186
Infectious diseases, 26, 29–31
Informed citizen input, 265, 268
Insurance, 10, 12, 45, 57, 58, 62, 64, 154, 187, 189, 266, 272, 273, 302, 308
Integration, 2, 9, 35, 83, 84, 88–90, 168, 170, 180, 183, 208, 279, 293, 370, 371, 377–380, 384, 387
Integrity, 243, 254–259, 261
Intelligence, 13, 14, 128, 198, 206, 219–233, 247
  quotient, 219
International community, 298
Intuition, 15, 201, 248, 249, 251–253, 255, 257–261
Invasiveness, 13, 167, 174

J
Jasanoff, S., 376, 377
Justice, 11, 79, 81, 82, 84–86, 89, 91, 181, 187, 220

K
Kass, L., 238, 257
K-12 education, 266, 272, 309
Kennedy, P.R., 132, 133, 223
Kepler's laws, 199, 200
Kinetra, 126
Kuiken, T.A., 136, 353, 355
Kulinowski, K., 271

L
Lab-on-a-chip, 28, 86
Legal, 13, 148, 159, 162, 170, 171, 179–189, 232, 239, 268, 271, 300, 304, 308, 342, 377
Legitimacy, 151, 155, 261, 373
Levodopa, 126
Locked-in syndrome, 129, 130
Lou Gehrig's disease (ALS), 129, 132, 133
Luddism, 108

M
Magnetic resonance imaging (MRI), 1, 71, 117, 140, 149
Magnetoencephalography (MEG), 117
Mann, S., 142
Mass media, 44, 50, 51, 63, 68–70, 74, 75
Media, 10, 44, 50, 51, 54, 60, 63, 67–70, 74–76, 105
Medtronic, 126
MEG. *See* Magnetoencephalography
Mehlman, M.J., 271
Memory and attention, 187
Mid-stream modulation, 372
Military uses, 272
Miniaturization, 6, 180, 181
Modafinil, 237, 239
Molecular binding sites, 33
Moniz, E., 122
Moral identity, 168, 171
MRI. *See* Magnetic resonance imaging

N
NAD. *See* National Association of the Deaf
Nagle, M., 133, 136, 169
Nanobots, 81, 84, 105
Nanocoatings, 36
Nanodialogues, 109
Nano-electrodes, 180
Nano-engagement, 98–101, 108, 110
Nanoethics, 2–6, 11, 38, 79, 80, 82, 249
NanoJury, 99–101, 109
Nanomedicine, 81, 86, 182, 183, 281
Nano-neuro, 1–35, 80, 86–87
Nanoparticles, 5, 11, 15, 28, 32, 68, 314–330, 343–346, 354, 362, 370, 375, 379, 382, 384
Nano-silver, 48, 360, 364, 382, 383
*Nanotalk*, 5, 101
Nanotechnology, biotechnology, information technologies and cognitive science (NBIC), 83, 85, 89, 90, 265–274, 278, 285–295, 300–306, 308
Nano $TiO_2$ (nano titanium dioxide), 324
Narrative, 70, 85, 101, 103, 107–109, 199, 219–233
National Association of the Deaf (NAD), 151, 154–156
National Citizens Technology Forum (NCTF), 15, 109, 265–282, 285–309
National Institutes of Health (NIH), 150, 168, 169, 377, 378

National Nanotechnology Initiative (NNI), 67, 69
National Science Foundation (NSF), 295
Naturally occurring nanoparticles, 343
NBIC. *See* Nanotechnology, biotechnology, information technologies and cognitive science
NCTF. *See* National Citizens Technology Forum
Nerve growth, 10, 34–36
Nerve regeneration, 10, 34–35
Nervous system, 1, 9, 15, 22–38, 67, 117, 122, 138, 140, 148, 171, 172, 251, 313–331
Neural-cognitive mapping, 196, 197, 215
*Neural nets*, 196
Neural Prosthesis Program, 168
Neural prosthetic, 6, 10, 21, 35–37, 141, 143, 180
Neural Signals Inc., 222
Neurite outgrowth, 26
Neurodegenerative diseases, 26, 33, 34, 181
Neuro-ethics, 38
Neuroimplantation, 86, 89, 90
Neurological enhancements, 5
Neurosurgery, 6, 181
Neurotoxicity, 68, 318, 320, 322–323
New energy sources, 276
Newspaper, 10, 67–76, 119, 269
Newton, I., 199, 200, 206
Nicolelis, M.A.L., 133
Nietzsche, 224
NIH. *See* National Institutes of Health
Nobel Prize, 75, 122
Nonlinearity, 203
NSF. *See* National Science Foundation

**O**
Oberdörster, E., 317, 318, 321
Off-label, 52, 61
Olds, J., 122, 123, 125
Olfactory nerve, 5, 345

**P**
P300, 120, 131
Pacemakers, 126, 128, 168–170, 172, 176
Pain receptors, 250, 253
*Parallel distributed processing*, 196, 211
Parkinson's, 9, 23, 26, 29, 34, 68, 126, 127, 169, 180, 181, 184, 185, 188
Patch clamp technique, 323–327
Paternalism, 186
Patient interest groups, 189
Personhood, 13, 167, 171–177
PET. *See* Positron emission tomography (PET)
Pharmacokinetic technologies, 27, 28
Phrenology, 222, 225
*Physical Control of the Mind*, 123
Physicians, 116, 237, 242, 243, 295
Plasticity, 140, 222
Playing God, 45, 59, 60, 64, 65
Positron emission tomography (PET), 117
Prey, 105, 274
Privacy, 53, 81–83, 87, 88, 266, 272, 289–293, 297, 300, 301, 307
Professions, 3, 242, 298
Professor the Lord Winston, 21
Prosthetic, 1, 6, 10, 12, 21, 30, 32, 35–38, 136, 138, 141, 143, 144, 164, 165, 168–170, 174, 180, 181, 237, 271, 283
Provigil, 61, 237
Psychometrics, 226
Psychopharmacology, 83, 84, 86, 88, 117
Psychosurgery, 86
Public attitude, 43–65, 68, 159, 164, 385
Public awareness, 46, 50, 69, 297, 355, 382, 385
Public deliberation, 7, 99, 289, 371
Public dialogue, 5, 8, 10, 102
Public engagement, 11, 15, 97–110, 370, 372, 373, 380–383, 385, 386
Public information, 16, 266, 274, 301, 350
Public opinion, attitudes, 10
Public value mapping, 8

**Q**
Quantum dot, 28, 32–34, 36, 37, 76, 86

**R**
Real-time technology assessment, 2, 6, 8–10, 80
Recombinant DNA, 374, 377, 378, 382
Red blood cells, 316
Reflexive engagement and integration, 9
Reflexive governance, 371, 372, 384
Regan, T., 252, 255, 260
Regularity, 196, 199–201
Regulatory adequacy, 272
Regulatory frameworks, 12, 378
Rehabilitation Institute of Chicago, 136
Rejeski, D., 270, 335, 355
Religion, 1, 54, 56, 60, 85, 97–110, 304
Religiosity, 44, 46, 54, 56

Religious perspectives, 11, 102, 104, 106, 107
Reputational cascades, 274
Research and innovation systems mapping, 8
Retinitis pigmentosa, 161
Reverse engineered, 197
Reversibility, 108, 302
Rights, 64, 184, 231, 252, 255, 267, 273, 275, 290, 293, 299, 300, 305, 307
Risk analysis, 3, 4, 335
Risk perception, 44, 50, 99
Ritalin, 14, 235–237, 241, 242
Robert, J.S., 1, 5, 8, 21, 82, 170, 271
Robo-rats, 125
Roco, M.C., 4, 21, 67, 81, 82, 86, 149, 187
Rollin, B., 248, 253

**S**

Sandøe, P., 248, 256
Scanning electron microscopy (SEM), 27, 315, 319, 324
Science citation index (SCI), 23, 24
Science fiction, 12, 115, 142–144, 173, 249
Scientific methodology, 195, 198, 207–208, 210
Screening, 82, 83, 86, 88, 155, 242, 279, 280, 316, 317
Second Sight, 160
SEM. *See* Scanning electron microscopy (SEM)
Sensory and motor nerves, 171
Sign language, 154
Silicon microelectrode, 180
Simon, T., 229, 230
Singer, P., 252, 254
Single photon emission computerized tomography (SPECT), 117
Single-wall carbon nanotubes (SWNTs), 180
Skin care products, 67
Slippery slope, 181
Small talk, 101–102, 109
Smart dust, 82
Social effects, 106, 274
Social equality, 188
Social justice, 11, 81, 91, 220
Sociotechnical, 48, 53, 219–233
Soldier, 52–54, 63, 83, 84, 88, 137, 140, 142, 239, 298, 299
Soletra, 126
Space exploration, 142, 274
SPECT. *See* Single photon emission computerized tomography
Speculative, 38, 49, 181, 274
Spinal cord injuries, 29, 33–35, 140
Sports, 54, 55, 64, 119, 238, 240, 242
SPR. *See* Surface plasmon resonance (SPR)
SQUID. *See* Superconducting quantum interface devices (SQUID)
Stem cell transplantation, 160
Stimoceiver, 124
Stroke, 23, 28, 29, 32, 129, 130, 132, 140, 141
Suffer, 106, 107, 123, 125, 128, 181, 186, 197, 232, 248, 250, 252–255, 257, 258, 260, 261, 291
Sun block, 44, 358
Superconducting quantum interface devices (SQUID), 27
Super soldiers, 84, 88
Surface plasmon resonance (SPR), 28, 33
Surveillance, 7, 82, 83, 88, 279–280, 293
SWNTs. *See* Single-wall carbon nanotubes

**T**

TEM. *See* Transmission electron microscopy (TEM)
*The Bell Curve*, 227, 229, 287
*The Diving Bell and the Butterfly*, 130
*The Mismeasure of Man*, 225
Therapy *vs.* enhancement, 14, 105, 187–188
*The Sound and Fury*, 151
*The Terminal Man*, 125
"The Wisdom of Repugnance," 257, 260
Thought Translation Device, 131, 132
Tillery, S.H., 271
Tissue printing, 277, 278
Toxicity, 11, 15, 68, 83, 88, 180, 189, 272, 309, 311, 313–329, 342, 344–346
Transcranial magnetic stimulation (TMS), 140
Transhumanism, 5, 84, 89, 90, 128, 143
Transhumanists, 5, 7, 115, 143, 144, 187
Transmission electron microscopy (TEM), 27, 315
Transmission pathways, 33
Transparency, 302, 373, 376, 386
Trauma, 34, 129, 141
Turing machine, 205
21st Century Nanotechnology Research and Development Act, 67

**U**

Ultrafine, 343–345
Uploading, 115

Upstream, 2, 7, 8, 11, 22, 97, 107, 147, 280, 373
U.S. Food and Drug Administration. *See* Food and Drug Administration (FDA)

**V**
Vader, D., 142
Vagus nerve stimulation, 128
Visually impaired persons, 159

**W**
Walter, W.G., 130, 133
Warwick, K., 143
Weapons for defense, 274
Web of Science (WOS), 22, 23
Weckert, J., 81, 82, 85
Wetmore, J.M., 11, 97, 154
White cane, 160
Willis, T., 116
Wolbring, G., 91, 98, 99, 102, 175
Wolpe, P.R., 86, 174
Woodrow Wilson International Center for Scholars, 270, 335, 355, 381
WOS. *See* Web of Science; Web of Science (WOS)

**X**
X-rays, 27, 117

Printed by Publishers' Graphics LLC